Systems and Synthetic Biotechnology for Production of Nutraceuticals

Long Liu • Jian Chen
Editors

Systems and Synthetic Biotechnology for Production of Nutraceuticals

Springer

Editors
Long Liu
School of Biotechnology
Jiangnan University
Wuxi, Jiangsu, China

Jian Chen
School of Biotechnology
Jiangnan University
Wuxi, China

ISBN 978-981-15-0445-7 ISBN 978-981-15-0446-4 (eBook)
https://doi.org/10.1007/978-981-15-0446-4

This Springer imprint is published by the registered company Springer Nature Singapore Pte Ltd.
The registered company address is: 152 Beach Road, #21-01/04 Gateway East, Singapore 189721, Singapore

Contents

Chapter 1
Nutraceuticals Definition, Kinds and Applications

Yanfeng Liu and Long Liu

1.1 Introduction

Nutraceuticals are important a class of important compounds with health-promoting and disease-preventing functions behind their nutritional value. Traditionally, nutraceuticals mainly refer to amino acids, vitamins, and nucleotide those have been widely used for promoting our health with long history. With the deeper understanding of the functionalities of various biomolecules, new nutraceuticals have been increasingly developed and commercialized, such as plant-derived nutraceuticals (terpenoid, flavonoid, plant oligosaccharide) and animal tissue-derived nutraceuticals (glucosamine, hyaluronic acid, and chondroitin sulfate) (Bian et al. 2017; Benavente-Garcıa et al. 1997; Weimer et al. 2014; Liu et al. 2011; Clegg et al. 2006). Because nutraceuticals are derived from food those have been intensively investigated for their functionality on health, two major characters of nutraceuticals are high safety and defined functionality.

Various nutraceuticals cover a broad range of functionalities, including regulating immunity, improve lipid metabolism, preventing aging-associated diseases, facilitating maintenance of blood glucose, improve brain development and health, maintaining joint health and preventing cardiovascular diseases (Das et al. 2012). Based on increasing numbers of commercially available nutraceuticals and the wide applications, global nutraceutical market reached over $230 billion in 2018. It is estimated that the global nutraceutical market will reach $336 billion by 2023 with a compound annual growth rate of 7.8% (BCC Research 2018).

Y. Liu · L. Liu (✉)
Key Laboratory of Carbohydrate Chemistry and Biotechnology, Ministry of Education, Jiangnan University, Wuxi, China

Key Laboratory of Industrial Biotechnology, Ministry of Education, Jiangnan University, Wuxi, China
e-mail: longliu@jiangnan.edu.cn

L. Liu, J. Chen (eds.), *Systems and Synthetic Biotechnology for Production of Nutraceuticals*, https://doi.org/10.1007/978-981-15-0446-4_1

In this chapter, we firstly discuss the definition of nutraceuticals. Next, functionality-based different kinds of nutraceuticals are summarized. Finally, the applications of important typical nutraceuticals are discussed, such as gamma-aminobutyric acid, alpha-ketoglutaric acid, hyaluronic acid, vitamin B12, folate, glutathione, carotenoids, and N-acetylglucosamine.

1.2 Nutraceutical Definition

The term nutraceutical is hybrid of 'nutrition' and 'pharmaceutical' proposed by Stephen L. DeFelice in 1989 (DeFelice 1995). Originally, nutraceutical referred to 'a food (or part of a food) that provides medical or health benefits, including the prevention and/or treatment of a disease' (DeFelice 1995). The definition of nutraceutical was lately changed to be 'a product isolated or purified from foods that is generally sold in medicinal forms not usually associated with food' (Pandey et al. 2010). Based on literature mining and analyzing, there are 25 definitions for nutraceutical with the majority of the definitions refer nutraceuticals to 'food, food components, or nutrients providing health benefits behind their nutritional value' (Andrew and Izzo 2017). Despite no satisfactory definition can be found to cover all the aspects of different existing definitions for nutraceutical, it is generally recognized that nutraceuticals are important a class of important compounds with health-promoting and disease-preventing functions behind their nutritional value (Aronson 2017).

1.3 Functionality-Based Different Kinds of Nutraceuticals

Based on the functionalities of various nutraceuticals, we can divide the commonly used nutraceuticals into the following five kinds: (1) cardiovascular disease-preventing nutraceuticals, (2) joint health-promoting nutraceuticals, (3) immunity health-promoting nutraceuticals, (4) diabetes-preventing nutraceuticals, and (5) brain health-promoting nutraceuticals (Table 1.1). The representative nutraceuticals of abovementioned functionality-based different kinds nutraceuticals were discussed.

Cardiovascular disease-preventing nutraceuticals are a series of molecules those can reduce cholesterol uptake or reducing cholesterol and homocysteine accumulation, which is of great importance for decreasing risk of cardiovascular disease. Folate and vitamin B12 are widely used compounds for reducing homocysteine abundance for preventing cardiovascular disease (Saini et al. 2016; Blancquaert et al. 2010; Fang et al. 2018). Nicotinic acid is an important compound for reducing cholesterol abundance for promoting cardiovascular health (Martin-Jadraque et al. 1996).

Table 1.1 Functionality-based different kinds of nutraceuticals

Nutraceutical kind	Representative nutraceutical	Specific functionality	Reference
Cardiovascular disease-preventing nutraceutical	Folate and vitamin B12	Reducing homocysteine abundance in blood	Saini et al. (2016), Blancquaert et al. (2010), Fang et al. (2018)
	Nicotinic acid	Reducing cholesterol abundance in blood	Martin-Jadraque et al. (1996)
Joint health-promoting nutraceutical	Glucosamine and chondroitin sulfate	Reducing joint flexibility and alleviating pain in join	Clegg et al. (2006), Anderson et al. (2005), Sawitzke et al. (2008), Lee et al. (2010)
Immunity health-promoting nutraceutical	β-glucan	Improving immunity response to external stimulus	Bashir and Choi (2017)
Diabetes-preventing nutraceutical	Polyphenols	Inhibiting carbohydrate metabolism for modulating blood glucose level	Xiao and Hogger (2015)
Brain health-promoting nutraceutical	Phosphatidylserine	Improving memory and preventing depression	Glade and Smith (2015)

Joint health-promoting nutraceuticals mainly consists of glucosamine and chondroitin sulfate, which are usually used in combination as nutraceuticals for increasing joint flexibility and alleviating pain in joint. Joint tissues contain high concentration of glucosamine, the hypothesis that glucosamine supplements can alleviate osteoarthritis was proposed nearly 40 years ago. This hypothesis has been tested by many clinical trials, and glucosamine supplements are widely used to relieve arthritic and maintain joint health (Anderson et al. 2005). Chondroitin sulfate is another important joint health-promoting nutraceutical, which is often used in combination with glucosamine to enhance the effectiveness of improve joint health (Clegg et al. 2006; Sawitzke et al. 2008; Lee et al. 2010).

Immunity health-promoting nutraceuticals can improve immune response to external stimulus.

As one of the important immunity health-promoting nutraceuticals, β-glucan can activate macrophage and neutrophil leucocyte, enhance leukin and cytokinin concentrations, which efficiently improves immune response to external stimulus and regulate the function of immunity (Bashir and Choi 2017). Diabetes-preventing nutraceutical mainly refers to polyphenols, which can inhibit carbohydrate metabolism for modulating blood glucose level (Xiao and Hogger 2015). Brain health-promoting nutraceutical is the fifth kind of functionality-based different kinds of nutraceuticals. Phosphatidylserine is the representative compound for brain health-promoting nutraceuticals with ranging applications, which can improve memory and preventing depression (Glade and Smith 2015).

1.4 Applications of Important Typical Nutraceuticals

Based on various defined functionalities, nutraceuticals have been widely applied into different aspect for promoting health and preventing disease. The typical application aspects of typical nutraceuticals are discussed, including amino acid and organic acid-based nutraceutical (gamma-aminobutyric acid and alpha-ketoglutaric acid), functional polysaccharide (hyaluronic acid), important vitamins (vitamin B12 and folate), antioxidant compounds (glutathione and carotenoids), and joint-health-promoting nutraceutical (glucosamine) (Table 1.2).

Gamma-aminobutyric acid and alpha-ketoglutaric acid are two typical amino acid and organic acid-based nutraceutical. Gamma-aminobutyric acid is the major inhibitory neurotransmitter in the central nervous system of human. It has been demonstrated that gamma-aminobutyric acid can be potentially used as a preventing-hypertensive and promoting relaxation compound (Boonstra et al. 2015). Therefore, this compound has recently been widely used as nutraceutical with an expected global market of $64 million by 2025 (Reports TM: Global GABA (CAS 56-12-2) 2018). Alpha-ketoglutaric acid is a primary metabolite from TCA cycle, which is also connected to central nitrogen metabolism and plays important functions for cell metabolism. Alpha-ketoglutaric acid supplement can enhance bone tissue formation

Table 1.2 Applications of important typical nutraceuticals

Nutraceutical name	Application	Reference
Gamma-aminobutyric acid	1) As nutraceutical for preventing hypertensive and promoting relaxation	Boonstra et al. (2015)
Alpha-ketoglutaric acid	1) As nutraceutical for enhancing bone tissue formation	Wu et al. (2016), Chin et al. (2014)
	2) Delaying age-related disease	
Hyaluronic acid	1) As nutraceutical for preventing cardiovascular disease	Fallacara et al. (2018)
	2) Delaying skin aging process	
Vitamin B12	1) As nutraceutical for preventing cardiovascular disease	Fang et al. (2018), Romain et al. (2016)
	2) Preventing neurological disorders	
Folate	1) As nutraceutical for preventing cardiovascular disease	Saini et al. (2016), Blancquaert et al. (2010)
	2) Preventing megaloblastic anemia and neural tube defects	
Glutathione	1) As antioxidant nutraceutical for modulating cell metabolism	Young et al. (2011)
	2) Delaying age-related disease	
Carotenoids	1) As antioxidant nutraceutical for modulating cell metabolism	Nishizaki et al. (2007), Yoshida et al. (2009)
Glucosamine	1) Promoting joint health	Clegg et al. (2006), Anderson et al. (2005), Sawitzke et al. (2008)

and potentially delay age-related disease, which are major application field for alpha-ketoglutaric acid (Wu et al. 2016; Chin et al. 2014).

Hyaluronic acid belongs to polysaccharide that is composed of disaccharide repeats of D-glucuronic acid and N-acetylglucosamine joined alternately by β-1,3 and β-1,4 glycosidic bonds with molecular weights distribution from 10^4 to10^7 Da. Hyaluronic acid has high abundance In the human body in the skin, umbilical cord, and vitreous humor. Due to its physiological, biological functions, and high moisturising retention capability with its lack of immunogenicity and toxicity, hyaluronic acid is used as nutraceutical for preventing cardiovascular disease and delaying skin aging process (Fallacara et al. 2018).

Folate and vitamin B12 are two typical vitamins those have been widely used for preventing cardiovascular diseases and neurological diseases. Both folate and vitamin B12 cannot be de novo synthetized in human body, therefore, additional supply via dietary intake for folate and vitamin B12 is of great importance for maintaining health. Folate is also an essential metabolite for vitamin B12 and vitamin B6 biosynthesis, which is related to numerous biological functions (Romain et al. 2016). Therefore, folate and vitamin B12 are applied as nutraceutical for facilitating our health.

Glutathione and carotenoids are important antioxidant compounds those have been used for nutraceuticals. Glutathione is a tripeptide formed by condensation of L-glutamate, L-cysteine, and L-glycine (Das et al. 2012; Young et al. 2011). Carotenoids, such as lycopene and β-carotene, are a large class of natural pigments synthetized by plants and microorganisms, which belong to terpenoids (Nishizaki et al. 2007; Yoshida et al. 2009). Both glutathione and carotenoids are important antioxidants, which are used as nutraceutical for modulating cell metabolism and delaying age-related disease. Based on the increasing demand for antioxidant compounds as nutraceutical, the global market for carotenoids is estimated to be \$2.0 billion by 2022 (McWilliams 2018).

Glucosamine helps to repair cartilage or promote new cartilage formation, which led glucosamine as a supplement that is thought to help the effective repair and regeneration of damaged cartilage in human joints. It has been demonstrated that glucosamine can delay the progression of knee osteoarthritis and promoting joint health by anti-inflammatory and chondro-protective effects. Glucosamine is widely used joint health-promoting nutraceutical and also available as an over-the-counter preparation in several countries, including European countries (Liu et al. 2013).

1.5 Concluding Remarks

In this chapter, the definition of nutraceuticals are discussed with the emphasis of their health-promoting and disease-preventing functions behind their nutritional value. Next, functionality-based different kinds nutraceuticals were summarized and discussed including (1) cardiovascular disease-preventing nutraceuticals, (2) joint health-promoting nutraceuticals, (3) immunity health-promoting nutraceuticals,

(4) diabetes-preventing nutraceuticals, and (5) brain health-promoting nutraceuticals. Finally, the applications of important typical nutraceuticals are discussed, including amino acid and organic acid-based nutraceutical (gamma-aminobutyric acid and alpha-ketoglutaric acid), functional polysaccharide (hyaluronic acid), important vitamins (vitamin B12 and folate), antioxidant compounds (glutathione and carotenoids), and joint-health-promoting nutraceutical (glucosamine).

References

Anderson J, et al. Glucosamine effects in humans: a review of effects on glucose metabolism, side effects, safety considerations and efficacy. Food Chem Toxicol. 2005;43(2):187–201.

Andrew R, Izzo AA. Principles of pharmacological research of nutraceuticals. Br J Pharmacol. 2017;174(11):1177–94.

Aronson JK. Defining ‘nutraceuticals’: neither nutritious nor pharmaceutical. Br J Clin Pharmacol. 2017;83(1):8–19.

Bashir KM, Choi J-S. Clinical and physiological perspectives of β-glucans: the past, present, and future. Int J Mol Sci. 2017;18(9):1906.

BCC Research. Nutraceuticals: global markets to 2023. 2018. https://www.bccresearch.com/market-research/food-and-beverage/nutraceuticals-global-markets.html.

Benavente-Garcıa O, et al. Uses and properties of citrus flavonoids. J Agric Food Chem. 1997;45:4505–15.

Bian G, et al. Strategies for terpenoid overproduction and new terpenoid discovery. Curr Opin Biotechnol. 2017;48:234–41.

Blancquaert D, et al. Folates and folic acid: from fundamental research toward sustainable health. Crit Rev Plant Sci. 2010;29(1):14–35.

Boonstra E, et al. Neurotransmitters as food supplements: the effects of GABA on brain and behavior. Front Psychol. 2015;6:1520.

Chin RM, et al. The metabolite α-ketoglutarate extends lifespan by inhibiting ATP synthase and TOR. Nature. 2014;510:397–401.

Clegg DO, et al. Glucosamine, chondroitin sulfate, and the two in combination for painful knee osteoarthritis. N Engl J Med. 2006;354(8):795–808.

Das L, et al. Role of nutraceuticals in human health. J Food Sci Technol. 2012;49(2):173–83.

DeFelice SL. The nutraceutical revolution: its impact on food industry R&D. Trends Food Sci Technol. 1995;6(2):59–61.

Fallacara A, et al. Hyaluronic acid in the third millennium. Polymers. 2018;10(7):701.

Fang H, et al. Metabolic engineering of Escherichia coli for de novo biosynthesis of vitamin B12. Nat Commun. 2018;9(1):4917.

Glade MJ, Smith K. Phosphatidylserine and the human brain. Nutrition. 2015;31(6):781–6.

Lee YH, et al. Effect of glucosamine or chondroitin sulfate on the osteoarthritis progression: a meta-analysis. Rheumatol Int. 2010;30(3):357–63.

Liu L, et al. Microbial production of hyaluronic acid: current state, challenges, and perspectives. Microb Cell Factories. 2011;10(1):99.

Liu L, et al. Microbial production of glucosamine and N-acetylglucosamine: advances and perspectives. Appl Microbiol Biotechnol. 2013;97(14):6149–58.

Martin-Jadraque R, et al. Effectiveness of low-dose crystalline nicotinic acid in men with low high-density lipoprotein cholesterol levels. Arch Intern Med. 1996;156(10):1081–8.

McWilliams A. 2018. https://www.bccresearch.com/market-research/food-and-beverage/the-global-market-for-carotenoids-fod025f.html.

Nishizaki T, et al. Metabolic engineering of carotenoid biosynthesis in Escherichia coli by ordered gene assembly in Bacillus subtilis. Appl Environ Microbiol. 2007;73(4):1355–61.

Pandey M, et al. Nutraceuticals: new era of medicine and health. Asian J Pharm Clin Res. 2010;3(1):11–5.

(Reports TM: Global GABA (CAS 56-12-2) Market Insights, Forecast to 2025. 2018). https://www.themarketreports.com/report/global-gaba-cas-56-12-52-market-insights-forecast-to-2025.

Romain M, et al. The role of Vitamin B12 in the critically ill—a review. Anaesth Intensive Care. 2016;44(4):447–52.

Saini RK, et al. Folates: chemistry, analysis, occurrence, biofortification and bioavailability. Food Res Int. 2016;89(Pt 1):1–13.

Sawitzke AD, et al. The effect of glucosamine and/or chondroitin sulfate on the progression of knee osteoarthritis: a report from the glucosamine/chondroitin arthritis intervention trial. Arthritis Rheum. 2008;58(10):3183–91.

Weimer S, et al. D-glucosamine supplementation extends life span of nematodes and of ageing mice. Nat Commun. 2014;5:3563.

Wu N, et al. Alpha-ketoglutarate: physiological functions and applications. Biomol Ther. 2016;24(1):1–8.

Xiao J, Hogger P. Dietary polyphenols and type 2 diabetes: current insights and future perspectives. Curr Med Chem. 2015;22(1):23–38.

Yoshida K, et al. Carotenoid production in Bacillus subtilis achieved by metabolic engineering. Biotechnol Lett. 2009;31(11):1789–93.

Young D, et al. Nutraceuticals and antioxidant function. In: Functional foods, nutraceuticals, and degenerative disease prevention; 2011. p. 75–112. https://doi.org/10.1002/9780470960844.

Chapter 2
Construction of Microbial Cell Factories by Systems and Synthetic Biotechnology

Yaokang Wu, Yang Gu, Rongzhen Tian, Guocheng Du, Jian Chen, and Long Liu

It is well known that microorganisms have been used for production of diverse chemicals with the benefits of sustainability and low environmental pressure (Gu et al. 2018; Nielsen et al. 2014). However, making microorganisms into effective cell factories is still challenging due to the extensive regulation and complex interaction of intracellular metabolism, the process will take great time, effort and investment (e.g., 50–300 person-years of work and up to several hundred million US dollars) (Lee and Kim 2015; Nielsen and Keasling 2016). The emergence of system biology and system biology has accelerated the construction process of cell factories (Choi et al. 2019; Stephanopoulos 2012). This chapter will summarize how to construct the efficient microbial cell factories by systems and synthetic metabolic engineering.

2.1 Design, Construction and Optimization of Metabolic Pathways for Bio-Chemicals Synthesis

With the depletion of traditional fossil resources, how to achieve the sustainability of society, economics, and environment has attracted much interest. Production of renewable resources by microbial fermentation is an available strategy to overcome

Y. Wu · Y. Gu · R. Tian · G. Du · L. Liu (✉)
Key Laboratory of Carbohydrate Chemistry and Biotechnology, Ministry of Education, Jiangnan University, Wuxi, China

Key Laboratory of Industrial Biotechnology, Ministry of Education, Jiangnan University, Wuxi, China
e-mail: longliu@jiangnan.edu.cn

J. Chen
Key Laboratory of Industrial Biotechnology, Ministry of Education, Jiangnan University, Wuxi, China

L. Liu, J. Chen (eds.), *Systems and Synthetic Biotechnology for Production of Nutraceuticals*, https://doi.org/10.1007/978-981-15-0446-4_2

the abovementioned challenge. In this part, we emphasize the essential process of constructing microbial cell factories, including the host strain selection, synthesis pathways design, biological synthesis pathways construction, and Scaffold-guided spatial organization pathway engineering.

2.1.1 The Host Strain Selection and Synthesis Pathways Design

2.1.1.1 Host Strain Selection

Selection of a host strain is the first and most important step in constructing microbial cell factories. Till date, microorganisms of most widely used for industrial production include *Saccharomyces cerevisiae* (Besada-Lombana et al. 2018; Lian et al. 2018), *Escherichia coli* (Pontrelli et al. 2018), *Bacillus species* (Gu et al. 2017; Harwood et al. 2013), *Clostridium species* (Charubin et al. 2018), *Pseudomonas species* (Nikel and de Lorenzo 2018), *Yarrowia lipolytic* (Abdel-Mawgoud et al. 2018), *Aspergillus niger* (Tong et al. 2019), and others (Becker et al. 2018; Wang et al. 2018). Among these microorganisms, three model microorganisms, namely *E. coli*, *B. subtilis* and *S. cerevisiae*, are often chosen as the preferred host strains for metabolic engineering due to the well-developed genetic manipulation tools and their clear inherited backgrounds (Gu et al. 2018), but in general, different microorganisms have diverse endogenous metabolisms and industrial production performances (Choi et al. 2019). Therefore, in some cases, some other microorganisms may be more suitable, such as amino acids production by *Corynebacterium sp.* (Hirasawa and Shimizu 2016), lipids and fatty acid production by *Y. lipolytic* (Qiao et al. 2017), lactate production by *lactobacillus sp.* (van Tilburg et al. 2019), succinic acid production by *Mannheimia succiniciproducens* (Ahn et al. 2018), malate production by *Aspergillus oryzae* (Ding et al. 2018), citric acid production by *A. niger* (Yu et al. 2018), and et al. In addition, other aspects of microorganisms need to be considered, including the capacity of utilizing the cheap carbon feedstock, the robustness in large-scale industrial fermentations and the safety (Y. Liu et al. 2017b). Thus, selection of a host strain needs to be according to the actual applications and requirements.

2.1.1.2 Synthesis Pathways Design

Once the host strain has been determined, the potential preferred bio-chemicals synthesis pathways in the strain need to be screened from the complex intracellular metabolism. The principle of designing the target biosynthesis pathway is assurance of high energy content (λ) of the final compound as well a competent synthesis route (Dugar and Stephanopoulos 2011). Energy content (λ) is the ratio of the reductance degrees of product to substrate. For example, the reductance degrees of glucose is 24 (C6H12O6 = 4 × 6 + 12 × 1 − 6 × 2). Here, we used the synthesis of acetate as a case to elaborate the adhibition. In microorganisms, acetate could be synthetized from glucose through glycolytic pathway via 10-steps reactions (Fig. 2.1a), generating two mole of acetate for each mole of internalized glucose. Additionally, three mole of acetate could be produced with the equivalent glucose by non-glycolytic pathway via 4-steps reactions (Fig. 2.1b). In this scenario, acetate synthesis in non-glycolytic pathway has higher energy content [$\lambda = (2 \times 4 + 4 - 2 \times 2) \times 3/(4 \times 6 + 12 \times 1 - 6 \times 2) = 1$] than that ($\lambda = (2 \times 4 + 4 - 2 \times 2) \times 2/(4 \times 6 + 12 \times 1 - 6 \times 2) = 0.667$) in glycolytic pathway. Here, to facilitate the assessment, we introduce an efficient and simple method by analyzing the reducing equivalents of synthesis pathways. Because cell metabolism needs to maintain redox neutrality, excess generated reducing equivalents, such as NAD(P)H and $FADH_2$, either participate in the synthesis of NAD(P)H-dependent byproducts or are oxidized to increase biomass (Dugar and Stephanopoulos 2011; Gu et al. 2019). As a result, synthesis pathways with generating excess reducing equivalents will lead to a low yield of the desired compound. For the convenience of readers to understand, we deduced the stoichiometry of glucose conversion to acetate of different pathways in the above-mentioned scenario. The stoichiometry of glucose conversion to acetate by glycolytic pathway is glucose + $4NAD^+$ + 4ADP → 4NADH + 2CO2 + 2acetate + 4ATP + 2H2O, and the stoichiometry of glucose conversion to acetate by non-glycolytic pathway is glucose + 2ADP → 3acetate + 2ATP. Obviously, the reducing equivalent

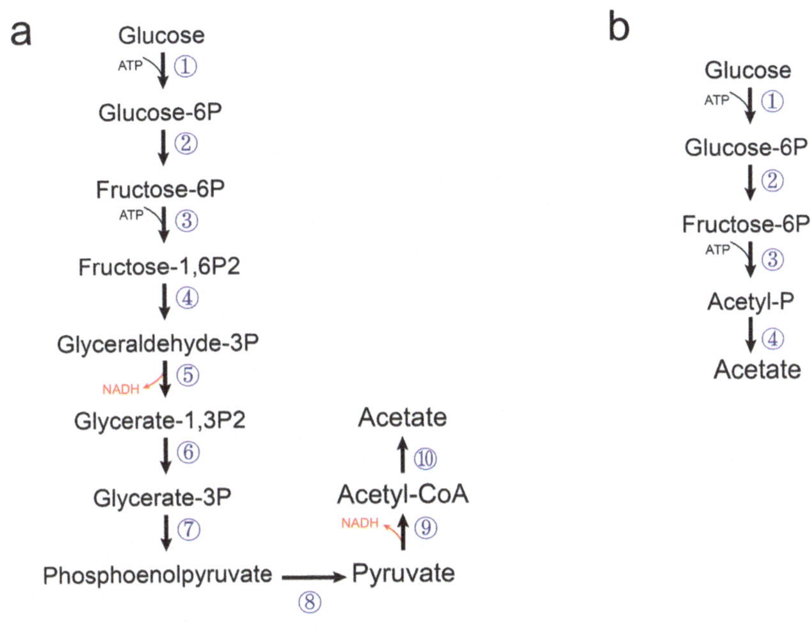

Fig. 2.1 The metabolic synthesis pathways of acetate from glucose in microorganisms

of glycolytic pathway producing acetate is higher than non-glycolytic pathway, which is consistent with our conclusion.

However, in some cases, there are no known pathways or enzymes in the host strain that can synthesize desired non-natural chemicals, such as pharmaceuticals and fuels (Biz et al. 2019). Literature mining is an effective method to explore potential synthetic pathways, because the forefront sources of newely-developed pathways and biology research have been recorded in the journal articles (Chen et al. 2017). Additionally, even though no enzyme is found for directly catalyzing the desired reaction, some ones that catalyze analogous reactions could be the candidates (Biz et al. 2019). With the advances in protein engineering and developments of synthetic biology, designing pathways with reactions not known in nature hitherto and elaborate control is now possible.

2.1.2 Construction of Biological Synthesis Pathways

The expressions of metabolism pathway genes are usual achieved by vector plasmids or genome integration. The methods of genome integration include site-specific recombination (SSR) systems, counter-selectable markers systems, and ultramodern CRISPR/Cas systems, which will be elaborated in the 3.3 part (Li et al. 2019). Here, we introduced multi-fragment DNA assembly methods used in constructing vector plasmids, such as Gibson assembly, Golden Gate assembly, BioBrick assembly, single strand assembly, ligase cycling reaction, uracil specific excision reagent (USER) cloning, and transformation-associated recombination (TAR) cloning. The commonly used methods of multi-fragment DNA assembly are Golden Gate assembly and Gibson assembly, therefore, we mainly focus on these two methods.

2.1.2.1 Golden Gate Assembly

Golden Gate assembly (Engler et al. 2008) belonging to restriction enzymes-based methods are established in the application of type IIS restriction enzymes and DNA ligases, which can successfully join multi-fragment DHA parts without the enzyme recognition sequences in the final constructed plasmids. Type IIS restriction enzymes, commonly used including *Bsa*I, *Bsm*BI, and *Bbs*I, are different from traditional type II restriction enzymes in the cleaving site, which cleave outside of the recognition sequences and generate four base flanking overhangs. Since the cleaving site is not part of the recognition sequence, it could obtain 256 potential different four base flanking overhangs theoretically. So, they can be customized to assembly of multiple DNA fragments with different flanking overhangs sequences by several thermal circulations between 37 °C (optimal for restriction enzymes) and 16 °C (optimal for DNA ligases). The detail cloning process has been shown in Fig. 2.2a with three DNA fragments assembly as an example.

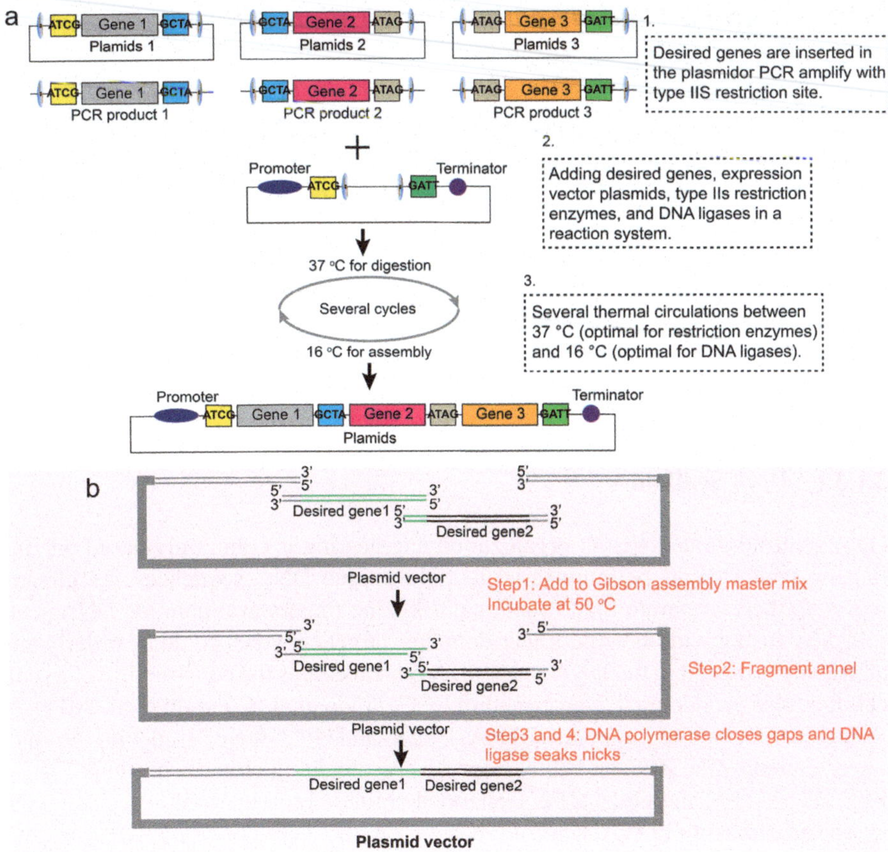

Fig. 2.2 The process of Golden Gate assembly and Gibson assembly

2.1.2.2 Gibson Assembly

Gibson assembly was first reported in 2009 (Gibson et al. 2008), which could simultaneously assemble up to 15 DNA fragments. In this method, DNA fragments to be assembled need 13–40 base pair overlap with adjacent DNA fragments. Additionally, three enzymes are required, including exonuclease, Taq DNA polymerase, and Taq DHA ligase. The process of Gibson assembly mainly includes four steps: (i) exonuclease chews back DNA from the 5′ end; (ii) generated single stranded regions on adjacent DNA fragments anneal; (iii) DNA polymerase recovers the gaps with nucleotides; (iv) DNA ligase covalently joins the DNA of adjacent segments and removes the nicks. The detail cloning process has been shown in Fig. 2.2b with two DNA fragments assembly as an example.

2.1.3 Scaffold-Guided Spatial Organization Pathway Engineering

Scaffold-guided spatial organization engineering could achieve the colocalization of pathway enzymes by structural scaffolds, such as DNA scaffolds (Conrado et al. 2012), RNA scaffolds (Delebecque et al. 2011) and protein scaffolds (Dueber et al. 2009). The potential benefits of scaffold-guided spatial organization engineering include: (i) convenient for optimizing the rations of pathway enzymes to improve the production efficiency; (ii) relieving the toxicity of excess intermediates for cell metabolism, and (iii) constructing artificial complexes of pathway enzymes and forming the substrate channel, and (iv) promoting the smooth progress of synthetic reactions.

2.1.3.1 DNA Scaffold

DNA scaffold-guided spatial organization engineering is commonly based on zinc fingers, which could be engineered to bind unique DNA sequences. As already reported, there are more than functional 700 zinc fingers available for DNA scaffolds. By fusing with the zinc-finger domains, target enzymes could be orderly and adjustably arranged in the DNA scaffold. A typical case is the construction of E. coli cell factories producing 1,2-propanediol by Conrado et al (Conrado et al. 2012). in their research, pathway enzymes MgsA, DkgA, and GldA were fused with the zinc-finger domain ZFa, ZFb, and ZFc, respectively, obtaining chimera proteins MgsA-ZFa, DkgA-ZFb, and GldA-ZFc. Generated chimera proteins could precisely bind to specific plasmid DNA scaffold basing on the corresponding ZF domains. Furthermore, optimizing the rations of various enzymes leaded a 4.5-fold higher titer of 1,2-propanediol than that without scaffold control. In general, DNA scaffolds have the stable, robust and configurable characteristics, thus, it is an alternative and effective method for constructing artificial complexes of pathway enzymes.

2.1.3.2 RNA Scaffold

RNA scaffolds are synthetic engineered non-coding RNA molecules, which could specifically recruit pathway enzymes in vivo. Different from DNA and protein scaffolds, RNA scaffolds could form complex multi-dimensional functional architectures due to 3D folding of RNA. In 2011, Delebecque et al. firstly described a protocol of design, expression and characteration of RNA scaffolds and the cognate proteins. The RNA scaffolds described in their research recruited pathway enzymes by harboring RNA aptamers as protein docking sites (Delebecque et al. 2011). Furthermore, this RNA scaffold could assemble into functional discrete, one-, and two-dimensional structures. And, they used this RNA scaffold to controll the spatial organization of the hydrpgen-producing pathway enzymes, which leaded to a remarkable increase in hydrogen output. This result indicated that RNA scaffolds are effective platforms that can be used in metabolic engineering and synthetic biology for increasing products yield and productivities.

2.1.3.3 Protein Scaffold

Protein scaffolds specifically recruit pathway enzymes by the interactions of cognate peptide ligands. Dueber et al. used protein-protein interaction ligands from metazoan cells to design protein scafolds for optimizing a three-enzyme pathway of mevalonate production from acetyl-CoA (Dueber et al. 2009), resulting in a 77-fold improvement in titer. Additionally, the same scaffold was applied in the synthesis of glucaric acid, which also got a significant outcome and the titer of glucaric acid reached 0.5 g/L. Generally, scaffold-guided spatial organization pathway engineering is an available engineering strategy for improving productivity and yield of desired products and relieving the toxicity of excess intermediates.

2.2 Optimization and Regulation of Metabolic Network

After a metabolic pathway was introduced into a microbial cell factory, further optimization and regulation will be needed to enhance synthetic efficiency, avoid metabolic burden, and keep balance of the metabolic network. For this purpose, modular metabolic engineering and dynamic regulation were often used.

2.2.1 Modular Metabolic Engineering

There are a lot of genes that need to be engineered in the process of building a cell factory. The multiple rounds-single gene modification method is time-consuming, and the interaction of different manipulation may be ignored by this way. However, the combination of each modification of each gene will produce too many phenotype spaces to test when an efficient high-throughput screening assay is lacking. Hence, modular metabolic engineering is often used in which genes was grouping together into modules based on the metabolic branch point or the properties of these enzymes (Biggs et al. 2014), and multiple modules with various expression levels can be assembled in a plug and play way to generate an appropriate landscape for the search of the optimal phenotype (Lu et al. 2019). The expression levels of different modules can be adjusted by change their copy number or transcription and translation strength. In addition, many synthetic biology tools were also developed and used for modular expression regulation.

2.2.1.1 Copy Number

The plasmids with different replication origin process corresponding copy numbers, and they are compatible in the same strain. Placing the different modules separately into different plasmids make it easy to change their expression levels and combine them together. For example, the fatty acids synthetic network was optimized in *E. coli* by expression the upstream GLY module (providing acetyl-CoA), the intermediary ACA module (converting acetyl-CoA to malonyl-ACP) and downstream FAS module (synthesizing fatty acid using malonyl-ACP) either on (h) high copy number plasmids (pETM6 or pRSM3), (m) medium copy number plasmid (pCDM4) or (l) low copy number plasmids (pACM4 or pCOM4). The result shows that expressing these three modules on the medium, low, and high copy number plasmid (mGLY-lACA-hFAS) respectively prevented the excessive accumulation of toxic intermediates acetyl-CoA/malonyl-ACP and resulted a high titer of fatty acid (Xu et al. 2013). This strategy was also used in the study of producing (2S)-pinocembrin from glucose, in which the (2S)-pinocembrin synthetic network was divided into four modular and were expression on the plasmids possessed copy numbers of 10, 20, 40, and 100, respectively (Wu et al. 2013).

However, this method is currently mainly used in *E. coli* because the lack of the compatible plasmids in the other strains. Besides, the metabolic burden of the high copy numbers plasmid is also an important factor to consider (Rozkov et al. 2004).

2.2.1.2 Transcription and Translation Strength

Except for change the copy number of different modular, change their transcription strength using different promoter is also a commonly used way to modulate their expression (Jones et al. 2015), and that is often combined with the change of copy number. For example, plasmids pSC101, p15A, and pBR322 were combined with promoters Trc, T5, and T7 in a taxadiene-producing *E. coli* cell factory to balance the upstream and downstream module so as to maximize the production with minimal accumulation of the toxic intermediate metabolite indole (Ajikumar et al. 2010). It is worth emphasizing that the strength of promoters could cover a larger span compared with the copy number of different plasmids, which offered more flexibility in the expression of modular. In addition, the synthetic promoter libraries (SPL), which have been successfully built for many organisms including *E. coli* (Alper et al. 2005), *B. subtilis* (Liu et al. 2018), *C. glutamicum* (Yim et al. 2013), *S. cerevisiae* (Redden and Alper 2015) and so on, also provided worthy toolboxes for the implement of this method.

In addition to changing copy number and promoter, RBS (ribosome binding sits) was also modulated to change the modular strength at translation level (Xu et al. 2013; Zelcbuch et al. 2013), and a tool called RBS calculator was often used to design RBSs as needed (Salis et al. 2009). For example, by choosing the most appropriate RBS combination of lower mevalonate pathway in an amorphadiene-producing strain, the production improved approximately fivefold with a large decrease in the accumulation of toxic metabolic intermediates (Nowroozi et al. 2014).

2.2.1.3 Synthetic Tools for Modular Regulation

Many synthetic biology tools like sRNA, RNAi, and CRISPRi were also developed and applied in the regulation of the strength of different modular (Man et al. 2011). The trans-acting sRNAs (small noncoding RNAs) that play an important role in the regulation of translation have been founded in many organisms (Gottesman 2004). Synthetic sRNAs could be designed rationally to modulate gene expression without the direct modification of chromosomal sequences modular control and is suitable for the regulation of the modular composed by the native pathways (Na et al. 2013; Yang et al. 2019). For instance, regulating glycolysis and peptidoglycan synthesis modules by sRNAs balanced the GlcNAc synthesis pathway and the primary metabolism for cell growth and resulted in a 3.3-fold improvement of GlcNAc yield on cell in an engineered *B. subtills* strain (Liu et al. 2014).

In eukaryote, RNA interference (RNAi) is often used to knockdown gene expression by the RNA-induced silencing complex (RISC) (Tomari and Zamore 2005). Recently, the RNAi system were restored in *S. cerevisiae* that lack of a functional RNAi pathway by expressing Dicer and Argonaute from *Saccharomyces castellii* (Crook et al. 2014), which provided an effective tool for modular regulation for this strain as sRNA did in *E. coli*.

In addition to synthetic sRNA and RNAi, the Clustered Regularly Interspaced Short Palindromic Repeats (CRISPR) interference (CRISPRi) systems were also developed both in prokaryote and eukaryote to repress gene expression using an inactive Cas9 (dCas9) protein, which can bind to a target gene and block the elongation of RNA polymerase (Gilbert et al. 2013, 2014). This system was used to regulate the three main competitive modular of production in a GlcNAc-producing *B. subtilis* strain, and 103.1 g/L GlcNAc was abstained in a 3-L fed-batch bioreactor after fine tuning the strength of these modular by using different sgRNAs (Wu et al. 2018).

2.2.2 *Dynamic Regulation*

The native metabolic network in a cell is keeping dynamic equilibrium through a series of complexity regulation mechanisms (Bervoets and Charlier 2019). The introduction of uncontrolled exogenous pathways certainly will break this balance, and lead to intermediate accumulation and cell viability impairment. Therefore, dynamic regulation has emerged as a promising strategy to solve this problem in which these exogenous pathways can be controlled with the help of the regulation mechanisms from the cell at transcription, post-transcription or protein level (Table 2.1) (Lalwani et al. 2018; Xu 2018).

2.2.2.1 Transcription Level

The control of gene expression at transcription level by dynamic promoters is often used by microorganisms, and the regulation of transcription, which is the first step of gene expression, is the most economical and direct way (Kochanowski et al. 2013). Induced promoters have been developed in many strains such as *E. coli, B. subtilis, S. cerevisiae,* and so on based on the repression or activation mechanism of transcription factor to promoter, and the switch of metabolic fluxes between cell growth and product synthetic can be achieved by constructing the open loop gene circuit using these promoters (Liu and Zhang 2018). For instant, the redirection of metabolic flux from TCA circle to isopropanol synthesis pathway was achieved by building a toggle switch using the lactose and tetracycline induced promoters (Soma et al. 2014).

Except for the open loop dynamic control of the metabolic network using the induced promoters, the natural ligand-responsive transcription factors (LRTFs) which can change gene expression by sensing the concentration of related Intracellular or extracellular metabolite were also used for dynamic regulation in a feedback way (Cress et al. 2015). The first example is that the use of the ntr regulon to redirect the excess glycolytic flux into lycopene biosynthesis pathway by sensing the intracellular overflow product acetyl phosphate, and applying this strategy significantly enhanced lycopene production while reducing the negative impact caused by metabolic imbalance (Farmer and Liao 2000). Since then, there have been many success cases in the construction of efficient microbial cell factories using this method (Dahl et al. 2013; Xu et al. 2014; Yang et al. 2018; Zhang et al. 2012). However, the availability of correlative LRTFs limited the application of this strategy in many metabolic pathways, but the development of synthetic biology makes it possible to redesign new LRTFs we needed by a rational or irrational way (Snoek et al. 2019; Tang and Cirino 2011).

Quorum sensing system was used by bacteria to sense their local population numbers and coordinate their behavior, which was achieved through the control of relate genes expression by some cell populations coupled signal molecules (Miller and Bassler 2001). For example, the QS system in *Vibrio fischeri* composed with acylhomoserine lactone (AHL) synthetase LuxI, AHL-dependent transcription activator LuxR, and a lux promote P_{lux}. When extracellular AHL reaches the threshold concentration caused by the plenty of quorum population, promoter P_{lux} will be activated by the AHL-LuxR complex. And this mechanism has been used to decouple cell growth and product synthesis dynamically. A typical case is the application

Table 2.1 Applications of dynamic regulation in microbial cell factories

Level	Regulation element	Inducer	Strain	Product	Ref.
Transcription	P_{suc2}	Sucrose	*S. cerevisiae*	GFP	Williams et al. (2015b)
	P_{lac}, P_{tet}	IPTG	*E. coli*	Isopropanol	Soma et al. (2014)
	Ntr regulon	Acetyl phosphate	*E. coli*	Lycopene	Farmer and Liao (2000)
	TF[①] FadR	Acyl-CoA	*E. coli*	FAEE	Zhang et al. (2012)
	TF FapR	Malonyl-CoA	*E. coli*	Fatty acids	Xu et al. (2014)
	TF CatR	Muconic acid	*E. coli*	Muconic acid	Yang et al. (2018)
	TF IpsA	Myo-inositol	*E. coli*	Glucaric acid	Doong et al. (2018)
	Stress-response promoters	Farnesyl pyrophosphate	*E. coli*	Amorphadiene	Dahl et al. (2013)
	lux QS system	High cell density	*E. coli*	Isopropanol	Soma and Hanai (2015)
	Esa QS system	High cell density	*E. coli*	Myo-inositol, Glucaric acid	Gupta et al. (2017)
	Esa QS system	High cell density	*E. coli*	Glucaric acid	Doong et al. (2018)
	P_L, P_R	Temperature	*E. coli*	D-lactate	Zhou et al. (2012)
	P_2, P_7	Temperature	*B. subtilis*	β-galactosidase	Li et al. (2007)
	P_{ohrB}	Environmental stresses	*B. subtilis*	Xylanase	Panahi et al. (2014)
	Lysine riboswitch	L-lysine	*C. glutamicum*	L-lysine	Zhou and Zeng (2015a)

(continued)

Table 2.1 (continued)

Level	Regulation element	Inducer	Strain	Product	Ref.
Post-transcription	Lysine riboswitch	L-lysine	*C. glutamicum*	L-lysine	Zhou and Zeng (Zhou and Zeng 2015a, b)
	glmS riboswitch	GlcN6P	*B. suntilis*	GlcNAc	Niu et al. (2018)
	sRNAs	Anhydrotetracycline	*E. coli*	gluconate	Solomon et al. (2012)
	sRNAs	IPTG	*E. coli*	Malonyl-CoA	Yang et al. (2015)
	sRNAs	None	*E. coli*	Tyrosine, Cadaverine	Na et al. (2013)
	sRNAs	Theophylline	*E. coli*	GFP	Qi et al. (2012)
	RNAi	High cell density	*S. cerevisiae*	PHBA	Williams et al. (2015a)
	RNAi	Sucrose	*S. cerevisiae*	GFP	Williams et al. (2015b)
	CRISPR	Anhydrotetracycline	*E. coli*	GFP, Mevalonate	Li et al. (2016)
	CRISPR	Lactose	*S. cerevisiae*	DV, V, PDV, PV	Zalatan et al. (2015)
	CRISPR	Xylose	*B. subtilis*	GlcNAc	Wu et al. (2018)
	tsRC9②	Temperature	*E. coli*	None	Richter et al. (2017)
Protein	HSD-S1	L-lysine	*C. glutamicum*	L-lysine	Chen et al. (2015)
	tmRNA②system	Anhydrotetracycline	*E. coli*	Myo-inositol	Brockman and Prather (2015a)
	tmRNA system	IPTG	*E. coli*	Medium-chain fatty acids	Torella et al. (2013)
	tmRNA system	Arabinose	*E. coli*	GFP, RFP	Cameron and Collins (2014)

① Transcript factor ② Transfer-messenger RNA

of the lux system from *Vibrio fischeri* in an isopropanol-producing strain, in which a cell density controlled metabolic toggle switch (MTS) was constructed to redirect metabolic flux from central metabolic pathways toward product synthetic pathway resulting a 3-fold and 2.3-fold improvement in the titer and yield of isopropanol, respectively (Soma and Hanai 2015). This dynamic regulation method is easier to adapted into different microbial cell factories compared with the LRTFs media manner because it is pathway-independent (Gupta et al. 2017).

2.2.2.2 Post-transcription Level

Control gene expression by the post-transcription regulation on mRNA is also a common way for many metabolic pathways in microbial (Oliva et al. 2015), where riboswitches that exited at the 5′ UTR of the mRNA and can affect downstream gene expression by conformations change in response to environmental signals may be the simplest one (Waters and Storz 2009; Winkler and Breaker 2005). For example, the *lysC* riboswitch of E. coli could inhibit gene expression by sheltering RBS and exposing RNase E cleavage sites in mRNA after sensing the high concentration lysine in the cell (Caron et al. 2012). And with the help of this mechanism, the metabolic flux was dynamically redirected from TCA into the synthetic pathway of L-lysine in *C. glutamicum* (Zhou and Zeng 2015a; b). Furthermore, it is easy to adapt various riboswitches into different species on account of their no protein required property (Serganov and Patel 2007). In addition to riboswitch, other regulation manner such as sRNA, RNAi, and CRISPRi talked above can also be applied for dynamic regulation at this level by controlling their expression using the induced or LRTFs-coupled promoter (Wu et al. 2018).

2.2.2.3 Protein Level

At protein level, the regulation on a pathway could be implemented by the change of activity or concentration of the pathway enzyme. There are many allosteric proteins in the metabolic pathways of microbial, which possess the active site and the allosteric site. The binding of the intracellular effector to the allosteric site will lead to reversible changes in the conformation of the active site, thus enhancing or reducing the catalytic activity of the enzyme (Motlagh et al. 2014). With that mechanism dynamic regulation can be achieved by the control of the activity of key enzyme in the metabolic network. For example, an artificial allosteric enzyme HSD-S1 that could responds to lysine inhibition was constructed by engineering the threonine binding site of *C. glutamicum* homoserine dehydrogenase (HSDH) into a lysine binding pock. And it was successfully applied in the dynamic control of growth-related byproduct formation pathway in a lysine-producing *C. glutamicum* cell factory (Chen et al. 2015).

Except for the control of activity of the enzyme, dynamic regulation at this level can also be achieved by the change of enzyme concentration. In bacteria, the nascent polypeptide chain will be removed by the transfer-messenger RNA (tmRNA) system though the SsrA peptide tag added into its C-terminus (Janssen and Hayes 2012). Based on this principle, the degradation rate can be adjusted by change the SsrA tag fused to the enzyme. For instance, the degradation of a key glycolytic enzyme phosphofructokinase-I (PFK-I) was controlled to redirect the metabolic flux from glycolysis into myo-inositol synthetic pathway resulting a two-fold improvement in yield and titers (Brockman and Prather 2015a). Protein degradation regulation is more sufficient than transcriptional or post-transcriptional regulation, because the expressed protein will remain stable for some time when transcription or translation was repressed.

2.3 Application of Systems Biology

Systems biology explains system-level cellular phenomena by using a wide range of experimental data and integrating a variety of high-throughput techniques and computational methods. In the construction of microbial cell factories, the most notable systems biology tools and strategies are the analysis of omics data and the simulation of metabolic models (Bordbar et al. 2014a; Chae et al. 2017; Choi et al. 2019; Dai and Nielsen 2015; Lee et al. 2012; Nielsen and Jewett 2008). Although it is achievable to develop microbial cell factories that can produce chemicals and biomaterials only through traditional metabolic engineering methods, there are still many challenges in obtaining industrially competitive strains (Lee et al. 2011; Ling et al. 2014; Nielsen et al. 2013; Peralta-Yahya et al. 2012). The complex metabolic pathways of cells and their associated regulatory networks are a major challenge (Zhu et al. 2012). Altering metabolic fluxes or introducing heterologous metabolic pathways in host strains often leads to metabolic competition, imbalances and inhibition (Biggs et al. 2014; Lee et al. 2012). In order to make a microbial cell factory more efficient, it usually takes a lot of time and cost. In order to overcome these obstacles, it is necessary to systematically understand the cellular metabolism and physiological characteristics of the cell factory. The biggest advantage of systems biology is that it can decode the details of microorganisms, and provide the required genome-scale targets for metabolic engineering more accurately and efficiently in a systematic and global way, which can speed up the construction of microbial cell factories (Lee et al. 2012; Nielsen and Jewett 2008). In addition, systems biology can help identify mutation targets for strains obtained by mutagenesis and screening or adaptive evolution, making the construction of cell factories more rational (Caspeta et al. 2014; Hong et al. 2011). At the same time, with a new generation of efficient genomic engineering tools, including oligonucleotide-mediated gene editing tools and CRISPR-based gene editing tools, systems biology is playing an increasingly important role in building efficient microbial cell factories.

2.3.1 Omics

Since the birth of genome and genomics, high-throughput analysis methods represented by genomics, transcriptomics, proteomics and metabolomics have been rapidly developed (Chen et al. 2017). Flux histology is also gaining more and more attention in the construction of cell factories, because optimizing flux distribution to produce products is the ultimate goal of metabolic engineering (Biz et al. 2019). The development of omics data has enabled the provision of valuable systematic information that comprehensively describes almost all components within a cell under a variety of genotypes and environmental conditions (Chae et al. 2017). In the process of building a cell factory, researchers can not only use a single omics information to purposefully solve problems in metabolic engineering, but also apply multi-omics methods to compensate for disadvantages of each omics by analyzing various omics data (Chae et al. 2017). The current technological developments make group data more and more accessible, allowing not only the production of specialized omics data through high-throughput experimental techniques, but also a large amount of omics data through publicly accessible databases, which makes its application prospects more and more extensive in the construction process of cell factories (Biz et al. 2019).

2.3.1.1 Identification of Enzymes and Metabolic Pathways

Due to the enormous advantages of microbial cell factories producing chemicals, more and more chemicals are expected to be biosynthesized, but the unknown metabolic pathways and the lack of key enzymes hinder the development of synthetic biology. In nature, after long-term evolution and selection, there are numerous synthetic pathways and efficient enzymes for the production of specific chemicals (Nielsen and Keasling 2016). However, in the traditional cell factory construction process, these huge and valuable natural resources are often not effectively exploited and applied. With the development of omics, especially the development of genomics and transcriptomics based on next-generation sequencing technology, more and more potentially valuable enzymes and metabolic pathways have been identified and has been successfully applied to the construction of microbial cell factories (Chae et al. 2017; Dai and Nielsen 2015; Keasling et al. 2016).

For example, by analyzing the transcriptome of glandular trichomes from female cannabis flowers, a major component of cannabinoid biosynthesis, olivetolic acid (OA) cyclase that catalyzes a C2–C7 intramolecular aldol condensation with carboxylate retention to form OA was found. The identification of OAC both cleaves the cannabinoid pathway and shows the evolutionary similarity between polyketide biosynthesis in plants and bacteria (Gagne et al. 2012). This has played an important role in metabolic engineering using microbial cell factories to produce cannabinoids and to develop treatments for various human health problems. There are also many important metabolic pathways for pharmaceutical compounds that are also obtained by mining omics data. The saponin biosynthetic genes was identified though monitored the expression of 18,695 transcript tags over on roots of methyl jasmonate (MeJA)-treated *Bupleurum falcatum* plants. After isolated and direct sequenced of 1,771 MeJA-responsive tags, CYP716Y1 was finally confirmed to be involved in the biosynthesis of triterpene saponin (Miller et al. 2008). In addition, six potential cytochrome P450 genes that convert miltiradiene to tanshinones (bioactive compounds from Chinese medicinal herb danshen) was fund through transcriptome analysis; Using *E. coli* for efficient biosynthesis of l-valine based on transcriptome analysis; The pathway of yeast resistance to high concentrations of ethanol was determined based on genomics, transcriptomics and metabolomics analyses (Caspeta et al. 2014; Guo et al. 2013; Park et al. 2007).

2.3.1.2 Optimization of Metabolic Pathways

The construction of a microbial cell factory that produces target chemicals is only the first step. The next step in optimizing the metabolic flux is very important, which determines whether the efficiency of the microbial cell factory is sufficient to compete with chemical synthesis. By optimizing the flux of metabolic pathways, the productivity and yield of microbial cell factories can be increased, making them both environmentally friendly and economically viable. Traditional metabolic engineering methods can make microbial factories to a satisfactory level, but often spend more time and costs. Therefore, it is particularly important to search and identify key points in metabolic pathways more accurately and predictably (Woolston et al. 2013). In other words, the key point lays on how to diagnose and find out where the problem lies in the product synthesis pathway. Based on traditional metabolic engineering methods, the easy solution is to use precursors to determine production bottlenecks. Examples include increasing the production of recombinant proteins in *Bacillus megaterium* (Korneli et al. 2012), Optimizing biodiesel production in marine Chlamydomonas sp. JSC4 (Ho et al. 2014). However, it is clear that this cannot meet the requirements of efficient genetic modification methods, and it is necessary to identify bottlenecks more rationally and accurately. In metabolic engineering, one solution is to diagnose and identify bottlenecks in the synthetic pathway through omics-type data (Liu et al. 2016). The omics data is a powerful tool for revealing metabolic flux problems, especially metabolomics data. The initial application of these omics data was mainly focused on steady-state analysis. Examples include rationally identifying new targets to improve riboflavin production; assessments and engineering microbial isopentenol production by analyzing proteomics and metabolomics; determining cellular physiology and behavior by combining multi-omics data (Brockman and Prather 2015b; George et al. 2014; Shi et al. 2009). With the rapid development of high-throughput metabolomics, measurement of dynamic metabolomics is becoming more and more available, and dynamic metabolomics analysis is becoming more and more important for target pathway scanning (Fuhrer and Zamboni 2015; Sévin et al. 2015; Zampieri et al. 2017). There are examples including the use of metabolic real-time analysis to rationally and rapidly derived an optimal blueprint to produce dihydroxyacetone phosphate (DHAP) and a real-time metabolome analysis of metabolic turnover between starvation and growth to increase substrate availability (Bujara et al. 2011; Link et al. 2015).

2.3.1.3 Analysis of the Genome

It is a very common method to obtain microbial cell factories by mutagenesis and screening among the conventional microbial cell factory construction methods (Brockman and Prather 2015b; Chatterjee and Walker 2017). Although the screening process requires a lot of time and labor costs, it is possible to obtain industrially superior strains. These strains often have mutations at key sites on the genome, but due to limitations in sequencing technology, changes in the genome are often masked. This not only hinders the analysis of important metabolic engineering mechanisms, but also fails to integrate dominant mutation sites to obtain more excellent strains. In addition, many bacteria that are potentially microbial cell factories may have different environmental tolerances and metabolic characteristics, and analysis of their genomes can also help us integrate the advantages of different strains. The situation begins to change due to the development of next-generation sequencing technology, sequencing of whole genomes has become cost-effective and efficient (van Dijk et al. 2018). At the same time, a variety of computational tools such as genomic alignment tools, coding sequence prediction tools, and gene function annotation tools have also developed rapidly (Clark et al. 2010; Kanehisa 2006), making our analysis of the dominant characteristics of target cells more and more available (Ashburner et al. 2000; Delcher 2002; Jones et al. 2014; Lagesen et al. 2007; Tatusov et al. 2000).

For the analysis of the genome, it has now developed into a specialized discipline, namely comparative genomics (Haft 2015). Mutations in the key genes *argF*, *argB*, *carB* were found by aligning the sequences of the genes of the L-arginine biosynthetic pathway of the AR0, AR1, AR2, AR3 and AR4 strains. By integrating the dominant genes, not only the cell growth rate is increased, but also the L-arginine titer is increased to 82 g/L with a yield of 0.35 g/g (Mattozzi et al. 2018). In addition, Mucor circinelloides pathogenesis was identified by comparative genomics (López-Fernández et al. 2018); a new strain that can eliminate of oil contamination from polluted environments (B. subtilis MJ01) was isolated from oil-contaminated soil based on comparative genomic analysis and genome annotation (Rahimi et al. 2018); 24 sequenced Bacillus velezensis strains were identified with the potential to degrade lignocellulose based on comparative genomic analysis and genome annotation (Chen et al. 2018); the microbial response of cells to acid stress during fermentation was revealed and the propionic acid yield increased to 10.31 ± 0.84 g/g DCW based on comparative genomic and transcriptomic analyses were conducted on wild-type and acid-tolerant Propionibacterium acidipropionici (Guan et al. 2018).

In addition to the applications described above, the largest application of multi-omics data is that it can be combined with a variety of computational modeling methods. The constructed genome-scale metabolic model can be used not only for the diagnosis and optimization of metabolic pathways, but also to efficiently predict metabolic pathways and find potential hosts, which will be described in detail in the next section.

2.3.2 Application of Genomics-Based Metabolic Model

In the construction of microbial cell factories, a variety of local and genome scale metabolic models have been developed for the need for systematic understanding of cells (Karr et al. 2012). These computational and simulation tools have become powerful tools for system-wide analysis and prediction of cellular metabolism and function. GEM has been developed for a variety of metabolic engineering strains, including *E. coli*, *B. subtilis*, *C. glutamicum*, *Ralstonia eutropha*, *C. acetobutyricum*, *M. succiniciproducens*, *S. cerevisiae*, etc (Kim et al. 2014; T. Y. Kim et al. 2012b)[54,55]. At the same time, a variety of simulation methods and computational methods have been used to construct scale metabolic models, including models based on kinetics and thermodynamics, models based on constraint-based flux analysis, and stoichiometric models (Liu et al. 2016; Stalidzans et al. 2018).

Although various algorithms have been developed to simulate the metabolic network of cells, due to the lack of large-scale data containing all components in the cell, these models can only be used to characterize intracellular metabolites but not enough to accurately simulate and predict the metabolic flux of cells. Simultaneously, as the number of omics-type data continues to increase, we need a better method to integrate these databases to understand complex dynamic processes which allows us to better solve the problems raised in metabolic engineering (Hao et al. 2018; O'Brien et al. 2015; Srinivasan et al. 2015), including diagnosis and identification of bottlenecks in metabolic pathways, prediction of the target metabolic pathway, screening for potential hosts, etc (Link et al. 2013). Therefore, the most promising approach is to combine computational models with metabolomics data (Bujara et al. 2011; Costa et al. 2015; Link et al. 2014). Currently, various tools for integrating genomic, transcriptomic, proteomic and metabolomic data with GEMs have been extensively developed, including gene inactivity moderated by metabolism and expression (GIMME), integrative metabolic analysis tool (iMAT), E-Flux, E-Flux2, probabilistic regulation of metabolism (PROM), transcriptomics-based strain optimization tool (tSOT), and gene inactivity moderated by metabolism and expression by proteome (GIMMEp) (Chae et al. 2017; Kim et al. 2014, 2016). The combination of GEM and multi-omics provide an important tool for us to better understand the complex processes in cellular metabolism and predict the effects of genetic alterations on cells, and can be used to accelerate the construction of cell factories (Chew et al. 2018; Hao et al. 2018). Most of the related algorithms have been given in detail, and multiple websites and software have been developed to further broaden the scope of GEM applications (Kim et al. 2014; O'Brien et al. 2015). Here we will present a series of examples detailing the contribution these models have made in the construction of microbial cell factories.

2.3.2.1 Prediction of Metabolic Ppathways

The analysis and selection of metabolic pathways is very important in the construction of cell factories. Building an efficient microbial cell factory requires not only enzymes that catalyze the biochemical reactions of each step, but also better metabolic properties of the target metabolic pathway, such as using as few reaction steps as possible, balanced cofactors, thermodynamic feasibility, and no allosteric regulatory enzymes, etc. (Bordbar et al. 2014b; Campodonico et al. 2014; McClymont and Soyer 2013; Pharkya 2004; Smanski et al. 2014; L. Wang et al. 2017b). If a complete or even optimal metabolic pathway can be predicted before the construction of the microbial cell factory, the time and labor required for the experimental screening will be greatly reduced. Therefore, many genome-scale pathway prediction tools have been developed to predict unknown biosynthetic pathways of various natural chemicals and to scan databases to screen for optimal metabolic pathways, including biochemical network integrated computational explorer (BNICE), RetroPath, GEM-Path, OptStrain and DESHARKY (Kim et al. 2014).

Because of limitations in computational methods and exponentially increasing genomic data, large-scale mining of RiPP data usually be very difficult. After combining hidden-Markov-model-based analysis, heuristic scoring, and machine learning, RODEO (Rapid ORF Description and Evaluation Online) was developed to identify biosynthetic gene clusters and predict RiPP precursor peptides, revealed more than 1,300 compounds (Tietz et al. 2017). This kind of genome-scale model provided a framework for future genome-mining efforts. Another successful examples the 1,4-Butanediol production in *E. coli*. Under the guide of a genome-scale metabolic model, 5 heterologous enzymes out of the 7 step pathway was screened from different microorganisms, followed by an increase in 1,4-Butanediol titer by over three orders of magnitude, to nearly 20 g/L (Yim et al. 2011).

2.3.2.2 Optimization of Metabolic Pathways

The host's own metabolic network often has a dual role for the target metabolic pathway: on the one hand, the host's own metabolic network can provide energy and redox cofactors for the target metabolic pathway, such as ATP, NADH or NADPH; on the other hand, the host's natural metabolic network usually has strong robustness and regulatory functions, which often hinder the excessive distribution of metabolic fluxes to the target metabolic pathway. Therefore, once the metabolic pathways for biosynthesis target chemicals are reconstituted, it is essential to reduce the barriers to the target metabolic pathway by the host's natural metabolic network, and to eliminate the rate-limiting steps that exist in the metabolic pathway itself.

These require the use of efficient genome-wide models to redistribute metabolic flux by identifying sites in the metabolic pathway that require knockout, overexpression, or sites that have allosteric regulation.

First, a variety of GEMs have been developed for genome-scale identification of gene knockout targets to increase microbial cell factory production efficiency. By using gene knockout simulation of the in silico genome-scale metabolic network, the *aceF*, *mdh*, and *pfkA* genes as knockout targets was identified, the titer of l-valine produced by *E. coli* was increased to 7.55 g/l with a high yield of 0.378 g of l-valine per gram of glucose, which suggest that based on GEM, an industrially competitive strain can be efficiently developed (Park et al. 2007). Besides, the synthesis of heterologous terpeniods was improved by identifying *S. cerevisiae* knockout targets using GEM (Sun et al. 2014); the computational method OptSwap could identify optimal modifications of the cofactor binding specificities of oxidoreductase enzyme and complementary reaction knockouts to predicts bioprocessing strain designs (King and Feist 2013). And recently, a Python Library for Computer Aided Metabolic Engineering and Optimization of Cell Factories, Cameo, was developed to enumerate and prioritize knockout, knock-in, overexpression, and down-regulation strategies (Cardoso et al. 2018).

There are also several examples of successful gene overexpression targets found through genome-scale metabolic models. When using metabolically engineered *E. coli* for the production of fumaric acid, one of the industrially important four carbon chemicals, in silico flux response analysis was used to identify genes that needed to be overexpressed. After plasmid-based overexpression of the native ppc gene, the titer of fumaric acid increased 2.8 fold (Song et al. 2013). Besides, FSEOF was developed to identify gene targets for overexpression of lycopene production, achieving a 2.7-fold increase in yield (Choi et al. 2010); FVSEOF was developed and successfully predicted *glk*, *acnA*, *acnB*, *ackA* and ppc genes as overexpression targets and the production of putrescine increased by 20.5% (H. U. Kim et al. 2012a).

In addition, it is also a key point to diagnose and identify of bottlenecks in target metabolic pathways in microbial cell factories, such as rate limiting sites, allosteric regulation. In the exploration of the mechanism, the kinetic model of the *B. subtilis* glycolytic pathway has been established based on the reactions and kinetic parameters of *B. subtilis*, and the allosteric regulation required for glycolytic flux reversal in *B. subtilis* was identified according to the simulated and experimental data (Buffing et al. 2018). In the application examples of metabolic engineering, the N-acetylneuramine (GlcNAc) synthetic pathway kinetic model has been established by simulation of kinetic parameters. Then, by comparing the kinetic model prediction data with the metabolomics data, an energy-dissipating futile cycle between N-acetylglucosamine 6-phosphate (GlcNAc6P) and GlcNAc was found, and the GlcNAc production was ultimately improved by metabolic engineering (Liu et al. 2016). After comparing the metabolic model prediction data with the metabolomics data, the kinetic bottleneck in the synthetic pathway can be found. This combination of metabolic models and metabolomics allows us to better understand the dynamics of chemicals biosynthesis, to more easily predict the effects of genetic changes on products, and to facilitate the diagnosis and identification of rate-limiting sites, allosteric regulations, substrate channels in metabolic pathways, which has great application potential in the construction of microbial cell factory.

2.3.2.3 Screening of Potential Dominant Hosts

For specific chemical biosynthesis, host selection often plays a decisive role in whether or not it is ultimately industrially competitive. Different hosts have different friendliness for different metabolic engineering modifications. This is because the metabolic pathways of interest often consume large amounts of carbon or nitrogen sources, energy and cofactors, and may produce toxic intermediate metabolites, while different hosts have unique advantages for the supply of different cofactors or different hosts may have different tolerances for different toxic intermediates. At the same time, part of the metabolic pathway of the target is often derived from the host itself. If the corresponding native enzymatic reaction is highly efficient, this will also facilitate further metabolic engineering. Therefore, it is necessary to use a genome-scale metabolic model to understand the metabolic network of cells on a genome scale (Magrane and Consortium 2011).

For examples, 245 unique synthetic pathways for 20 large volume compounds were predicted by the GEM-Path algorithm, which integrated with genome-scale model and a novel approach to address reaction promiscuity. GEM-Path not only characterizes the potential for *E. coli* to produce commodity chemicals, but also provide a model for selecting potential hosts (Campodonico et al. 2014). Further, an integrated metabolic model that included native *E. coli* reactions and known heterologous reactions was developed to systematically evaluated *E. coli*'s potential ability to produce different chemicals. 1777 non-native products could theoretically be produced in *E. coli*, of which 279 non-native products have commercial applications including pharmaceuticals, food, cosmetics and perfume, agriculture, manufacturing, and others (Zhang et al. 2016).

2.3.3 *Genome-Scale Engineering Tools*

In the construction of microbial cell factories, thanks to the rapid development of omics technology and genome-scale metabolic models, screening of gene targets has become easier 3,75,80. These targets include multiple knockout targets, multiple overexpression targets, and multiple allosteric regulatory targets on the genome scale that were previously difficult to identify. However, the engineering of these gene targets and the construction of microbial cell factories are often complex and time consuming (Ronda et al. 2016). Homologous recombination (Datsenko and Wanner 2000; Sharan et al. 2009; Yu et al. 2002; Zhang et al. 1998). However, these methods are usually inefficient and antibiotic-dependent. In order to removed selection markers in genome, Cre-Lox recombinase-based and FLP flippase-based genome editing method were developed (Enyeart et al. 2014; Sukhija et al. 2012).

But both these enzymes leave genetic scars in the genome which may increase the risk of internal chromosomal rearrangements. Therefore, genome-scale engineering tools are needed, and two main methods have been developed, including oligo-mediated genome engineering tools and CRISPR-based genome engineering tools (Esvelt and Wang 2014)1.

Oligo-mediated genome engineering tools are represented by multiplex automated genome engineering (MAGE) (Chao et al. 2017; Singh and Braddick 2015). MAGE is an automated, fast and efficient tool designed to modify multiple targeted genes in a single cell or whole cell population. Its key point is to introduce directing ssDNA or oligonucleotides (oligos) into cells to modify target genes through multiple rounds of electroporation (Wang et al. 2009). This method allows for the modification of many targets ranging from nucleotides to genome lengths for different purposes, including knockout, overexpression or point mutation of multiple genes. Initially, it was applied to improve the biosynthesis of lycopene. In *E. coli*, nearly 15 billion genetic variants were obtained by random optimization of RBS or silencing of 24 target genes on genomes using MAGE. By screening, the yield of lycopene increased to 9000 p.p.m. (μg per g dry cell weight), which was better than documented yields (Wang et al. 2009). Currently, oligo-mediated genome engineering tools have been widely used in the construction of microbial cell factories. Methods of combining omics data or computational tools with MAGE have also been developed. By combining MAGE with conjugative assembly genome engineering (CAGE), all 314 TAG stop codons on the E. coli genome were replaced with TAA (Isaacs et al. 2011). Trackable multiplex recombineering (TRMR) is a method for gene-trait mapping which creates simulated knockdown and overexpression mutants for virtually all genes in the E. coli genome (Mansell et al. 2013). By combining directed evolution and TRMR, a broad range of mutations (>25 growth-enhancing mutations confirmed), which improved growth rate 10–200% for several different conditions was successfully identified on laboratory timescales (Sandoval et al. 2012). In addition, co-selection MAGE (CoS-MAGE) was developed to optimize biosynthesis of aromatic amino acid derivatives (Wang et al. 2012); microarray-oligonucleotide (MO)-MAGE was developed to perturb thousands of genomic sites simultaneous (Microarray-derived and Church 2014); BioDesignER, a high-fidelity genome engineering strain was developed to enable high-efficiency recombineering with a low basal mutagenesis rate (Egbert et al. 2019). To simplify the cumbersome and time-consuming design of oligonucleotides for recombinant engineering and MAGE, some automated design tools such as the MAGE oligo design tool (MODEST) have also been developed (Bonde et al. 2014). The development of these Oligo-mediated genome engineering tools will help to better understand the host's metabolic mechanisms and develop more efficient microbial cell factories.

Another type of efficient CRISPR-based genomic scale engineering tools has also developed rapidly (Doudna and Charpentier 2014; Knott and Doudna 2018; Zhang et al. 2014). The CRISPR-Cas system not only enables genome-scale gene editing, but also enables dynamic regulation of multi-gene expression during cell plant construction (Jakočiunas et al. 2016). For genome-scale gene editing, an easy-to-use and efficient tool has been developed in E. coli that allows simultaneous editing (insertion or deletion) of three targeted genes with a maximum efficiency of 100% (Ao et al. 2018; Jiang et al. 2015; Zhang et al. 1998). At the same time, the combination of CRISPR/Cas9 and λ Red recombineering based MAGE technology (CRMAGE) not only enables efficient and rapid genome editing, but also opens up new possibilities for the automation of genome-scale gene editing (Ronda et al. 2016). And CRISPR-enabled trackable genome engineering (CREATE) that integrates multiplex genome editing using CRISPR–Cas9 with barcode-enabled tracking of mutations in cell populations was also developed (Garst et al. 2017). Through the CRISPR-Cas system, it is even allowed to eliminate targeted chromosomes (Zuo et al. 2017).

Genome-scale gene expression regulation based on CRISPR has also been applied to the construction of microbial cell factories. An orthogonal tri-functional CRISPR system was developed to realize transcriptional activation, transcriptional interference, and gene deletion (CRISPR-AID) in the yeast Saccharomyces cerevisiae (Lian et al. 2017). Similarly, the CRISPR/Cas9-facilitated multiplex pathway optimization (CFPO) technique was developed to simultaneously modulate the expression of multiple genes on the genome (Zhu et al. 2017). This strategy can not only achieve the regulation of large-scale metabolic networks in a high-throughput manner, but also realize the conversion and regulation of intracellular metabolic pathways at different times, which is of great significance for the centralized and efficient use of resources by microbial cell factories.

2.4 Conclusions and Perspectives

Metabolic engineering has undergone rapid development since it was proposed in 1991 (Bailey 1991; Stephanopoulos and Vallino 1991), and lots of microbial cell factories have been constructed for the production of chemicals (Becker and Wittmann 2016; Cordova and Alper 2016; Lee et al. 2011), biofuel (C. Wang et al. 2017a; Zhou et al. 2018), nutraceuticals (L. Liu et al. 2017a), pharmaceuticals (Hirasawa and Shimizu 2016) and so on. The development of synthetic and system biology has dramatically improved our ability in metabolic pathways building and metabolic networks optimization, and the emerging tools and strategies in these fields will provide us with new ideas in the construction of more efficient microbial cell factories.

Many TF-based or riboswitch-based biosensors that can sense specific intracellular or extracellular metabolites have been constructed and applied in the dynamic regulation of metabolic network. However, there is no established model for how to efficiently apply these elements in the building of microbial cell factories. Recently studies about the design, construction, and testing of programmable genetic circuits provide a good idea for their further improvement (Bashor et al. 2019; Hoynes-O'Connor and Moon 2015). In addition, advances in measurement technique also help us to overcome the disparities in strain construction and product detection. For example, the optically guided matrix-assisted laser desorption/ionization mass spectrometry was used in the profiling of microbial colonies for high-throughput engineering of multistep enzymatic reactions (Si et al. 2017). Furthermore, the progress on chemically synthesized genomes give us a glimpse of the future, in which microbial cell factory could be built by a bottom-up design manner so as to save multifarious gene engineering operation (Cai et al. 2017).

References

Abdel-Mawgoud AM, Markham KA, Palmer CM, Liu N, Stephanopoulos G, Alper HS. Metabolic engineering in the host Yarrowia lipolytica. Metab Eng. 2018;50:192–208. https://doi.org/10.1016/j.ymben.2018.07.016.

Ahn JH, Lee JA, Bang J, Lee SY. Membrane engineering via trans-unsaturated fatty acids production improves succinic acid production in Mannheimia succiniciproducens. J Ind Microbiol Biotechnol. 2018;45:555–66. https://doi.org/10.1007/s10295-018-2016-6.

Ajikumar PK, Xiao W-H, Tyo KEJ, Wang Y, Simeon F, Leonard E, Mucha O, Phon TH, Pfeifer B, Stephanopoulos G. Isoprenoid pathway optimization for taxol precursor overproduction in Escherichia coli. Science. 2010;330:70–4. https://doi.org/10.1126/science.1191652.

Alper H, Fischer C, Nevoigt E, Stephanopoulos G. Tuning genetic control through promoter engineering. Proc Natl Acad Sci. 2005;102:12678–83. https://doi.org/10.1073/pnas.0504604102.

Ao X, Yao Y, Li T, Yang T-T, Dong X, Zheng Z-T, Chen G-Q, Wu Q, Guo Y. A multiplex genome editing method for Escherichia coli based on CRISPR-Cas12a. Front Microbiol. 2018;9:1–13. https://doi.org/10.3389/fmicb.2018.02307.

Ashburner M, Ball CA, Blake JA, Botstein D, Butler H, Cherry JM, Davis AP, Dolinski K, Dwight SS, Eppig JT, Harris MA, Hill DP, Issel-Tarver L, Kasarskis A, Lewis S, Matese JC, Richardson JE, Ringwald M, Rubin GM, Sherlock G. Gene ontology: tool for the unification of biology. Nat Genet. 2000;25:25–9. https://doi.org/10.1038/75556.

Bailey J. Toward a science of metabolic engineering. Science. 1991;252:1668–75. https://doi.org/10.1126/science.2047876.

Bashor CJ, Kondev J, Khalil AS. Complex signal processing in synthetic circuits using cooperative regulatory assemblies. Science. 2019;8287:1–11.

Becker J, Wittmann C. Systems metabolic engineering of Escherichia coli for the heterologous production of high value molecules. Curr Opin Biotechnol. 2016;42:178–88. https://doi.org/10.1016/j.copbio.2016.05.004.

Becker J, Rohles CM, Wittmann C. Metabolically engineered Corynebacterium glutamicum for bio-based production of chemicals, fuels, materials, and healthcare products. Metab Eng. 2018:1–20. https://doi.org/10.1016/j.ymben.2018.07.008.

Bervoets I, Charlier D. Diversity, versatility and complexity of bacterial gene regulation mechanisms: opportunities and drawbacks for applications in synthetic biology. FEMS Microbiol Rev. 2019; https://doi.org/10.1093/femsre/fuz001.

Besada-Lombana PB, McTaggart TL, Da Silva NA. Molecular tools for pathway engineering in Saccharomyces cerevisiae. Curr Opin Biotechnol. 2018;53:39–49. https://doi.org/10.1016/j.copbio.2017.12.002.

Biggs BW, De Paepe B, Santos CNS, De Mey M, Kumaran Ajikumar P. Multivariate modular metabolic engineering for pathway and strain optimization. Curr Opin Biotechnol. 2014;29:156–62. https://doi.org/10.1016/j.copbio.2014.05.005.

Biz A, Proulx S, Xu Z, Siddartha K, Indrayanti AM, Mahadevan R. Systems biology based metabolic engineering for non-natural chemicals. Biotechnol Adv. 2019; https://doi.org/10.1016/j.biotechadv.2019.04.001.

Bonde MT, Klausen MS, Anderson MV, Wallin AIN, Wang HH, Sommer MOA. MODEST: a web-based design tool for oligonucleotide-mediated genome engineering and recombineering. Nucleic Acids Res. 2014;42:W408–15. https://doi.org/10.1093/nar/gku428.

Bordbar A, Monk JM, King ZA, Palsson BO. Constraint-based models predict metabolic and associated cellular functions. Nat Rev Genet. 2014a;15:107–20. https://doi.org/10.1038/nrg3643.

Bordbar A, Nagarajan H, Lewis NE, Latif H, Ebrahim A, Federowicz S, Schellenberger J, Palsson BO. Minimal metabolic pathway structure is consistent with associated biomolecular interactions. Mol Syst Biol. 2014b;10:737. https://doi.org/10.15252/msb.20145243.

Brockman IM, Prather KLJ. Dynamic knockdown of E. coli central metabolism for redirecting fluxes of primary metabolites. Metab Eng. 2015a;28:104–13. https://doi.org/10.1016/j.ymben.2014.12.005.

Brockman IM, Prather KLJ. Dynamic metabolic engineering: new strategies for developing responsive cell factories. Biotechnol J. 2015b;10:1360–9. https://doi.org/10.1002/biot.201400422.

Buffing MF, Link H, Christodoulou D, Sauer U. Capacity for instantaneous catabolism of preferred and non-preferred carbon sources in Escherichia coli and Bacillus subtilis. Sci Rep. 2018;8:11760. https://doi.org/10.1038/s41598-018-30266-3.

Bujara M, Schümperli M, Pellaux R, Heinemann M, Panke S. Optimization of a blueprint for in vitro glycolysis by metabolic real-time analysis. Nat Chem Biol. 2011;7:271–7. https://doi.org/10.1038/nchembio.541.

Cai Y, Huang CLV, Richardson SM, Stracquadanio G, Mitchell LA, Lee D, DiCarlo JE, Chandrasegaran S, Yang K, Dymond JS, Bader JS, Boeke JD. Design of a synthetic yeast genome. Science. 2017;355:1040–4. https://doi.org/10.1126/science.aaf4557.

Cameron DE, Collins JJ. Tunable protein degradation in bacteria. Nat Biotechnol. 2014;32:1276–81. https://doi.org/10.1038/nbt.3053.

Campodonico MA, Andrews BA, Asenjo JA, Palsson BO, Feist AM. Generation of an atlas for commodity chemical production in Escherichia coli and a novel pathway prediction algorithm, GEM-Path. Metab Eng. 2014;25:140–58. https://doi.org/10.1016/j.ymben.2014.07.009.

Cardoso J, Jensen K, Lieven C, Hansen ASL, Galkina S, Beber ME, Özdemir E, Herrgard M, Redestig H, Sonnenschein N. Cameo: a python library for computer aided metabolic engineering and optimization of cell factories. ACS Synth Biol. 2018:acssynbio.7b00423. https://doi.org/10.1021/acssynbio.7b00423.

Caron M-P, Bastet L, Lussier A, Simoneau-Roy M, Masse E, Lafontaine DA. Dual-acting riboswitch control of translation initiation and mRNA decay. Proc Natl Acad Sci. 2012;109:E3444–53. https://doi.org/10.1073/pnas.1214024109.

Caspeta L, Chen Y, Ghiaci P, Feizi A, Baskov S, Hallström BM, Petranovic D, Nielsen J. Altered sterol composition renders yeast thermotolerant. Science. 2014;346:75–8. https://doi.org/10.1126/science.1258137.

Chae TU, Choi SY, Kim JW, Ko Y-S, Lee SY. Recent advances in systems metabolic engineering tools and strategies. Curr Opin Biotechnol. 2017;47:67–82. https://doi.org/10.1016/j.copbio.2017.06.007.

Chao R, Mishra S, Si T, Zhao H. Engineering biological systems using automated biofoundries. Metab Eng. 2017;42:98–108. https://doi.org/10.1016/j.ymben.2017.06.003.

Charubin K, Bennett RK, Fast AG, Papoutsakis ET. Engineering clostridium organisms as microbial cell-factories: challenges & opportunities. Metab Eng. 2018; https://doi.org/10.1016/j.ymben.2018.07.012.

Chatterjee N, Walker GC. Mechanisms of DNA damage, repair, and mutagenesis. Environ Mol Mutagen. 2017;58:235–63. https://doi.org/10.1002/em.22087.

Chen Z, Rappert S, Zeng AP. Rational design of allosteric regulation of homoserine dehydrogenase by a nonnatural inhibitor l -lysine. ACS Synth Biol. 2015;4:126–31. https://doi.org/10.1021/sb400133g.

Chen X, Gao C, Guo L, Hu G, Luo Q, Liu J, Nielsen J, Chen J, Liu L. DCEO biotechnology: tools to design, construct, evaluate, and optimize the metabolic pathway for biosynthesis of chemicals. Chem Rev. 2017:acs.chemrev.6b00804. https://doi.org/10.1021/acs.chemrev.6b00804.

Chen L, Gu W, Xu H y, Yang GL, Shan XF, Chen G, Kang Y h, Wang CF, Qian AD. Comparative genome analysis of Bacillus velezensis reveals a potential for degrading lignocellulosic biomass. 3 Biotech. 2018;8:253. https://doi.org/10.1007/s13205-018-1270-7.

Chew YH, Goldberg AP, Sekar JAP, Roth YD, Karr JR, Szigeti B, Chew YH, Sekar JAP, Roth YD, Karr JR. Emerging whole-cell modeling principles and methods. Curr Opin Biotechnol. 2018;51:97–102. https://doi.org/10.1016/j.copbio.2017.12.013.

Choi HS, Lee SY, Kim TY, Woo HM. In silico identification of gene amplification targets for improvement of lycopene production. Appl Environ Microbiol. 2010;76:3097–105. https://doi.org/10.1128/AEM.00115-10.

Choi KR, Jang WD, Yang D, Cho JS, Park D, Lee SY. Systems metabolic engineering strategies: integrating systems and synthetic biology with metabolic engineering. Trends Biotechnol. 2019:1–21. https://doi.org/10.1016/j.tibtech.2019.01.003.

Clark TA, Olivares EC, Travers KJ, Webster DR, Lee JH, Turner SW, Korlach J, Flusberg BA. Direct detection of DNA methylation during single-molecule, real-time sequencing. Nat Methods. 2010;7:461–5. https://doi.org/10.1038/nmeth.1459.

Srinivasan S, Cluett WR, Mahadevan R. Constructing kinetic models of metabolism at genome-scales: a review. Biotechnol J. 2015;10:1345–59. https://doi.org/10.1002/biot.201400522.

Conrado RJ, Wu GC, Boock JT, Xu H, Chen SY, Lebar T, Turnek J, Tomšič N, Avbelj M, Gaber R, Koprivnjak T, Mori J, Glavnik V, Vovk I, Beninča M, Hodnik V, Anderluh G, Dueber JE, Jerala R, Delisa MP. DNA-guided assembly of biosynthetic pathways promotes improved catalytic efficiency. Nucleic Acids Res. 2012;40:1879–89. https://doi.org/10.1093/nar/gkr888.

Cordova LT, Alper HS. Central metabolic nodes for diverse biochemical production. Curr Opin Chem Biol. 2016;35:37–42. https://doi.org/10.1016/j.cbpa.2016.08.025.

Costa RS, Hartmann A, Vinga S. Kinetic modeling of cell metabolism for microbial production. J Biotechnol. 2015;219:126–41. https://doi.org/10.1016/j.jbiotec.2015.12.023.

Cress BF, Trantas E a, Ververidis F, Linhardt RJ, Koffas M a G. Sensitive cells: enabling tools for static and dynamic control of microbial metabolic pathways. Curr Opin Biotechnol. 2015;36:205–14. https://doi.org/10.1016/j.copbio.2015.09.007.

Crook NC, Schmitz AC, Alper HS. Optimization of a yeast RNA interference system for controlling gene expression and enabling rapid metabolic engineering. ACS Synth Biol. 2014;3:307–13. https://doi.org/10.1021/sb4001432.

Dahl RH, Zhang F, Alonso-Gutierrez J, Baidoo E, Batth TS, Redding-Johanson AM, Petzold CJ, Mukhopadhyay A, Lee TS, Adams PD, Keasling JD. Engineering dynamic pathway regulation using stress-response promoters. Nat Biotechnol. 2013;31:1039–46. https://doi.org/10.1038/nbt.2689.

Dai Z, Nielsen J. Advancing metabolic engineering through systems biology of industrial microorganisms. Curr Opin Biotechnol. 2015;36:8–15. https://doi.org/10.1016/j.copbio.2015.08.006.

Datsenko KA, Wanner BL. One-step inactivation of chromosomal genes in Escherichia coli K-12 using PCR products. Proc Natl Acad Sci U S A. 2000;97:6640–5. https://doi.org/10.1073/pnas.120163297.

Delcher A. Improved microbial gene identification with GLIMMER. Nucleic Acids Res. 2002;27:4636–41. https://doi.org/10.1093/nar/27.23.4636.

Delebecque CJ, Lindner AB, Silver PA, Aldaye FA. Organization of intracellular reactions with rationally designed RNA assemblies. Science. 2011;333:470–4. https://doi.org/10.1126/science.1206938.

Ding Q, Luo Q, Zhou J, Chen X, Liu L. Enhancing L-malate production of Aspergillus oryzae FMME218-37 by improving inorganic nitrogen utilization. Appl Microbiol Biotechnol. 2018;102:8739–51. https://doi.org/10.1007/s00253-018-9272-2.

Doong SJ, Gupta A, Prather KLJ. Layered dynamic regulation for improving metabolic pathway productivity in Escherichia coli. Proc Natl Acad Sci. 2018:201716920. https://doi.org/10.1073/pnas.1716920115.

Doudna JA, Charpentier E. The new frontier of genome engineering with CRISPR-Cas9. Science. 2014;346:1258096. https://doi.org/10.1126/science.1258096.

Dueber JE, Wu GC, Malmirchegini GR, Moon TS, Petzold CJ, Ullal AV, Prather KLJ, Keasling JD. Synthetic protein scaffolds provide modular control over metabolic flux. Nat Biotechnol. 2009;27:753–9. https://doi.org/10.1038/nbt.1557.

Dugar D, Stephanopoulos G. Relative potential of biosynthetic pathways for biofuels and bio-based products. Nat Biotechnol. 2011;29:1074–8. https://doi.org/10.1038/nbt.2055.

Egbert RG, Rishi HS, Adler BA, McCormick DM, Toro E, Gill RT, Arkin AP. A versatile platform strain for high-fidelity multiplex genome editing. Nucleic Acids Res. 2019;47:3244–56. https://doi.org/10.1093/nar/gkz085.

Engler C, Kandzia R, Marillonnet S. A one pot, one step, precision cloning method with high throughput capability. PLoS One. 2008;3:e3647. https://doi.org/10.1371/journal.pone.0003647.

Enyeart PJ, Chirieleison SM, Dao MN, Perutka J, Quandt EM, Yao J, Whitt JT, Keatinge-Clay a T, Lambowitz a M, Ellington a D. Generalized bacterial genome editing using mobile group II introns and Cre-lox. Mol Syst Biol. 2014;9:685. https://doi.org/10.1038/msb.2013.41.

Esvelt KM, Wang HH. Genome-scale engineering for systems and synthetic biology. Mol Syst Biol. 2014;9:641. https://doi.org/10.1038/msb.2012.66.

Farmer WR, Liao JC. Improving lycopene production in Escherichia coli by engineering metabolic control. Nat Biotechnol. 2000;18:533–7. https://doi.org/10.1038/75398.

Fuhrer T, Zamboni N. ScienceDirect High-throughput discovery metabolomics. Curr Opin Biotechnol. 2015;31:73–8. https://doi.org/10.1016/j.copbio.2014.08.006.

Gagne SJ, Stout JM, Liu E, Boubakir Z, Clark SM, Page JE. Identification of olivetolic acid cyclase from Cannabis sativa reveals a unique catalytic route to plant polyketides. Proc Natl Acad Sci. 2012;109:12811–6. https://doi.org/10.1073/pnas.1200330109.

Garst AD, Bassalo MC, Pines G, Lynch SA, Halweg-Edwards AL, Liu R, Liang L, Wang Z, Zeitoun R, Alexander WG, Gill RT. Genome-wide mapping of mutations at single-nucleotide resolution for protein, metabolic and genome engineering. Nat Biotechnol. 2017;35:48–55. https://doi.org/10.1038/nbt.3718.

George KW, Chen A, Jain A, Batth TS, Baidoo EEK, Wang G, Adams PD, Petzold CJ, Keasling JD, Lee TS. Correlation analysis of targeted proteins and metabolites to assess and engineer microbial isopentenol production. Biotechnol Bioeng. 2014;111:1648–58. https://doi.org/10.1002/bit.25226.

Gibson DG, Benders G a, Axelrod KC, Zaveri J, Algire M a, Moodie M, Montague MG, Venter JC, Smith HO, Hutchison C a. One-step assembly in yeast of 25 overlapping DNA fragments to form a complete synthetic Mycoplasma genitalium genome. Proc Natl Acad Sci. 2008;105:20404–9. https://doi.org/10.1073/pnas.0811011106.

Gilbert LA, Larson MH, Morsut L, Liu Z, Brar GA, Torres SE, Stern-Ginossar N, Brandman O, Whitehead EH, Doudna JA, Lim WA, Weissman JS, Qi LS. CRISPR-mediated modular RNA-guided regulation of transcription in eukaryotes. Cell. 2013;154:442. https://doi.org/10.1016/j.cell.2013.06.044.

Gilbert LA, Horlbeck MA, Adamson B, Villalta JE, Chen Y, Whitehead EH, Guimaraes C, Panning B, Ploegh HL, Bassik MC, Qi LS, Kampmann M, Weissman JS. Genome-scale CRISPR-mediated control of gene repression and activation. Cell. 2014;159:647–61. https://doi.org/10.1016/j.cell.2014.09.029.

Gottesman S. The small RNA regulators of *Escherichia coli* : roles and mechanisms. Annu Rev Microbiol. 2004;58:303–28. https://doi.org/10.1146/annurev.micro.58.030603.123841.

Gu Y, Deng J, Liu Y, Li J, Shin HD, Du G, Chen J, Liu L. Rewiring the glucose transportation and central metabolic pathways for overproduction of N-acetylglucosamine in Bacillus subtilis. Biotechnol J. 2017;12:1700268. https://doi.org/10.1002/biot.201700020.

Gu Y, Xu X, Wu Y, Niu T, Liu Y, Li J, Du G, Liu L. Advances and prospects of Bacillus subtilis cellular factories: from rational design to industrial applications. Metab Eng. 2018;50:109–21. https://doi.org/10.1016/j.ymben.2018.05.006.

Gu Y, Lv X, Liu Y, Li J, Du G, Chen J, Rodrigo LA, Liu L. Synthetic redesign of central carbon and redox metabolism for high yield production of N-acetylglucosamine in Bacillus subtilis. Metab Eng. 2019;51:59–69. https://doi.org/10.1016/j.ymben.2018.10.002.

Guan N, Du B, Li J, Shin HD, Chen RR, Du G, Chen J, Liu L. Comparative genomics and transcriptomics analysis-guided metabolic engineering of Propionibacterium acidipropionici

for improved propionic acid production. Biotechnol Bioeng. 2018;115:483–94. https://doi.org/10.1002/bit.26478.
Guo J, Zhou YJ, Hillwig ML, Shen Y, Yang L, Wang Y, Zhang X, Liu W, Peters RJ, Chen X, Zhao ZK, Huang L. CYP76AH1 catalyzes turnover of miltiradiene in tanshinones biosynthesis and enables heterologous production of ferruginol in yeasts. Proc Natl Acad Sci. 2013;110:12108–13. https://doi.org/10.1073/pnas.1218061110.
Gupta A, Reizman IMB, Reisch CR, Prather KLJ. Dynamic regulation of metabolic flux in engineered bacteria using a pathway-independent quorum-sensing circuit. Nat Biotechnol Adv. 2017; https://doi.org/10.1038/nbt.3796.
Haft DH. Using comparative genomics to drive new discoveries in microbiology. Curr Opin Microbiol. 2015;23:189–96. https://doi.org/10.1016/j.mib.2014.11.017.
Hao T, Wu D, Zhao L, Wang Q, Wang E, Sun J. The genome-scale integrated networks in microorganisms. Front Microbiol. 2018;9:296. https://doi.org/10.3389/fmicb.2018.00296.
Harwood CR, Pohl S, Smith W, Wipat A. Bacillus subtilis. Model Gram-Positive Synthetic Biology Chassis. In: Methods in microbiology. Copyright {©} 2013 Elsevier Ltd. All rights reserved. 1st ed; 2013. https://doi.org/10.1016/B978-0-12-417029-2.00004-2.
Hirasawa T, Shimizu H. Recent advances in amino acid production by microbial cells. Curr Opin Biotechnol. 2016;42:133–46. https://doi.org/10.1016/j.copbio.2016.04.017.
Ho S-H, Nakanishi A, Ye X, Chang J-S, Hara K, Hasunuma T, Kondo A. Optimizing biodiesel production in marine Chlamydomonas sp. JSC4 through metabolic profiling and an innovative salinity-gradient strategy. Biotechnol Biofuels. 2014;7:97. https://doi.org/10.1186/1754-6834-7-97.
Hong K-K, Vongsangnak W, Vemuri GN, Nielsen J. Unravelling evolutionary strategies of yeast for improving galactose utilization through integrated systems level analysis. Proc Natl Acad Sci. 2011;108:12179–84. https://doi.org/10.1073/pnas.1103219108.
Hoynes-O'Connor A, Moon TS. Programmable genetic circuits for pathway engineering. Curr Opin Biotechnol. 2015;36:115–21. https://doi.org/10.1016/j.copbio.2015.08.007.
Isaacs FJ, Carr PA, Wang HH, Lajoie MJ, Sterling B, Kraal L, Tolonen AC, Gianoulis TA, Goodman DB, Reppas NB, Emig CJ, Bang D, Hwang SJ, Jewett MC, Jacobson JM, Church GM. Precise manipulation of chromosomes in vivo enables genome-wide codon replacement. Science. 2011;333:348–53. https://doi.org/10.1126/science.1205822.
Jakočiunas T, Jensen MK, Keasling JD. CRISPR/Cas9 advances engineering of microbial cell factories. Metab Eng. 2016;34:44–59. https://doi.org/10.1016/j.ymben.2015.12.003.
Janssen BD, Hayes CS. The tm RNA ribosome-rescue system. Adv Protein Chem Struct Biol. 2012;86:151–91. https://doi.org/10.1016/B978-0-12-386497-0.00005-0.
Jiang Y, Chen B, Duan C, Sun B, Yang J, Yang S. Multigene editing in the Escherichia coli genome via the CRISPR-Cas9 system. Appl Environ Microbiol. 2015;81:2506–14. https://doi.org/10.1128/aem.04023-14.
Jones P, Binns D, Chang H-Y, Fraser M, Li W, McAnulla C, McWilliam H, Maslen J, Mitchell A, Nuka G, Pesseat S, Quinn AF, Sangrador-Vegas A, Scheremetjew M, Yong S-Y, Lopez R, Hunter S. InterProScan 5: genome-scale protein function classification. Bioinformatics. 2014;30:1236–40. https://doi.org/10.1093/bioinformatics/btu031.
Jones JA, Vernacchio VR, Lachance DM, Lebovich M, Fu L, Shirke AN, Schultz VL, Cress B, Linhardt RJ, Koffas MAG. EPathOptimize: a combinatorial approach for transcriptional balancing of metabolic pathways. Sci Rep. 2015;5:1–10. https://doi.org/10.1038/srep11301.
Kanehisa M. From genomics to chemical genomics: new developments in KEGG. Nucleic Acids Res. 2006;34:D354–7. https://doi.org/10.1093/nar/gkj102.
Karr JR, Sanghvi JC, Macklin DN, Gutschow MV, Jacobs JM, Bolival B, Assad-Garcia N, Glass JI, Covert MW. A whole-cell computational model predicts phenotype from genotype. Cell. 2012;150:389–401. https://doi.org/10.1016/j.cell.2012.05.044.
Keasling JD, Chubukov V, Mukhopadhyay A, Petzold CJ, Keasling JD, Martín HG. Synthetic and systems biology for microbial production of commodity chemicals. NPJ Syst Biol Appl. 2016;2:16009. https://doi.org/10.1038/npjsba.2016.9.

Kim HU, Kim TY, Kim WJ, Lee SY, Park HM, Park JM. Flux variability scanning based on enforced objective flux for identifying gene amplification targets. Bmc Syst Biol. 2012a;6:106. https://doi.org/10.1186/1752-0509-6-106.

Kim TY, Sohn SB, Kim YB, Kim WJ, Lee SY. Recent advances in reconstruction and applications of genome-scale metabolic models. Curr Opin Biotechnol. 2012b;23:617–23. https://doi.org/10.1016/j.copbio.2011.10.007.

Kim B, Kim WJ, Kim DI, Lee SY. Applications of genome-scale metabolic network model in metabolic engineering. J Ind Microbiol Biotechnol. 2014;42:339–48. https://doi.org/10.1007/s10295-014-1554-9.

Kim MK, Lane A, Kelley JJ, Lun DS. E-Flux2 and SPOT: validated methods for inferring intracellular metabolic flux distributions from transcriptomic data. PLoS One. 2016;11:e0157101. https://doi.org/10.1371/journal.pone.0157101.

King ZA, Feist AM. Optimizing cofactor specificity of oxidoreductase enzymes for the generation of microbial production strains—OptSwap. Ind Biotechnol. 2013;9:236–46. https://doi.org/10.1089/ind.2013.0005.

Knott GJ, Doudna JA. CRISPR-Cas guides the future of genetic engineering. Science. 2018; https://doi.org/10.1126/science.aat5011.

Kochanowski K, Sauer U, Chubukov V. Somewhat in control-the role of transcription in regulating microbial metabolic fluxes. Curr Opin Biotechnol. 2013;24:987–93. https://doi.org/10.1016/j.copbio.2013.03.014.

Korneli C, Bolten CJ, Godard T, Franco-Lara E, Wittmann C. Debottlenecking recombinant protein production in Bacillus megaterium under large-scale conditions-targeted precursor feeding designed from metabolomics. Biotechnol Bioeng. 2012;109:1538–50. https://doi.org/10.1002/bit.24434.

Lagesen K, Hallin P, Rødland EA, Staerfeldt H-H, Rognes T, Ussery DW. RNAmmer. Nucleic Acids Res. 2007;35:3100–8. https://doi.org/10.1093/nar/gkm160.

Lalwani MA, Zhao EM, Avalos JL. Current and future modalities of dynamic control in metabolic engineering. Curr Opin Biotechnol. 2018;52:56–65. https://doi.org/10.1016/j.copbio.2018.02.007.

Lee SY, Kim HU. Systems strategies for developing industrial microbial strains. Nat Biotechnol. 2015;33:1061–72. https://doi.org/10.1038/nbt.3365.

Lee JW, Kim HU, Choi S, Yi J, Lee SY. Microbial production of building block chemicals and polymers. Curr Opin Biotechnol. 2011;22:758–67. https://doi.org/10.1016/j.copbio.2011.02.011.

Lee JW, Na D, Park JM, Lee J, Choi S, Lee SY. Systems metabolic engineering for natural and non-natural chemicals. Nat Chem Biol. 2012;8:536–46. https://doi.org/10.1038/nchembio.970.

Li W, Li HX, Ji SY, Li S, Gong YS, Yang MM, Chen YL. Characterization of two temperature-inducible promoters newly isolated from B. subtilis. Biochem Biophys Res Commun. 2007;358:1148–53. https://doi.org/10.1016/j.bbrc.2007.05.064.

Li S, Jendresen CB, Grünberger A, Ronda C, Jensen SI, Noack S, Nielsen AT. Enhanced protein and biochemical production using CRISPRi-based growth switches. Metab Eng. 2016;38:274–84. https://doi.org/10.1016/j.ymben.2016.09.003.

Li L, Liu X, Wei K, Lu Y, Jiang W. Synthetic biology approaches for chromosomal integration of genes and pathways in industrial microbial systems. Biotechnol Adv. 2019; https://doi.org/10.1016/j.biotechadv.2019.04.002.

Lian J, HamediRad M, Hu S, Zhao H. Combinatorial metabolic engineering using an orthogonal tri-functional CRISPR system. Nat Commun. 2017;8:1688. https://doi.org/10.1038/s41467-017-01695-x.

Lian J, Mishra S, Zhao H. Recent advances in metabolic engineering of Saccharomyces cerevisiae: new tools and their applications. Metab Eng. 2018;50:85–108. https://doi.org/10.1016/j.ymben.2018.04.011.

Ling H, Teo W, Chen B, Leong SSJ, Chang MW. Microbial tolerance engineering toward biochemical production: from lignocellulose to products. Curr Opin Biotechnol. 2014;29:99–106. https://doi.org/10.1016/j.copbio.2014.03.005.

Link H, Kochanowski K, Sauer U. Systematic identification of allosteric protein-metabolite interactions that control enzyme activity in vivo. Nat Biotechnol. 2013;31:357–61. https://doi.org/10.1038/nbt.2489.

Link H, Christodoulou D, Sauer U. Advancing metabolic models with kinetic information. Curr Opin Biotechnol. 2014;29:8–14. https://doi.org/10.1016/j.copbio.2014.01.015.

Link H, Fuhrer T, Gerosa L, Zamboni N, Sauer U. Real-time metabolome profiling of the metabolic switch between starvation and growth. Nat Methods. 2015;12 https://doi.org/10.1038/nmeth.3584.

Liu D, Zhang F. Metabolic feedback circuits provide rapid control of metabolite dynamics. ACS Synth Biol. 2018;7:347–56. https://doi.org/10.1021/acssynbio.7b00342.

Liu Y, Zhu Y, Li J, Shin HD, Chen RR, Du G, Liu L, Chen J. Modular pathway engineering of Bacillus subtilis for improved N-acetylglucosamine production. Metab Eng. 2014;23:42–52. https://doi.org/10.1016/j.ymben.2014.02.005.

Liu Y, Link H, Liu L, Du G, Chen J, Sauer U. A dynamic pathway analysis approach reveals a limiting futile cycle in N-acetylglucosamine overproducing Bacillus subtilis. Nat Commun. 2016;7:11933. https://doi.org/10.1038/ncomms11933.

Liu L, Guan N, Li J, Shin H, Du G, Chen J. Development of GRAS strains for nutraceutical production using systems and synthetic biology approaches: advances and prospects. Crit Rev Biotechnol. 2017a;37:139–50. https://doi.org/10.3109/07388551.2015.1121461.

Liu Y, Li J, Du G, Chen J, Liu L. Metabolic engineering of Bacillus subtilis fueled by systems biology: recent advances and future directions. Biotechnol Adv. 2017b;35:20–30. https://doi.org/10.1016/j.biotechadv.2016.11.003.

Liu D, Mao Z, Guo J, Wei L, Ma H, Tang Y, Chen T, Wang Z, Zhao X. Construction, model-based analysis, and characterization of a promoter library for fine-tuned gene expression in Bacillus subtilis. ACS Synth Biol. 2018;7:1785–97. https://doi.org/10.1021/acssynbio.8b00115.

López-Fernández L, Sanchis M, Navarro-Rodríguez P, Nicolás FE, Silva-Franco F, Guarro J, Garre V, Navarro-Mendoza MI, Pérez-Arques C, Capilla J. Understanding Mucor circinelloides pathogenesis by comparative genomics and phenotypical studies. Virulence. 2018;9:707–20. https://doi.org/10.1080/21505594.2018.1435249.

Lu H, Villada JC, Lee PKH. Modular metabolic engineering for biobased chemical production. Trends Biotechnol. 2019;37:152–66. https://doi.org/10.1016/j.tibtech.2018.07.003.

Magrane M, Consortium U. UniProt Knowledgebase: a hub of integrated protein data. Database. 2011;2011:bar009. https://doi.org/10.1093/database/bar009.

Man S, Cheng R, Miao C, Gong Q, Gu Y, Lu X, Han F, Yu W. Artificial trans-encoded small non-coding RNAs specifically silence the selected gene expression in bacteria. Nucleic Acids Res. 2011;39:e50. https://doi.org/10.1093/nar/gkr034.

Mansell TJ, Warner JR, Gill RT. Trackable multiplex recombineering for gene-trait mapping in E. coli. In: Alper HS, editor. Methods in molecular biology. Totowa: Humana Press; 2013. p. 223–46. https://doi.org/10.1007/978-1-62703-299-5_12.

Mattozzi M, Zang Y, Gupta M, Wu X, Plassmeier J, Clarkson S, Zha J, Koffas MAG. Metabolic engineering of Corynebacterium glutamicum for anthocyanin production. Microb Cell Fact. 2018;17:1–13. https://doi.org/10.1186/s12934-018-0990-z.

McClymont K, Soyer OS. Metabolic tinker: an online tool for guiding the design of synthetic metabolic pathways. Nucleic Acids Res. 2013;41:e113. https://doi.org/10.1093/nar/gkt234.

Microarray-derived TU, Church G. Direct mutagenesis of thousands of genomic. ACS Synth Biol. 2014;4:1–10. https://doi.org/10.1021/sb5001565.

Miller MB, Bassler BL. Quorum sensing in bacteria. Annu Rev Microbiol. 2001;55:165–99. https://doi.org/10.1146/annurev.micro.55.1.165.

Miller JR, Delcher AL, Koren S, Venter E, Walenz BP, Brownley A, Johnson J, Li K, Mobarry C, Sutton G. Aggressive assembly of pyrosequencing reads with mates. Bioinformatics. 2008;24:2818–24. https://doi.org/10.1093/bioinformatics/btn548.

Motlagh HN, Wrabl JO, Li J, Hilser VJ. The ensemble nature of allostery. Nature. 2014;508:331–9. https://doi.org/10.1038/nature13001.

Na D, Yoo SM, Chung H, Park H, Park JH, Lee SY. Metabolic engineering of Escherichia coli using synthetic small regulatory RNAs. Nat Biotechnol. 2013;31:170–4. https://doi.org/10.1038/nbt.2461.
Nielsen J, Jewett MC. Impact of systems biology on metabolic engineering of Saccharomyces cerevisiae. FEMS Yeast Res. 2008;8:122–31. https://doi.org/10.1111/j.1567-1364.2007.00302.x.
Nielsen J, Keasling JD. Engineering cellular metabolism. Cell. 2016;164:1185–97. https://doi.org/10.1016/j.cell.2016.02.004.
Nielsen J, Larsson C, van Maris A, Pronk J. Metabolic engineering of yeast for production of fuels and chemicals. Curr Opin Biotechnol. 2013;24:398–404. https://doi.org/10.1016/j.copbio.2013.03.023.
Nielsen J, Fussenegger M, Keasling J, Lee SY, Liao JC, Prather K, Palsson B. Engineering synergy in biotechnology. Nat Chem Biol. 2014;10:319–22. https://doi.org/10.1038/nchembio.1519.
Nikel PI, de Lorenzo V. Pseudomonas putida as a functional chassis for industrial biocatalysis: From native biochemistry to trans-metabolism. Metab Eng. 2018;0–1 https://doi.org/10.1016/j.ymben.2018.05.005.
Niu T, Liu Y, Li J, Koffas M, Du G, Alper HS, Liu L. Engineering a glucosamine-6-phosphate responsive glmS Ribozyme switch enables dynamic control of metabolic flux in Bacillus subtilis for overproduction of N-acetylglucosamine. ACS Synth Biol. 2018;7:2423–35. https://doi.org/10.1021/acssynbio.8b00196.
Nowroozi FF, Baidoo EEK, Ermakov S, Redding-Johanson AM, Batth TS, Petzold CJ, Keasling JD. Metabolic pathway optimization using ribosome binding site variants and combinatorial gene assembly. Appl Microbiol Biotechnol. 2014;98:1567–81. https://doi.org/10.1007/s00253-013-5361-4.
O'Brien EJ, Monk JM, Palsson BO. Using genome-scale models to predict biological capabilities. Cell. 2015;161:971–87. https://doi.org/10.1016/j.cell.2015.05.019.
Oliva G, Sahr T, Buchrieser C. Small RNAs, 5' UTR elements and RNA-binding proteins in intracellular bacteria: Impact on metabolism and virulence. FEMS Microbiol Rev. 2015;39:331–49. https://doi.org/10.1093/femsre/fuv022.
Panahi R, Vasheghani-Farahani E, Shojaosadati SA, Bambai B. Induction of Bacillus subtilis expression system using environmental stresses and glucose starvation. Ann Microbiol. 2014;64:879–82. https://doi.org/10.1007/s13213-013-0719-5.
Park JH, Lee KH, Kim TY, Lee SY. Metabolic engineering of Escherichia coli for the production of L-valine based on transcriptome analysis and in silico gene knockout simulation. Proc Natl Acad Sci. 2007;104:7797–802. https://doi.org/10.1073/pnas.0702609104.
Peralta-Yahya PP, Zhang F, Del Cardayre SB, Keasling JD. Microbial engineering for the production of advanced biofuels. Nature. 2012;488:320–8. https://doi.org/10.1038/nature11478.
Pharkya P. OptStrain: a computational framework for redesign of microbial production systems. Genome Res. 2004;14:2367–76. https://doi.org/10.1101/gr.2872004.
Pontrelli S, Chiu TY, Lan EI, Chen FYH, Chang P, Liao JC. Escherichia coli as a host for metabolic engineering. Metab Eng. 2018;0–1 https://doi.org/10.1016/j.ymben.2018.04.008.
Qi L, Lucks JB, Liu CC, Mutalik VK, Arkin AP. Engineering naturally occurring trans-acting non-coding RNAs to sense molecular signals. Nucleic Acids Res. 2012;40:5775–86. https://doi.org/10.1093/nar/gks168.
Qiao K, Wasylenko TM, Zhou K, Xu P, Stephanopoulos G. Lipid production in Yarrowia lipolytica is maximized by engineering cytosolic redox metabolism. Nat Biotechnol. 2017;35:173–7. https://doi.org/10.1038/nbt.3763.
Rahimi T, Niazi A, Deihimi T, Taghavi SM, Ayatollahi S, Ebrahimie E. Genome annotation and comparative genomic analysis of Bacillus subtilis MJ01, a new bio-degradation strain isolated from oil-contaminated soil. Funct Integr Genomics. 2018;18:533–43. https://doi.org/10.1007/s10142-018-0604-1.
Redden H, Alper HS. The development and characterization of synthetic minimal yeast promoters. Nat Commun. 2015;6:7810.
Richter F, Fonfara I, Gelfert R, Nack J, Charpentier E. Switchable Cas9. Curr Opin Biotechnol. 2017;48:119–26. https://doi.org/10.1016/j.copbio.2017.03.025.

Ronda C, Pedersen LE, Sommer MOA, Nielsen AT. CRMAGE: CRISPR optimized MAGE recombineering. Sci Rep. 2016;6:19452. https://doi.org/10.1038/srep19452.

Rozkov A, Avignone-Rossa CA, Ertl PF, Jones P, O'Kennedy RD, Smith JJ, Dale JW, Bushell ME. Characterization of the metabolic burden on Escherichia coli DH1 cells imposed by the presence of a plasmid containing a gene therapy sequence. Biotechnol Bioeng. 2004;88:909–15. https://doi.org/10.1002/bit.20327.

Salis HM, Mirsky E a, Voigt C a. Automated design of synthetic ribosome binding sites to control protein expression. Nat Biotechnol. 2009;27:946–50. https://doi.org/10.1038/nbt.1568.

Sandoval NR, Kim JYH, Glebes TY, Reeder PJ, Aucoin HR, Warner JR, Gill RT. Strategy for directing combinatorial genome engineering in Escherichia coli. Proc Natl Acad Sci. 2012;109:10540–5. https://doi.org/10.1073/pnas.1206299109.

Serganov A, Patel DJ. Ribozymes, riboswitches and beyond: regulation of gene expression without proteins. Nat Rev Genet. 2007;8:776–90. https://doi.org/10.1038/nrg2172.

Sévin DC, Kuehne A, Zamboni N, Sauer U. Biological insights through nontargeted metabolomics. Curr Opin Biotechnol. 2015;34:1–8. https://doi.org/10.1016/j.copbio.2014.10.001.

Sharan SK, Thomason LC, Kuznetsov SG, Court DL. Recombineering: A homologous recombination-based method of genetic engineering. Nat Protoc. 2009;4:206–23. https://doi.org/10.1038/nprot.2008.227.

Shi S, Chen T, Zhang Z, Chen X, Zhao X. Transcriptome analysis guided metabolic engineering of Bacillus subtilis for riboflavin production. Metab Eng. 2009;11:243–52. https://doi.org/10.1016/j.ymben.2009.05.002.

Si T, Li B, Comi TJ, Wu Y, Hu P, Wu Y, Min Y, Mitchell DA, Zhao H, Sweedler JV. Profiling of microbial colonies for high-throughput engineering of multistep enzymatic reactions via optically guided matrix-assisted laser desorption/ionization mass spectrometry. J Am Chem Soc. 2017;139:12466–73. https://doi.org/10.1021/jacs.7b04641.

Singh V, Braddick D. Recent advances and versatility of MAGE towards industrial applications. Syst Synth Biol. 2015;9:1–9. https://doi.org/10.1007/s11693-015-9184-8.

Smanski MJ, Bhatia S, Zhao D, Park Y, B A Woodruff L, Giannoukos G, Ciulla D, Busby M, Calderon J, Nicol R, Gordon DB, Densmore D, Voigt C a. Functional optimization of gene clusters by combinatorial design and assembly. Nat Biotechnol. 2014;32:1241–9. https://doi.org/10.1038/nbt.3063.

Snoek T, Chaberski EK, Ambri F, Kol S, Bjørn SP, Pang B, Barajas JF, Welner DH, Jensen MK, Keasling JD. Evolution-guided engineering of small-molecule biosensors. Biorxiv. 2019; https://doi.org/10.1101/601823.

Solomon KV, Sanders TM, Prather KLJ. A dynamic metabolite valve for the control of central carbon metabolism. Metab Eng. 2012;14:661–71. https://doi.org/10.1016/j.ymben.2012.08.006.

Soma Y, Hanai T. Self-induced metabolic state switching by a tunable cell density sensor for microbial isopropanol production. Metab Eng. 2015;30:7–15. https://doi.org/10.1016/j.ymben.2015.04.005.

Soma Y, Tsuruno K, Wada M, Yokota A, Hanai T. Metabolic flux redirection from a central metabolic pathway toward a synthetic pathway using a metabolic toggle switch. Metab Eng. 2014;23:175–84. https://doi.org/10.1016/j.ymben.2014.02.008.

Song CW, Kim DI, Choi S, Jang JW, Lee SY. Metabolic engineering of Escherichia coli for the production of fumaric acid. Biotechnol Bioeng. 2013;110:2025–34. https://doi.org/10.1002/bit.24868.

Stalidzans E, Seiman A, Peebo K, Komasilovs V, Pentjuss A. Model-based metabolism design: constraints for kinetic and stoichiometric models. Biochem Soc Trans. 2018;46:261–7. https://doi.org/10.1042/BST20170263.

Stephanopoulos G. Synthetic biology and metabolic engineering. ACS Synth Biol. 2012;1:514–25. https://doi.org/10.1021/sb300094q.

Stephanopoulos G, Vallino J. Network rigidity and metabolic engineering in metabolite overproduction. Science. 1991;252:1675–81. https://doi.org/10.1126/science.1904627.

Sukhija K, Pyne M, Ali S, Orr V, Abedi D, Moo-Young M, Chou CP. Developing an extended genomic engineering approach based on recombineering to knock-in heterologous genes to Escherichia coli genome. Mol Biotechnol. 2012;51:109–18. https://doi.org/10.1007/s12033-011-9442-2.

Sun Z, Meng H, Li J, Wang J, Li Q, Wang Y, Zhang Y. Identification of novel knockout targets for improving terpenoids biosynthesis in saccharomyces cerevisiae. PLoS One. 2014;9:e112615. https://doi.org/10.1371/journal.pone.0112615.

Tang SY, Cirino PC. Design and application of a mevalonate-responsive regulatory protein. Angew Chemie Int Ed. 2011;50:1084–6. https://doi.org/10.1002/anie.201006083.

Tatusov RL, Natale DA, Garkavtsev IV, Tatusova TA, Shankavaram UT, Rao BS, Kiryutin B, Galperin MY, Fedorova ND, Koonin EV. The COG database. Nucleic Acids Res. 2000;29:22–8.

Tietz JI, Schwalen CJ, Patel PS, Maxson T, Blair PM, Tai HC, Zakai UI, Mitchell DA. A new genome-mining tool redefines the lasso peptide biosynthetic landscape. Nat Chem Biol. 2017;13:470–8. https://doi.org/10.1038/nchembio.2319.

Tomari Y, Zamore PD. Machines for RNAi. Genes Dev. 2005;19:517–29. https://doi.org/10.1101/gad.1284105.Box.

Tong Z, Zheng X, Tong Y, Shi YC, Sun J. Systems metabolic engineering for citric acid production by Aspergillus niger in the post-genomic era. Microb Cell Fact. 2019;18:28. https://doi.org/10.1186/s12934-019-1064-6.

Torella JP, Ford TJ, Kim SN, Chen AM, Way JC, Silver PA. Tailored fatty acid synthesis via dynamic control of fatty acid elongation. Proc Natl Acad Sci. 2013;110:11290–5. https://doi.org/10.1073/pnas.1307129110.

van Dijk EL, Jaszczyszyn Y, Naquin D, Thermes C. The third revolution in sequencing technology. Trends Genet. 2018;34:666–81. https://doi.org/10.1016/j.tig.2018.05.008.

van Tilburg AY, Cao H, van der Meulen SB, Solopova A, Kuipers OP. Metabolic engineering and synthetic biology employing Lactococcus lactis and Bacillus subtilis cell factories. Curr Opin Biotechnol. 2019;59:1–7. https://doi.org/10.1016/j.copbio.2019.01.007.

Wang HH, Isaacs FJ, Carr P a, Sun ZZ, Xu G, Forest CR, Church GM. Programming cells by multiplex genome engineering and accelerated evolution. Nature. 2009;460:894–8. https://doi.org/10.1038/nature08187.

Wang HH, Kim H, Cong L, Jeong J, Bang D, Church GM. Genome-scale promoter engineering by coselection MAGE. Nat Methods. 2012;9:591–3. https://doi.org/10.1038/nmeth.1971.

Wang C, Pfleger BF, Kim SW. Reassessing Escherichia coli as a cell factory for biofuel production. Curr Opin Biotechnol. 2017a;45:92–103. https://doi.org/10.1016/j.copbio.2017.02.010.

Wang L, Dash S, Ng CY, Maranas CD. A review of computational tools for design and reconstruction of metabolic pathways. Synth Syst Biotechnol. 2017b;2:243–52. https://doi.org/10.1016/j.synbio.2017.11.002.

Wang X, He Q, Yang Y, Wang J, Haning K, Hu Y, Wu B, He M, Zhang Y, Bao J, Contreras LM, Yang S. Advances and prospects in metabolic engineering of Zymomonas mobilis. Metab Eng. 2018;50:57–73. https://doi.org/10.1016/j.ymben.2018.04.001.

Waters LS, Storz G. Regulatory RNAs in bacteria. Cell. 2009;136:615–28. https://doi.org/10.1016/j.cell.2009.01.043.

Williams TC, Averesch NJ, Plan M, Winter G, Vickers CE, Nielsen LK, Krömer JO, Lekieffre N, Winter G, Vickers CE, Nielsen LK, Krömer JO. Quorum-sensing linked RNAi for dynamic pathway control in Saccharomyces cerevisiae. Metab Eng. 2015a;29:124–34. https://doi.org/10.1016/j.ymben.2015.03.008.

Williams TC, Espinosa MI, Nielsen LK, Vickers CE. Dynamic regulation of gene expression using sucrose responsive promoters and RNA interference in Saccharomyces cerevisiae. Microb Cell Fact. 2015b;14:43. https://doi.org/10.1186/s12934-015-0223-7.

Winkler WC, Breaker RR. Regulation of bacterial gene expression by riboswitches. Annu Rev Microbiol. 2005;59:487–517. https://doi.org/10.1146/annurev.micro.59.030804.121336.

Woolston BM, Edgar S, Stephanopoulos G. Metabolic engineering: past and future. Annu Rev Chem Biomol Eng. 2013;4:259–88. https://doi.org/10.1146/annurev-chembioeng-061312-103312.

Wu J, Du G, Zhou J, Chen J. Metabolic engineering of Escherichia coli for (2S)-pinocembrin production from glucose by a modular metabolic strategy. Metab Eng. 2013;16:48–55. https://doi.org/10.1016/j.ymben.2012.11.009.

Wu Y, Chen T, Liu Y, Lv X, Li J, Du G, Ledesma-Amaro R, Liu L. CRISPRi allows optimal temporal control of N-acetylglucosamine bioproduction by a dynamic coordination of glucose and xylose metabolism in Bacillus subtilis. Metab Eng. 2018;49:232–41. https://doi.org/10.1016/j.ymben.2018.08.012.

Xu P. Production of chemicals using dynamic control of metabolic fluxes. Curr Opin Biotechnol. 2018;53:12–9. https://doi.org/10.1016/j.copbio.2017.10.009.

Xu P, Gu Q, Wang W, Wong L, Bower AG, Collins CH, Koffas MA. Modular optimization of multi-gene pathways for fatty acids production in E. coli. Nat Commun. 2013;4:1409. https://doi.org/10.1038/ncomms2425.

Xu P, Li L, Zhang F, Stephanopoulos G, Koffas M. Improving fatty acids production by engineering dynamic pathway regulation and metabolic control. Proc Natl Acad Sci. 2014;111:11299–304. https://doi.org/10.1073/pnas.1406401111.

Yang Y, Lin Y, Li L, Linhardt RJ, Yan Y. Regulating malonyl-CoA metabolism via synthetic antisense RNAs for enhanced biosynthesis of natural products. Metab Eng. 2015;29:217–26. https://doi.org/10.1016/j.ymben.2015.03.018.

Yang Y, Lin Y, Wang J, Wu Y, Zhang R, Cheng M, Shen X, Wang J, Chen Z, Li C, Yuan Q, Yan Y. Sensor-regulator and RNAi based bifunctional dynamic control network for engineered microbial synthesis. Nat Commun. 2018;9:1–10. https://doi.org/10.1038/s41467-018-05466-0.

Yang D, Yoo SM, Gu C, Ryu JY, Lee JE, Lee SY. Expanded synthetic small regulatory RNA expression platforms for rapid and multiplex gene expression knockdown. Metab Eng. 2019; https://doi.org/10.1016/j.ymben.2019.04.003.

Yim H, Haselbeck R, Niu W, Pujol-Baxley C, Burgard A, Boldt J, Khandurina J, Trawick JD, Osterhout RE, Stephen R, Estadilla J, Teisan S, Schreyer HB, Andrae S, Yang TH, Lee SY, Burk MJ, Van Dien S. Metabolic engineering of Escherichia coli for direct production of 1, 4-butanediol. Nat Chem Biol. 2011;7:445–52. https://doi.org/10.1038/nchembio.580.

Yim SS, An SJ, Kang M, Lee J, Jeong KJ. Isolation of fully synthetic promoters for high-level gene expression in corynebacterium glutamicum. Biotechnol Bioeng. 2013;110:2959–69. https://doi.org/10.1002/bit.24954.

Yu D, Ellis HM, Lee E-C, Jenkins NA, Copeland NG, Court DL. An efficient recombination system for chromosome engineering in Escherichia coli. Proc Natl Acad Sci. 2002;97:5978–83. https://doi.org/10.1073/pnas.100127597.

Yu B, Zhang X, Sun W, Xi X, Zhao N, Huang Z, Ying Z, Liu L, Liu D, Niu H, Wu J, Zhuang W, Zhu C, Chen Y, Ying H. Continuous citric acid production in repeated-fed batch fermentation by Aspergillus niger immobilized on a new porous foam. J Biotechnol. 2018;276–277:1–9. https://doi.org/10.1016/j.jbiotec.2018.03.015.

Zalatan JG, Lee ME, Almeida R, Gilbert LA, Whitehead EH, La Russa M, Tsai JC, Weissman JS, Dueber JE, Qi LS, Lim WA. Engineering complex synthetic transcriptional programs with CRISPR RNA Scaffolds. Cell. 2015;160:339–50. https://doi.org/10.1016/j.cell.2014.11.052.

Zampieri M, Sekar K, Zamboni N, Sauer U. Frontiers of high-throughput metabolomics. Curr Opin Chem Biol. 2017;36:15–23. https://doi.org/10.1016/j.cbpa.2016.12.006.

Zelcbuch L, Antonovsky N, Bar-Even A, Levin-Karp A, Barenholz U, Dayagi M, Liebermeister W, Flamholz A, Noor E, Amram S, Brandis A, Bareia T, Yofe I, Jubran H, Milo R. Spanning high-dimensional expression space using ribosome-binding site combinatorics. Nucleic Acids Res. 2013;41 https://doi.org/10.1093/nar/gkt151.

Zhang Y, Buchholz F, Muyrers JPP, Francis Stewart A. A new logic for DNA engineering using recombination in Escherichia coli. Nat Genet. 1998;20:123–8. https://doi.org/10.1038/2417.

Zhang F, Carothers JM, Keasling JD. Design of a dynamic sensor-regulator system for production of chemicals and fuels derived from fatty acids. Nat Biotechnol. 2012;30:354–9. https://doi.org/10.1038/nbt.2149.

Zhang F, Wen Y, Guo X. CRISPR/Cas9 for genome editing: progress, implications and challenges. Hum Mol Genet. 2014;23:R40–6. https://doi.org/10.1093/hmg/ddu125.

Zhang X, Tervo CJ, Reed JL. Metabolic assessment of E. coli as a Biofactory for commercial products. Metab Eng. 2016;35:64–74. https://doi.org/10.1016/j.ymben.2016.01.007.

Zhou LB, Zeng AP. Exploring Lysine Riboswitch For Metabolic Flux Control And Improvement of l -lysine synthesis in Corynebacterium glutamicum. ACS Synth Biol. 2015a;4:729–34. https://doi.org/10.1021/sb500332c.

Zhou LB, Zeng AP. Engineering a lysine-ON riboswitch for metabolic control of lysine production in Corynebacterium glutamicum. ACS Synth Biol. 2015b;4:1335–40. https://doi.org/10.1021/acssynbio.5b00075.

Zhou L, Niu DD, Tian KM, Chen XZ, Prior BA, Shen W, Shi GY, Singh S, Wang ZX. Genetically switched d-lactate production in Escherichia coli. Metab Eng. 2012;14:560–8. https://doi.org/10.1016/j.ymben.2012.05.004.

Zhou YJ, Kerkhoven EJ, Nielsen J. Barriers and opportunities in bio-based production of hydrocarbons. Nat Energy. 2018; https://doi.org/10.1038/s41560-018-0197-x.

Zhu L, Zhu Y, Zhang Y, Li Y. Engineering the robustness of industrial microbes through synthetic biology. Trends Microbiol. 2012;20:94–101. https://doi.org/10.1016/j.tim.2011.12.003.

Zhu X, Zhao D, Qiu H, Fan F, Man S, Bi C, Zhang X. The CRISPR/Cas9-facilitated multiplex pathway optimization (CFPO) technique and its application to improve the Escherichia coli xylose utilization pathway. Metab Eng. 2017;43:37–45. https://doi.org/10.1016/j.ymben.2017.08.003.

Zuo E, Huo X, Yao X, Hu X, Sun Y, Yin J, He B, Wang X, Shi L, Ping J, Wei Y, Ying W, Wei W, Liu W, Tang C, Li Y, Hu J, Yang H. CRISPR/Cas9-mediated targeted chromosome elimination. Genome Biol. 2017;18:224. https://doi.org/10.1186/s13059-017-1354-4.

Chapter 3
Microbial Production of Functional Organic Acids

Xueqin Lv, Jingjing Liu, Xian Yin, Liuyan Gu, Li Sun, Guocheng Du, Jian Chen, and Long Liu

3.1 Introduction

The petrochemical industry has influenced the aspects of daily life for many years. Since the beginning of the industry revolution, the demand for valuable chemicals that depend on fossil resources continued to grow. It brought a series of problems, such as excess utilization of petrochemical resource, emission of greenhouse gases, and accumulation of wastes, etc. (Becker et al. 2015). The awareness of necessity to take measures to alleviate the above consequences is raising, mainly in development and utilization of renewable biological resources for production of chemicals (Arslan et al. 2012).

Organic acids are low-molecular-weight organic compounds, their acidity originate from acidic groups such as carboxyl, sulfonic, alcohol, and thiol groups. As an intermediate metabolite of biological cells, they have become mature products after years of research and are widely used in food, pharmaceutical, cosmetics, detergents, polymers and textiles industries (Becker and Wittmann 2015). The growing market demand and emergence of novel applications have resulted in different biosynthetic organic acids (Becker et al. 2015). In 2004, the US Department of Energy's report "Top value added chemicals from biomass: Volume I-Results of screening for potential candidates from sugars and synthesis gas" included four-carbon dicarboxylic acids (succinic acid, fumaric acid, and malic acid), 2,5-furandicarboxylic acid,

X. Lv · J. Liu · X. Yin · L. Gu · L. Sun · G. Du · L. Liu (✉)
Key Laboratory of Carbohydrate Chemistry and Biotechnology, Ministry of Education, Jiangnan University, Wuxi, China

Key Laboratory of Industrial Biotechnology, Ministry of Education, Jiangnan University, Wuxi, China
e-mail: longliu@jiangnan.edu.cn

J. Chen
Key Laboratory of Industrial Biotechnology, Ministry of Education, Jiangnan University, Wuxi, China

L. Liu, J. Chen (eds.), *Systems and Synthetic Biotechnology for Production of Nutraceuticals*, https://doi.org/10.1007/978-981-15-0446-4_3

3-hydroxypropionic acid, glucose diacid, itaconic acid, levulinic acid, etc. (Werpy et al. 2004). In recent years, public concerns over alternative and renewable chemical products, which are expected to reduce dependence on oil reserves and carbon dioxide emissions to the environment. Most organic acids are intermediates in the metabolic pathways that occur naturally in microorganisms (Yin et al. 2015), can be used as substitutes for these chemical products.

Natural organic acids mainly can be found in fruits, such as grapes, apples, peaches and so on. Then the original extracting method is used for extracting organic acid. Nowadays, the methods for producing organic acids include chemical synthesis, enzymatic conversion and microbial fermentation. However, chemical production processes usually lead to unwanted environmental consequences although they are competitive in price. With the development of industrial biotechnology, microbial production has the potential to achieve economic benefit in industry and at the same time reduce toxic waste, alleviate environmental pollution and save energy.

Multiple types of microorganisms can be used in the production of organic acids, such as aspergillus, yeast and bacteria. Some strains for certain organic acid production have natural advantages. For example, the industrial production bacterium for citric acid is *Aspergillus niger* (Xu et al. 2016), *Actinobacillus succinogenes* and *Anaerobiospirillum succiniciproducens* are capable of producing higher production of succinic acid (Carvalho et al. 2016). Yeast is the simplest single-cell eukaryotic model organism. It can tolerate high osmotic pressure and low pH. Due to the clear genetic background and mature genetic manipulation tools, it is the ideal host for the production of organic acids (Sitepu et al. 2014). Although yeast has many advantages as a host, it is only suitable for the production of some organic acids and the yield is low or the production intensity is far from the industrial demand. Filamentous fungi, mainly food-safe *A. oryzae* and *A. niger*, are natural strains with high organic acid production (Yang et al. 2013). *Escherichia coli* and *Bacillus subtilis* are also used for the heterologous expression of organic acid synthesis pathways because they are well-established industrial production hosts (Hossain et al. 2014).

If microorganisms are compared to "surgical objects", industrial biotechnologies can be called "scalpels" to achieve the goal of obtaining strains with high organic acid production. Strain improvement is crucial for reducing the fermentation costs and this is meaningful for expanding potential market of organic acids. Traditional optimization of fermentations focuses on the conditions for the growth of strains. But now the strategies like directed evolution, random mutagenesis, systems biology and synthetic biology are efficient tools for improving the production of organic acids (Chen et al. 2015; Knuf et al. 2014; Zhou et al. 2012). These strategies combine data and omics with metabolic engineering. Results can be predicted in silico by modelling and then be tested in the wet lab (Thakker et al. 2015).

This chapter introduces seven organic acids and summarizes the microbial production of these functional organic acids. Strategies for increasing the production of organic acids in past studies are also summarized.

3.2 Citric Acid

Citric acid (CA), which is commercially produced by microbial fermentation process and is the world's largest consumed organic acid, is extensively used as food additive and commonly used in pharmaceutics and cosmetics industry because of its property as non-toxic acidulant, buffering agent and metallic-ion chelator with pleasant fruit flavor and sour taste (Show et al. 2015). As an intermediate of tricarboxylic acid cycle, CA universally exists in all living cells from microorganisms, animals to plants. Molds (such as *Aspergillus niger*) and yeast strains can produce high level CA in certain condition. It is crucial to optimize CA production by improving the parameters for more economical and environmentally friendly methods and genetic manipulation of microbes employed (Sawant 2018) (Fig. 3.1).

Microbial fermentation was carried out mainly through submerged fermentation and a few through solid substrate fermentation and surface fermentation (Kumar Gupta et al. 2015). Nutritional conditions synergistically affect the CA production including carbon source concentration, dissolved oxygen, hydrogen ions and suboptimal concentrations of phosphate and trace metals, such as manganese (Show et al. 2015).

3.2.1 *Enhanced Citric Acid Production in* Aspergillus niger

For metabolic engineering of *A. niger* to make CA production more effective, the genetic background of the species need to be clear, and several strains either for high enzyme production or for CA production were investigated. Nevertheless, gene examination of *A. niger* ATCC 1015, a CA producing strain, did not point to any unique gene leading to CA hyper-production, when comparing to the genome of *A. niger* 513.88, which is an enzyme producing strain (Andersen et al. 2011). Comparative transcriptome analysis of these two strains grow on glucose-based minimal medium only showed differences of higher expression of alternative oxidative pathway in ATCC 1015. Another comparative genome analysis of wild *A. niger*, which can produce CA and oxalic acid, and its mutant, which cannot secrete large amount organic acids, shows a putative methyltransferase-domain protein (LaeA) is

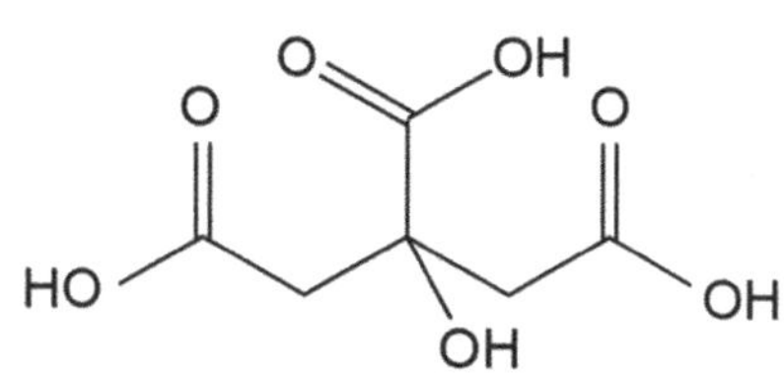

Fig. 3.1 The structure of CA

required for CA production (Niu et al. 2015). The further transcriptome analysis however provide new insight into some details on gene expression of the response by *A. niger* to different culture conditions. The CA fermentation condition of *A. niger* required extreme low pH. As *A. niger* has the ability to grow on low pH medium, the response of *A. niger* to ambient pH on transcriptome level were studied. The pH-dependent cis-acting promoter elements were identified and all steps of the pal/pacC signaling pathway were identified (Andersen et al. 2009). In addition, according to the transcriptome analysis of 4 time-point during CA fermentation of *A. niger* H915–1 (Yin et al. 2017), a CA producing strain, several candidates of low-pH-inducible promoters were studied, and P*gas* was found to be induced by low pH, which can coordinate with the extreme low pH condition during fermentation. The promoter was succeeded used for dynamic control of gene expression during CA fermentation (Ruijter et al. 1999). The low pH required for CA producing is provided mainly by oxalic acid production, but when the CA begins to be produced, the oxalic acid pathway is reduced and the oxalic acid becomes a by-product finally. The oxaloacetate acetylhydrolase (OAH) of *A. niger* was identified and the strain lacking OAH and glucose oxidase could produce CA at pH 5, in which condition the main fermentation product usually was oxalic acid (Ruijter et al. 1999).

The CA fermentation condition of *A. niger* requires high dissolved oxygen, and the alternative oxidation pathway plays an important role. The alternative oxidase (AOX) expression pattern is studied and the results indicate only one copy of aox1 exist on the *A. niger* WU-2223 L, which is another CA producing strain and the activity of *aox1* is comparable to the level of *aox1* transcription (Hattori et al. 2009). The role of AOX performance was detected through deletion and overexpression of AOX in *A. niger.* The *aox1* gene disruption strain reduced CA production from 158.9 g/L to 125.6 g/L and loose mycelial pellets were formed. Moreover the *aox1* overexpression strains had more oxygen consumption, which lead to higher energy metabolism to produce higher NADH, and CA yield increased by 13.5% (Hou et al. 2018).

The citrate acid transport system for high secretion of CA has some breakthrough these years. A CA exporter was identified through engineering *A. niger* by overexpression and disruption the transporter gene. Through homology searching using an itaconic acid transporter from *Ustilago maydis* as template, a major facilitator superfamily protein, CexA, was identified. Knockout the CexA makes the strain secret mainly oxalic acid and abolishes CA secretion completely. Expression of CexA in *Saccharomyces cerevisiae* provided the yeast the ability to secret CA. Furthermore, overexpression of CexA in *A. niger* significantly improved the CA production and employing the inducible expression system, tet-on system, the production increased by fivefolds (Steiger et al. 2019). The experimental evidence of the CA export process provide important element for scientist to improve *A. niger* physical characteristics for more efficient CA production.

As TCA cycle exists in mitochondria, it depends on a malate-citrate shuttle transporter to facilitate the CA being transformed to cytosol and a CA secreting protein to transform CA further outside the *A. niger*. Deletion a putative mitochondrial citrate transport protein (CTP) in *A. niger* WU-2223 L changed cell growth in early-log phase, and the CA production was reduced maybe resulted from growth inhibition

(Kirimura et al. 2016). The function of mitochondrial citrate transporters CtpA and hmA from *A. kawachii*, which were homologous to Ctp1 and Yhm2 from *S. cerevisiae* were studied. The purified CtpA and YhmA were reconstituted into liposome and both of them can transport CA with counter substrates. Disruption of either gene caused deficient hyphal growth and CA production because of acetyl-CoA deficient (Kadooka et al. 2019). When comparing the spore germination rate and growth characteristics of CA high-yield strain *A. niger* CGMCC 5751 and *A. niger* ATCC 1015, the CA high-yield strain was more sensitive to antimycin, and its energy metabolism system was weaker than those of ATCC 1015, suggesting excess ATP as an inhibitor for CA accumulation (Wang et al. 2015).

The glycolysis and TCA cycle are strictly controlled by intermediate metabolite feedback. One of the key enzymes, whose activity was stringent controlled, is 6-phosphofructo-1-kinase (PFK1). A modified PFK1, which is resistant to CA inhibition, was obtained through expression a shorter form of PFKq with a modified threonine residue to glutamic acid (Capuder et al. 2009). The effect of the shorter enzyme on CA producing is still unknown. Nevertheless, Overexpression of whole length PFK1 and pyruvate kinase in *A. niger* have no influence on CA production and even not affect intermediary metabolite levels (Ruijter et al. 1997). Some effects have made to decrease the hexokinase inhibitor, trehalose-6-phosphate, through disruption of encoding gene, *ggsA*. The resulted strain accumulated CA earlier, whereas the multicopy strain showed the reverse effect (Arisan-Atac et al. 1996). The citrate synthase (*citA*) of *A. niger* was also identified through overexpression, the CA level did not improved though the CitA activity increased by tenfolds (Ruijter et al. 2000). The cytosolic ATP: citrate lyase (*acl*) may conduct a futile cycle to hydrolyze cytosolic CA transformed from mitochondria before secretion. The *acl* gene deletion *A. niger* was studied for the organic acid producing pattern, and the production changed to the direction of succinic acid (Meijer et al. 2009). The cytosolic Acl contains 2 subunits, Acl1 and Acl2, and loss of either the subunit lead to loss of enzyme activity. Deletion of acl1 or acl2 gene in *A. niger* resulted in a dramatic decrease of acetyl-CoA and CA level (Chen et al. 2014a).

The mycelial morphology of *A. niger* also plays important role in CA fermentation. It influences the oxygen and mass transfer efficiency. The cell wall of *A. niger* consists of 80–90% polysaccharides, and chitin is one of the compounds in the cell wall. The role of chitin synthase gene, *chsC*, in morphogenesis and CA production was studied through RNAi to silence *chsC*. The compactness of the mycelial pellets decreased, as a result reduced viscosity of the medium and improved solute transfer efficiency, and finally improved the CA production (Sun et al. 2018). The influence of manganese on CA production was also studied. Fourteen ppb or higher level of manganese switches the morphology from pellet to filamentous hypha. Twenty-two genes were identified to be responsive to manganese through suppression subtractive hybridization (Dai et al. 2004).

During CA fermentation, the non-fermentable isomaltose was produced as by-product and as a result inhibit the product yield. The alpha-glucosidase (*agdA*) gene was disrupted and the resulted strain overexpressed glucoamylase and the alpha glucoamylase activity was reduced by 62.5%. The *agdA* deleted strain could not accumulate isomaltase and the final CA production increased by 16.87% (Wang et al. 2016).

Several genes involved in the reductive branch of the TCA cycle, for instance malate dehydrogenase, cytosolic targeted fumarase and fumarate reductase from *S. cerevisiae*, were inserted into genome of *A. niger*, and the transformant strains produced CA with higher yield and productivity than that of wild strain. Moreover, the strain, which overexpressed both fumarase and fumarate reductase resulted in a maximum yield of 0.9 g/g glucose (de Jongh and Nielsen 2008).

Fermentation conditions for CA production by *A. niger* are well studied and the yield often exceed 70% on the carbon source (Papagianni 2007). Though monosaccharides and disaccharides can be more rapidly metabolized, polysaccharides processed decomposition can be used to achieving economic purpose, and are more commercially used by CA producing company (Yin et al. 2017). For instance, the sweet potato starch hydrolysate can serve as carbon source for CA fermentation with maximum product concentration of 83 g/L (Betiku and Adesina 2013). Nevertheless, many studies focused on searching for novel and economical sources for fermentation. A variety of wastes produced from agriculture and oil industry were served as cost effective substrate for CA fermentation (Sawant 2018). When oil palm empty fruit bunches were used for citrate fermentation, 6.4% (w/w) of sucrose and 9% (v/w) of minerals addition with 15.5% (v/w) of inoculum provided maximum CA production (Bari et al. 2009). Banana peel as a substrate required moisture of 70%, temperature of 28 degrees C, initial pH of 3 and inoculum of 10(8) spores/mL as the most suitable condition for *A. niger* to produce CA (Karthikeyan and Sivakumar 2010). As a waste of apple processing industries, apple pomace sludge can also serve as inexpensive substrate for CA producing. The condition of 25 g/L initial total solids with 3% (v/v) methanol can conduct 44.9 g/100 g dry substrate for 144 h fermentation (Dhillon et al. 2011). The CA production by *A. niger* on cassava peel substrate can reach to 88.73 g/L (Adeoye et al. 2015). Mixing Spanish-style green olive processing waste with equal quantity of white grape pomace, a satisfactory amount of carbon sources were provided. Various nutrients and fermentation conditions were investigated and a final CA production of 85 g/L was obtained (Papadaki and Mantzouridou 2019). The solid-state fermentation condition for CA production of *A. niger* ATCC 9142 to utilize the corn distillers dried grains, which was an ethanol fermentation co-product, was studied and the phosphate-treated grains increased CA production (Xie and West 2009). *A. niger* AA120, a mutant strain tolerant to tannin, was used for fermentation on acorn starch medium containing 20 g/L tannin, and the CA production can reach 130.8 g/L with biomass at 32.9 g/L (Zhang et al. 2018).

CA fermentation by immobilized *A.niger* was also investigated. Conidiaspores were entrapped in Ca-alginate beads, and a starting concentration of 0.05 g/L nitrogen and 140 g/L sucrose addition with 4.0 mL methanol and 3.0 mL ethanol was found provide maximum CA production (Demirel et al. 2005).

3.2.2 *Citric Acid Production by Yeast*

Several kinds of yeast were also employed for CA production. Comparing to mold, yeast has also good production of CA and is tolerance to high substrate concentrations, and it does not need to control the filamentous morphology, which is essential for oxygen transfer and mixing efficiency and complicated fermentation process by *A. niger* (Carsanba et al. 2019). Nevertheless, the drawback for CA production by yeast is a high amount of iso-CA as by-product (Cavallo et al. 2017). The Fermentation conditions are related to the types of bacteria, carbon sources, C/N ratio, pH, temperature and oxygen conversion rate. For *Yarrowia lipolytica*, the initial C/N molar ratio of 367 provides CA titer of 73.3 g/L (Carsanba et al. 2019). Expression of INU1 gene to immobilize inulinase on *Kluyveromyces marxianue* CBS 6556 provided the transformants ability to hydrolyze 9.3% inulin within 10 h and CA production reached 77.9 g/L (Liu et al. 2010a). Based on the inulinase gene recombinant *Y. lipolytica* strain SWJ-1b, knock-out of the ATP-citrate lyase genes (ACL1) and overexpression of isocitrate lyase gene (ICL1) increased the yield of CA to 89.6% when using inulin as sole carbon source (Liu et al. 2013a).

The CA accumulation in *Y. lipolytica* required a suitable N/C ratio at around 0.021 Nmol/Cmol. If the N/C ratio arrived to 0.085 Nmol/Cmol, the yeast will increase lipid accumulation (Ochoa-Estopier and Guillouet 2014). The medium constituents for *Y. lipolytica* to use pineapple waste as substrate in solid state fermentation were optimized according to Plackett-Burman design and further central composite design, and the CA production reached above 200 g/kg dried pineapple waste (Imandi et al. 2008). The olive-mill wastewater were also used as based media for CA production, and the unsaturated fatty acids synthesis was improved simultaneously (Papanikolaou et al. 2008). During batch cultivation, *Y. lipolytica* produced CA at 112 g/L with a yield of 0.6 g/g and productivity of 0.71 g/L/h in the medium containing glycerol waste of biodiesel industry. Moreover, the productivity reached to 1.42 g/L/h with little changes in production and yield when CA biosynthesis prolonged up to 300 h through cell recycling (Rymowicz et al. 2010). The addition of Triton X-100, which is a surfactant with effects upon cell membrane permeability, increased the CA production by 40–80% (Mirbagheri et al. 2011).

Overexpression of pyruvate carboxylase gene from *Penicillium rubens* in yeast *Y. lipolytica* SWJ-1b enhanced CA production. The final concentration of CA reached 111 g/L and the yield was 0.93 g/g (Fu et al. 2016). Gene-dose-dependent overexpression of isocitrate lyase (ICL1) in *Y. lipolytica* resulted in a strong metabolite shift to form CA and synthesis less iso-CA (Förster et al. 2017). The influence of aconitase (ACO) overexpression on CA and iso-CA fermentation in *Y. lipolytica* was also studied. The organic acid production pattern changed into the direction of higher iso-CA production on either sunflower oil medium or glycerol medium (Holz et al. 2009) (Fig. 3.2).

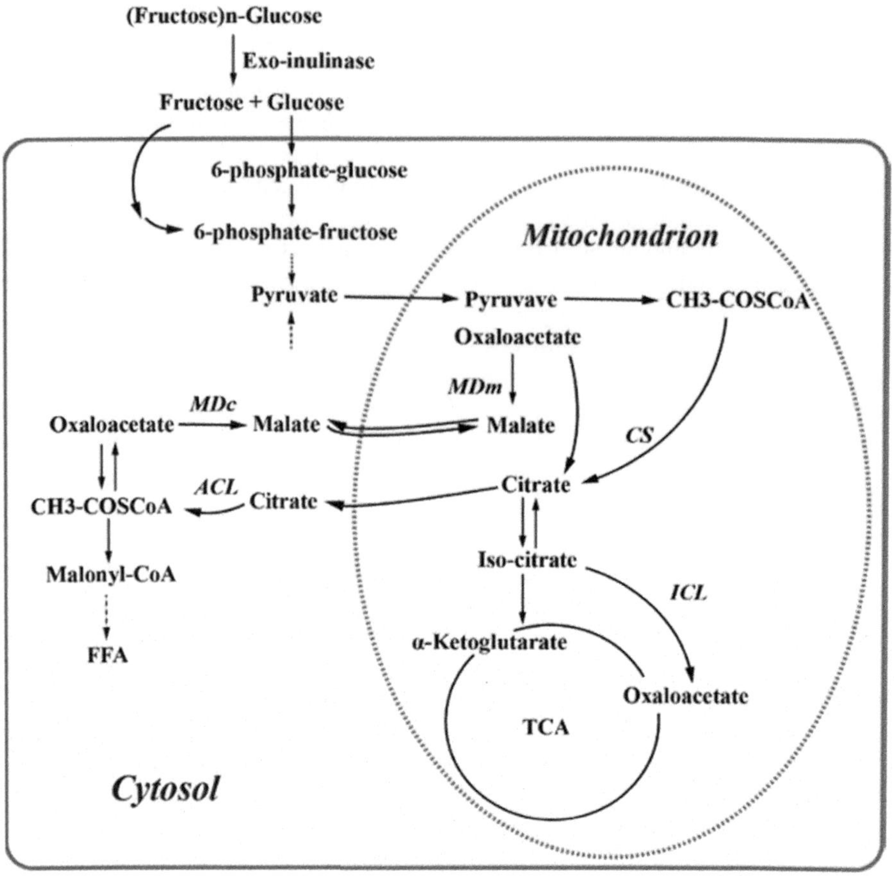

Fig. 3.2 The pathway of inulin hydrolysis and citric acid biosynthesis. *(Fructose)n-Gluocse* inulin, *CS* citrate synthase, *ICL* iso-citrate lyase, *FFA* free fatty acid, *ACL* ATP-citrate lyase, *MDc* malate dehydrogenase (cytoplasmic), *MDm* malate dehydrogenase (mitochondria)

3.3 Alpha-Ketoglutaric Acid

Alpha-ketoglutaric acid (α-KG) is intermediate in the TCA cycle and amino acid metabolism (Yin et al. 2015; Brunhuber et al. 2000). It could be synthesized into polymers for tissue scaffolds and therapeutic delivery (Barrett and Yousaf 2008) (Fig. 3.3).

Currently, α-KG is chemically synthesized from succinic acid and oxalic acid diethyl esters with cyanohydrines or by hydrolysis of acyl cyanides (Stottmeister et al. 2005). This chemical strategy for α-KG is not only environmentally unfriendly but also generates toxic waste. Microbial fermentation, enzymes and whole-cell

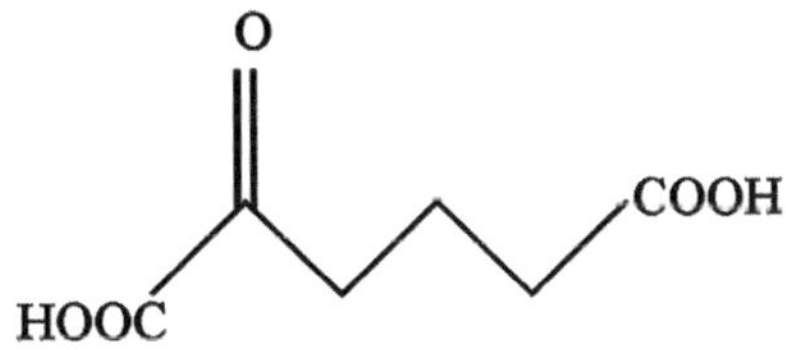

Fig. 3.3 The structure of α-KG

biotransformation are effective strategies to replace chemical production ofα-KG (Finogenova et al. 2005). However, these strategies have different advantages and disadvantages in the production of alpha-KG.

3.3.1 α-KG Production by Microbial Fermentation

In 1946, Lockwood and Stodola firstly developed the synthesis of α-KG by cultivating *Pseudomonas fluorescens*. The cheap carbon resource glucose was excess and this was important for production of α-KG. After 112 h fermentation in bioreactor, α-KG reached 16–17 g/100 g of glucose (Lockwood and Stodola 1946). Koepsell et al. used synthetic medium and 48 g α-KG/100 g of glucose was obtained after 263 h by *P. fluorescens* (Koepsell et al. 1952). Asai et al. found *Serratia marcescens* no.18 can produce 18.2 g/L α-KG (Asai et al. 1955). Latter, they found that *Bacillus megatherium*, *Bacillus mesentericus*, *Bacterium succinicum* and so on could also produce α-KG. In addition, Tsugawa et al. described *Candida lipolytica* AJ5004, which was later renamed *Y. lipolytica*. It has been reported that 195 g/L was produced from *Y. lipolytica* H355 using n-paraffins (C12-C18) as substrates. However, due to the high price and limited supply of n-paraffins, as well as the difficulty of fermentation operation, the α-KG production using paraffins has not reached the industrial scale. Although using ethanol as carbon source, the amount of α-KG produced by *Y. lipolytica* N1 can reach 49 g/L (Chernyavskaya et al. 2000; Il'Chenko et al. 2002), the high cost of ethanol also makes it difficult to carry out large-scale production. *Candida paludigena* BKM Y-2443 and *Pichia inositovora* BKM Y-2494 can produce 17.1 g/L and 21.9 g/L of α-KG on ethanol. However, thiamine was the production limitation of *Candida* and *Pichia* strains (Chernyavskaya et al. 1997). The production of α-KG increased to 186 g/L by using renewable carbon source raw glycerol (Yovkova et al. 2014).

Therefore, efforts to find a better source of substrates are still under way in order to make this process economically viable. Although microbial fermentation produces less environmental pollution than chemical synthesis, it is still limited to laboratory research due to the high production cost and a large number of by-products (Liu et al. 2013b).

3.3.2 α-KG Production by Biotransformation

Recombinant L-glutamate oxidase was used to transform L-glutamic acid into α-KG and the titer could reach 104.7 g/L (Niu et al. 2014), but catalase, together with oxidase, was required to eliminate large amounts of H_2O_2 produced. It has been reported that L-amino acid deaminase (L-AAD) from *Proteus* species was used as whole-cell biocatalyst to produce alpha-KG in different expression systems without producing H_2O_2 as a by-product (Molla et al. 2017). In addition, *Bacillus subtilis* expressing an L-amino acid deaminase (L-AAD) from *Proteus mirabilis* KCTC 2566 was selected to as a whole-cell biocatalyst for the production of α-KG from L-glutamic acid with a yield of about 4.65 g/L (Hossain et al. 2014).

Whole-cell biotransformation is suitable for the production ofα-KG with the advantages of simple separation process and high bioconversion rate (95.18%), and it is an immobilized biocatalyst, which can be used in large-scale production (Liu et al. 2013b; Niu et al. 2014). However, the whole-cell biocatalysis showed some disadvantages, like substrate uptake limitation and product inhibition possible.

3.3.3 Use of Metabolic Engineering and Other Novel Methods to Improve α-KG Production

The culture medium optimization is the basic strategy because a good culture medium is the key part in both laboratory and industry. In *T. glabrata* CCTCC M202019, α-KG concentration increased to 43.7 g/L by changing the concentrations of thiamine, biotin and Ca^{2+} (Liu et al. 2007).

Metabolic engineering is a powerful tool for the improvement of α-KG. By optimizing the existing pathways of introducing pathway components, the metabolism of the strain is modified and the production of α-KG was improved. To enhance the pyruvate carboxylation pathway and redistribute the carbon flux from pyruvic acid (PA) to α-KG, the pyruvate carboxylase genes *ScPYC1* from *S. cerevisiae* and *RoPYC2* from *Rhizopus oryzae* were overexpressed. The yields of α-KG increased by 24.5 and 35.3%, respectively (Yin et al. 2012). By overexpressing genes encoding $NADP^+$-dependent isocitrate dehydrogenase (*IDP1*) and pyruvate carboxylase (*PYC1*) in *Y. lipolytica*, the amount of secreted α-KG was increased and by-product PA was significantly reduced. Simultaneous overexpression of *IDP1* and *PYC1* made a shift of the carbon flux from glycolysis to oxaloacetate in tricarboxylic acid cycle (Yovkova et al. 2014).

To solve the degradation of α-KG in *B. subtilis* 168, the *sucA* gene encoding α-KG dehydrogenase was deleted. Error-prone PCR is a way of introducing mutation and confirming the sites influencing the catalytic efficiency. Then directed evolution, which is an effective way to engineer enzymes, can accomplished in the laboratory test tubes. Combining the above deletion with evolution of L-ADD, α-KG production was increased from 4.65 g/L to 12.21 g/L (Hossain et al. 2014).

The integration of error-prone PCR and gene shuffling was shown to be an effective method. To solve the low substrate solubility and low α-KG production, eight rounds of error-prone PCR and four rounds of gene shuffling were applied in *P. mirabilis* L-AAD KCTC 2566. The titer of α-KG reached 89.11 g/L and was seven times than before (Hossain et al. 2016).

Novel tools like CRISPR/Cas9 editing technology developed for metabolic engineering has great potential for improvement for α-KG production. The biosensor constructed by combining enzyme with carbon fiber can show a rapid response to dynamic changes in the α-KG concentrations, which may be easier to detect α-KG in further studies (Poorahong et al. 2011). Traditional metabolic engineering combined with new biotechnology creates more potential of strains.

3.4 Succinic Acid

C4 dicarboxylic acids, including malic acid (MA), fumaric acid (FA), and succinic acid (SA), are essential intermediates of cell metabolism. In 2004, the U.S. Department of Energy identified C4 dicarboxylic acids as one of the top 12 biomass-derived building block chemicals, which could be produced from renewable carbohydrates by chemical or biological conversion (Werpy et al. 2004). At present, C4 dicarboxylic acids have been widely used in food, agriculture, pharmaceutical, fine chemical industry and other fields (Thakker et al. 2015).

SA is an intermediate of the TCA cycle and one of the final products of anaerobic metabolism. Among organisms, a variety of fungi and bacteria have been recognized as suitable hosts for the efficient SA production, such as *Issatchenkia orientalis*, *Y. lipolytica*, *S. cerevisiae*, *Actinobacillus succinogenes*, *E. coli*, *Anaerobiospirillum succiniciproducens*, and *Mannheimia succiniciproducens*, and much efforts including strain selection, metabolic engineering approaches and process optimization has been exerted to develop processes for SA production from renewable resources (Ahn et al. 2016) (Fig. 3.4).

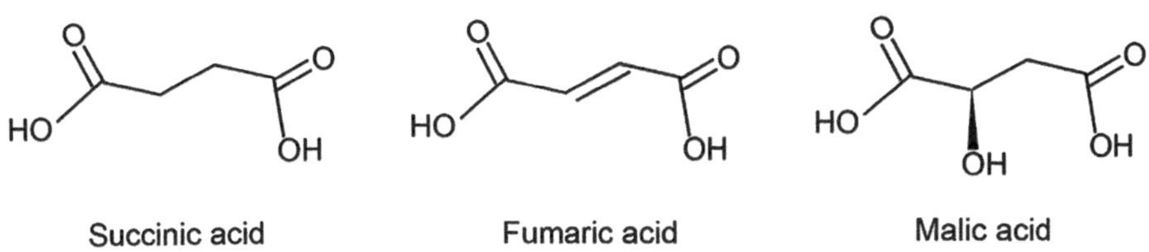

Fig. 3.4 The structure of C4 dicarboxylic acids

3.4.1 *Succinic Acid Production by* Actinobacillus succinogenes

A. succinogenes is considered to be the best natural SA producer (Dessie et al. 2018). Early in 1999, *A. succinogenes* was isolated from bovine rumen, it is an osmotolerant, facultative, anaerobic gram-negative bacterium, can accumulate more than 70 g/L SA (Guettler et al. 1999). Except production capacity, as a producer, *A. succinogenes* has more advantages including non-pathogenic, extensive use of cheap carbon sources, scalable biorefinery flow performance (Bradfield et al. 2015; Salvachúa et al. 2016), tolerance to high concentrations of glucose and SA (Guettler et al. 1999; Song and Lee 2015), fixation and consumption of CO_2 (Van der Werf et al. 1997) and so on.

Generally, there are three pathways for SA production in microbes: reductive TCA (rTCA) branch, oxidative TCA cycle, and glyoxylate shunt (Fig. 3.5). Because of the lack of glyoxylate and Entner-Doudoroff pathways, citrate synthase and isocitrate dehydrogenase, wild *A. succinogenes* could not produce SA through glyoxylate shunt and oxidative TCA cycle, only via rTCA pathway. Under anaero-

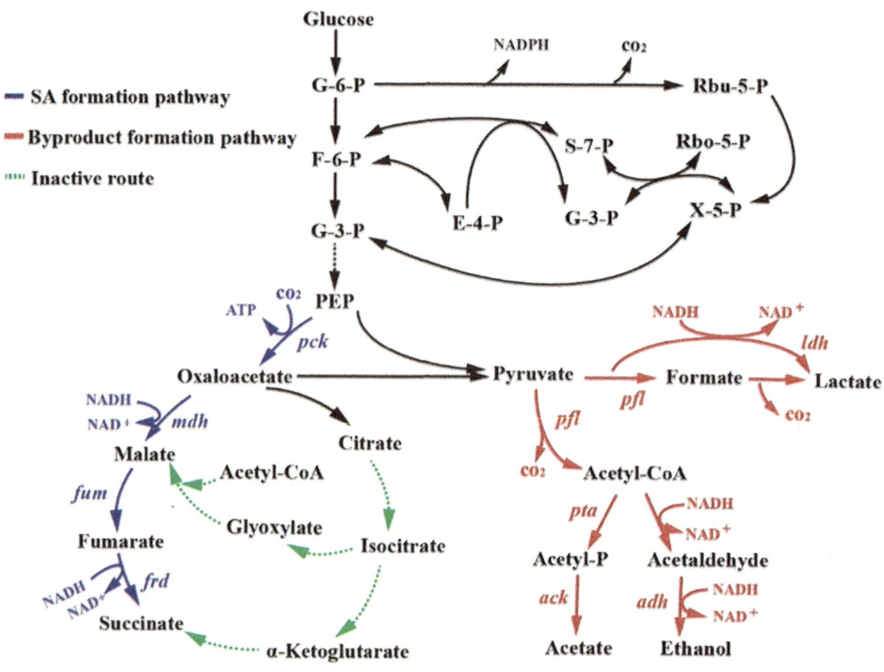

Fig. 3.5 Pathways related to succinate production in *A. succinogenes*. G-6-P glucose-6-phosphate, F-6-P fructose-6-phosphate, G-3-P glyceraldehyde-3-phosphate, PEP phosphoenolpyruvate, Rbu-5-P ribulose-5-phosphate, Rbo-5-P ribose-5-phosphate, S-7-P sedoheptulose-7-phosphate, X-5-P xylulose-5-phosphate, E-4-P erythrose-4-phosphate, *pck* PEP carboxykinase, *mdh* malate dehydrogenase, *fum* fumarase, *frd* fumarate reductase, *ldh* lactate dehydrogenase, *pfl* pyruvate-formate lyase, *pta* phosphotransacetylase, *ack* acetyl-kinase, *adh* acetaldehyde dehydrogenase

bic conditions, the non-oxidative rTCA pathway is ATP neutral and can fix 1 mol CO_2/mol SA, resulting in a maximum theoretical yield of 2 mol SA/mol glucose. However, the synthesis of 1 mol succinate requires 2 mol NADH, while 1 mol glucose can only provide 2 mol NADH through glycolytic pathway. Therefore, assuming that all carbon fluxes go through anaerobic fermentation only, the yield of SA is limited to 1 mol SA/mol glucose (Dessie et al. 2018; Vuoristo et al. 2016).

Due to the immaturity of genetic tools, some chemical mutation methods were used for strain evolution, such as adaptation to fluoroacetate (Guettler et al. 1999), ammoniumion (Ye et al. 2013a), and NaCl (Aramaki et al. 2002). A genome shuffling technique was applied to improve fermentative production of SA by *A. succinogenes*, a high SA-producing strain was obtained (95.6 g/L), due to increased activities of enzymes about SA production and NADH synthesis, and decreased acetic acid formation (Zheng et al. 2013).

Although these mutation methods can increase SA production, they are also uncontrollable, because it is generally difficult to test the exact pattern of mutagenesis, and even cause irreversible damage to the genome. Therefore, it is very important and crucial to design systematic metabolic engineering strategies to achieve expected genotype or phenotypic traits to improve the performance of strains (Dessie et al. 2018). SA can be produced by *A. succinogenes* from a variety of carbon sources, including cane molasses (Liu et al. 2008), glycerol (Vieille 2014), xylose (Bradfield and Nicol 2016), corn steep liquor (Guarnieri et al. 2017), etc. Guarnieri et al. enhanced SA flux by overexpressing SA biosynthetic machinery in rTCA branch and removal of competitive carbon pathways (acetate and formate synthesis) leads to the production of SA with higher purity, but also to the production of new by-products (lactic acid) (Guarnieri et al. 2017).

Moreover, cell immobilization technology can promote the separation of products, and the immobilized cells can be reused with high stability. *A. succinogenes* can be efficiently attached on plastic composite supports (Urbance et al. 2004), Poraver® (Maharaj et al. 2014), cotton fiber (Yan et al. 2014b), porous polyurethane filler (Shi et al. 2014), microfiber membrane (Chen et al. 2017a) and luffa sponge matrices (Cao et al. 2018). Compared with batch culture, repeated batch culture can significantly increase the yield of SA, but the production of SA with sodium hydroxide as neutralizer has the problems of prolonged cell cycle and blocked cell attachment (Cao et al. 2018).

3.4.2 *Exploring Succinic Acid Production by Engineered Yeast*

Yeast has remarkable advantages in organic acid production because of its good resistance to low pH and high osmotic pressure (Sitepu et al. 2014). Due to the lack of reduction pathway in yeast, constructing rTCA pathway in yeast can induce cells to produce high concentration of C4 dicarboxylic acid. For example, when PYC, MDH, FRD and Fum wrer overexpressed in the cytoplasm and GPD1 was knocked out, the SA production of engineering yeast can reach 12.97 ± 0.42 g/L under optimal conditions (Yan et al. 2014a).

Y. lipolytica is an unconventional and strictly aerobic yeast, it can produce considerable amounts of SA at low pH (Yuzbashev et al. 2010). In *Y. lipolytica*, the defects of succinate dehydrogenase subunit (SDH1 or 2) inhibited the growth on glucose while the mutants grew on glycerol and produced succinate in the presence of the buffer $CaCO_3$. The strain can produce more than 45 g/L succinate in shaking flask with buffering or 17.5 g/L without buffering after deletion of SDH2 (Yuzbashev et al. 2010). During fermentation, if no additional pH adjustment procedure is helpful to decrease the downstream cost in industrial application. The evolutional strain from long-term cultivation in chemostat culture was isolated and can utilize glucose, high titer SA (45 g/L) was accumulated without any pH control (Yuzbashev et al. 2016). Knockout of SDH5 subunit in wild *Y. lipolytica* led to SA accumulation and secretion significantly. In combination with optimization of media, which resulted in 43 g/L SA production from crude glycerol. Using the fed-batch strategy in 2.5 L fermenter, up to 160 g/L SA was yielded (Gao et al. 2016). For elimination of by-product after inactivation of SDH5, the source of acetate be identified and CoA-transferase gene *Ylach* be deleted. Overexpression of phosphoenolpyruvate carboxykinase (ScPCK) from *S. cerevisiae* and endogenous succinyl-CoA synthase beta subunit (YlSCS2), the engineered strain produced 110.7 g/L SA with a yield of 0.53 g/g glycerol without pH control in fed-batch fermentation, this high efficiency with great industrial prospects for SA production.

3.4.3 Metabolic Engineering of Escherichia coli *for Succinic Acid Production*

The production of SA by *E. coli* has been widely studied and applied, and the production strategy is generally dual-phase fermentation, which consists of aerobic growth and anaerobic fermentation (Isar et al. 2006; Chen et al. 2014b; Wu et al. 2012). Engineering *E. coli* constructed by knocking out lactate dehydrogenase A (LdhA) and pyruvate formate lyase (PflB) or further deleting phosphotransacetylase acetate kinase (PtsG) and PPC was usually used as the original producer. Considering that *E. coli* can produce SA from various cheap renewable carbon sources, using mixed carbohydrates and improving saccharifying metabolism as the main strategies to improve fermentation performance (Liu et al. 2013c; Zheng et al. 2009a). Furthermore, in order to improve ATP supply during *E. coli* fermentation to increase SA production, it is an effective method to overexpress the PEP carboxykinase (PCK) in strains with the deletion of *ppc*, *ldhA*, *pflB* and *ptsG*. For example, the concentration of SA produced from bagasse hydrolysate increased to 39.3 g/L after 120 h of batch fed-batch fermentation with the recombinant bacteria mentioned above (Liu et al. 2013c). In addition, the repeated production of SA by *E. coli* was realized by simultaneously consuming glucose and xylose. The final SA titer was 83 g/L and the yield was 0.87 g/g sugar mixture (Liang et al. 2013). A new engineering strain was constructed by overexpressing the magnesium transporter gene *mgtA* in *E. coli* mutant strain, which could use $Mg(OH)_2$ and $NH_3{\cdot}H_2O$ instead of $MgCO_3$

as alkaline neutralizer (Wang et al. 2014). Additionally, methanol as an auxiliary carbon source was benefit for SA production in methylotrophic *E. coli* (Zhang et al. 2018).

3.5 Malic Acid

As an acidulant and flavor enhancer, malic acid (MA) is widely used in the food and beverage industries. In addition, MA is also used for alkyl and unsaturated polyester resins and coatings (Liu et al. 2017a). According to the analysis of MA demand in the global market, the annual demand for MA is 200,000 tons, while its actual annual output is only 40,000 tons, which is far from meeting the requirements (Chi et al. 2016a). Due to the advantages of low cost, low energy and high yield of MA biosynthesis, many studies on MA biosynthesis and enzyme catalysis have been carried out in the past two or three decades, and two-step or one-step fermentation technology of microorganisms has been developed for this purpose (Chi et al. 2016a).

Compared with conventional chemical synthesis, MA synthesized by microbial fermentation has high purity (no D-MA pollution), mild reaction conditions and high production efficiency. And microbial fermentation can also produce MA from multiple substrates (Liu et al. 2017a). In addition, several strains have been reported for MA production, such as *Aspergillus flavus*, *E. coli*, *Zygosaccharomyces rouxii*, *Rhizopus delemar*, *R. oryzae*, *Ustilago trichophora*, *S. cerevisiae*, *A. oryzae*, and *Aureobasidium pullulans*.

Some natural strains, including *A. flavus* and *Z. rouxii*, accumulate a large amount of L-MA in the medium containing high concentration of glucose, inorganic salts, a certain amount of nitrogen and $CaCO_3$ (Battat et al. 1991; Taing and Taing 2006). With the development of systems biology, synthetic biology and other technologies, many new metabolic engineering strategies have been used in the design and construction of cell factories.

3.5.1 Engineering Cytosolic rTCA Pathway and Transporter for Malic Acid Production

There are four metabolic pathways for MA production (Fig. 3.6) (Brown et al. 2013). Reductive tricarboxylic acid (rTCA) pathway is the first and the most important pathway, in which pyruvate carboxylates pyruvate to oxaloacetic acid (OAA) via pyruvate carboxylase (PYC), and then malic dehydrogenase (MDH) reduces OAA to MA. The maximum theoretical yield of this process is 2 mol MA/mol glucose (Yin et al. 2015). The second pathway is the TCA cycle, and the theoretically maximum yield of MA is limited to 1 mol/mol glucose. The other two pathways of MA formation are cyclic and noncyclic glyoxylate cycles. The maximum MA yield from glucose is 1 mol MA/mol glucose due to the oxidative decarboxyl-

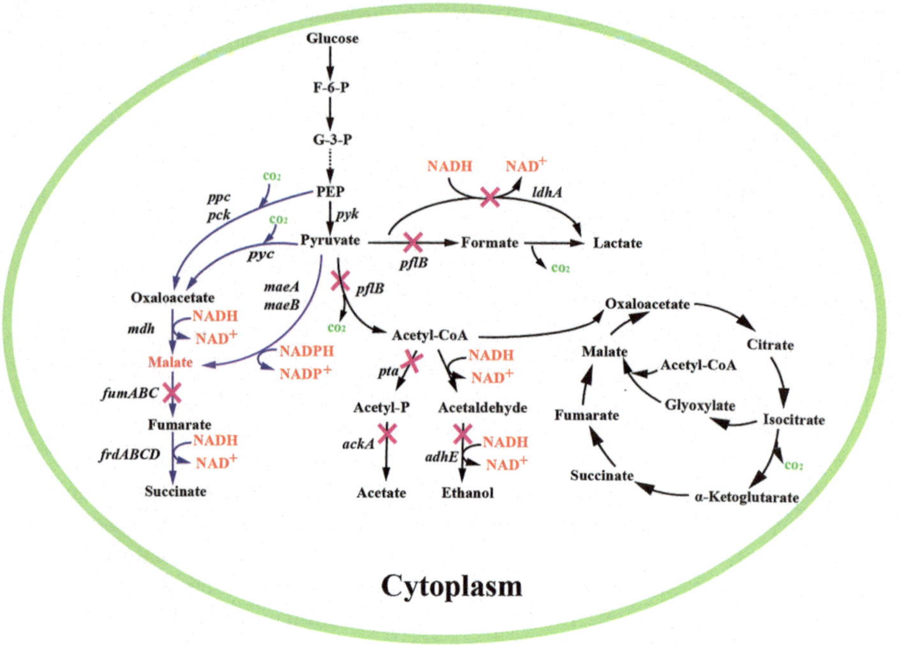

Fig. 3.6 Engineering strategies for malic acid production. The rTCA pathway is represented by purple lines. *pyk* pyruvate kinase, PEP phosphoenolpyruvate, *pck* phosphoenolpyruvate carboxykinase, *ppc* phosphoenolpyruvate carboxylase, *mdh* malic dehydrogenase, pyc pyruvate carboxylase, *fumABC* fumarase, *frdABCD* fumaric reductase, *aspA* aspartase, *pflB* formate acetyltransferase, *ldhA* lactate dehydrogenase, *pta* phosphate acetyltransferase, *adhE* alcohol dehydrogenase, *ackA* acetate kinase

ation of acetyl-Co A by pyruvate in cyclic glyoxylate cycle. However, pyruvate carboxylation can supplement OAA in the noncyclic glyoxylate cycle route, resulting in a theoretical maximum yield of 1.33 mol MA/mol glucose.

Modular pathway reorganization of intracellular product synthesis and transport through metabolic engineering is an effective way to improve microbial production capacity (Qin et al. 2015). Because rTCA pathway has a high carbon conversion rate, which leads to net fixation rather than release of carbon dioxide, overexpression or construction of rTCA pathway has always been the preferred rational strategy. Overexpression of natural or heterologous C4 dicarboxylic acid transporters is a common method to improve MA transport. In addition, regulation of the coenzyme regeneration and the evolution of malic enzyme catalyzing the production of MA from pyruvate are also used to produce MA (Ye et al. 2013b; Zheng et al. 2009b).

Yeast is tolerant to low pH and high osmotic pressure, so it has certain advantages in organic acid production (Taing and Taing 2006). So far, the use of *S. cerevisiae* for MA production has been widely studied. In the engineered yeast, overexpression of the *FUMI* gene which encodes fumarase in *S. cerevisiae*, was proved to be an efficient process for converting fumarate acid into MA (Peleg et al.

1990). And an ideal engineering system for MA production was obtained by strengthening the MA transport and optimizing the rTCA pathway. In order to enhance the transportability of MA, the C4T318 gene of *A. oryzae* and the *mael* gene encoding C4 dicarboxylate transporter of *Schizosaccharomyces pombe* were overexpressed, respectively (Brown et al. 2013; Grobler et al. 1995). Overexpression of *MDH3, PYC2* and *mae1* genes in yeast showed that the MA titer could reach 59 g/L and the yield was 0.42 mol/mol glucose (Zelle et al. 2008). Under the optimum conditions (10 mM $CaCl_2$, 15% CO_2-enriched air, oxygen concentration reduced to 3%, and pH 6.8), the MA yield of the engineering strain was 0.48 mol/mol glucose in the bioreactor (Zelle et al. 2010). Similarly, when *mae1* gene and heterologous genes AfPYC, ROPYC and ROMDH were co-expressed in multi-vitamin auxotrophic *E. coli* and *T. glabrata*, the cumulative amount of MA was 30.25 g/L and 8.5 g/L, respectively (Chen et al. 2013, 2017b). Moreover, genetically engineered *E. coli* for succinate and fumarate production can also be used as a MA producer after knocking out the fumarate reductase (*frdBC*) in the rTCA pathway (Zhang et al. 2011). In addition, by metabolizing more than 2000 generations and deleting central anaerobic fermentation genes related to NAD+ regeneration, including *ldhA*, *pfl*B, *ackA, adhE, focA* and *mgsA*, an engineering *E. coli* for the production of succinic acid and MA was constructed. Under the control of anaerobic stirring and pH, the yield of MA by the recombinant strain can reach 516 mM with inorganic salts as raw materials (Jantama et al. 2008).

The difference between *A. oryzae* and *A. flavus* is that the former is GRAS and does not produce aflatoxin, which has been used in food industry for a long time. *A. oryzae* has great potential for MA production. By overexpressing natural cytoplasmic alleles such as pyc, MDH and C4T318, the MA titer of 154 g/L was achieved in 164 h with a yield of 1.38 mol/mol glucose (Brown et al. 2013). It is known that in *E. coli,* phosphoenolpyruvate carboxylase (PPC) with high substrate affinity can convert PEP into OAA without ATP production, while phosphoenolpyruvate carboxylase (PCK) has low substrate affinity and ATP is produced during the transformation process (Moon et al. 2008). Liu et al. further improvedthe MA production by *A. oryzae*. The recombinant *A. oryzae* produced 165 g/L MA in 3-L fed-batch culture with a productivity of 1.38 g/L/h by heterologous expression of PCK and PPC from *E. coli* and 6-phosphofructokinase coding gene and *mae1* gene to construct the oxaloacetate anaplerotic reactions (Liu et al. 2017a).

3.5.2 Malic Acid Production by Engineering One-Step Pathway

As a catalyst for reversible carboxylation, malic enzyme can catalyze pyruvate to MA accompanied by oxidation of NAD(P)H to $NAD(P)^+$ (Stols and Donnelly 1997). It has been pointed out that one-step synthesis of high-efficiency MA using pyruvate as raw material by evolutionary malic enzyme is an effective and promising method for CO_2 immobilization and MA production (Ye et al. 2013a; Dong et al. 2017).

The formation of MA catalyzed by malic enzyme is thermodynamically disadvantageous (Stols and Donnelly 1997). Considering that the reaction only takes place in the direction of negative Gibbs energy, combined with the second law of thermodynamics, the most key breakthrough is to clarify the thermodynamic feasibility of carboxylation reaction (Ye et al. 2013a). To solve the above problems, coupling with NAD(P)H supply (such as overexpression of glucose-6-phosphate dehydrogenase) can be considered, which is thermodynamically advantageous (Zheng et al. 2009b; Ohno et al. 2008). The direct conversion of glucose to MA can be achieved through synthetic biology and metabolic engineering strategies (Ye et al. 2013a). It has been proved that the thermodynamic feasibility of redirecting reversible carboxylation via Embden-Meyerhof pathway formed by non-ATP, in which the catalytic action of malic enzyme from *Thermococcus kodakarensis* is NAD(P)H-dependent ($\Delta G = +7.3$ kJ/mol). A new balance between consumption and regeneration of redox cofactors is beneficial to MA production (glucose +2 $HCO_3^- + 2H \rightarrow 2$ MA + 2 H_2O; $\Delta G = -121.4$ kJ/mol). Therefore, by increasing HCO_3^- concentration, glucose was directly converted to MA with a molar yield of 60%.

In addition, the directed evolution of malic enzyme has also been studied to change its specific activity from pyruvate to MA or its cofactor preference from NADP(H) to NAD(H) (Dong et al. 2017; Morimoto et al. 2014). For instance, the overexpression of the NADH kinase and mutant malic enzyme in mutant *E. coli* with multiple deletions increased MA production to 21.65 g/L in a 5 L bioreactor with the yield of 0.36 g/g (Dong et al. 2017), indicating that this one-step pathway can maximize carbon flux to MA production.

3.5.3 Formation of Malic Acid by Hydrolysis of PMA

Because MA has a significant effect on TCA cycle and inhibits ABC-type transport system and glycolysis (down-regulation of fructose-bisphosphate aldolase and 6-phosphofructokinase), it significantly inhibits the growth of cells (Vasco-Cardenas et al. 2013). Therefore, direct MA production by microbial fermentation is usually limited by the low product yield, titer and productivity, because of the end-product inhibition (Zou et al. 2013).

As a linear anionic C4-polyester, PMA is composed of MA monomeric units, and is generally made of carbohydrates and $CaCO_3$ by one-step fermentation by fungi (Chi et al. 2016b). Recently, many researchers have attempted to obtain MA from hydrolysis of PMA (Zhang et al. 2011; Zou et al. 2013). Using silico analysis technique and genome-scale metabolic model, a yeast-like fungus *A. pullulans* was reconstructed for PMA production (Feng et al. 2017). 87.6 g/L MA (hydrolysis of 76.2 g/L PMA) was obtained in free-cell fermentation of *A. pullulans*, and the productivity was as high as 0.61 g/L/h. In addition, for the immobilized cells, the titer of fed batch fermentations in a fibrous-bed bioreactor (FBB) was 144.2 g/L MA (hydrolysis of 123.7 g/L PMA), and the productivity increased to 0.74 g/L/h. The inhibition of the product was not found in the process of acid hydrolysis, which provided a promising way for the industrialized production of MA from glucose and other substrates (Zou et al. 2013).

3.5.4 *Production of Malic Acid from Low-Value Feedstock*

If the renewable materials such as starch, crude glycerol and corn straw can be directly used as carbon source to produce MA, the production cost will be greatly reduced compared with the use of glucose as carbon source. It has been reported that *A. niger* ATCC 10577 has the highest yield of 19 g/L using thin stillage as carbon source (West 2011). From contaminated citrate fermentation medium, a mutant of *R. delemar* was isolated, and further study found that MA was a main product, besides, by-products such as succinate, fumarate and ethanol, were also produced. Combined with metabolic pathway analysis and metabolic network regulation, more than 120 g/L MA can be produced from 125 g/L biomass hydrolate within 60 h in a pilot-scale jar by simultaneously utilizing 6C sugar glucose and 5C sugar xylose (Li et al. 2014), which overcomes the key technical bottleneck of MA biotransformation to a certain extent. In *E.coli*, firstly, the biosynthetic pathway of malic acid was constructed and the consumption pathway was eliminated. Secondly, the conversion of glycolic acid to malic acid was strengthened. Then the overexpression of catalase HPII was used to decompose H_2O_2 to reduce its toxicity. Finally, the MA titer of the recombinant bacteria was 5.90 g/L, and the yield was 0.80 g/g xylose.

L-malic acid can be produced directly from corn starch as raw material by liquefaction-saccharification and fermentation in *A. oryzae* (Liu et al. 2017b). To reduce the concentration of by-products and further improve the L-malate titer, in the cytosol and mitochondria, Liu et al. synergistically engineered the redox metabolism and the carbon metabolism, and found that 117.2 g/L L-malate and 3.8 g/L succinate were produced in the engineered *A. oryzae* strain, and the yield of L-malate was 0.9 g/g corn starch with the1.17 g/L/h productivity (Liu et al. 2018).

It has been reported that the highest titer of MA was produced by microbial fermentation of crude glycerol by non-engineering *U. trichophora* (Zambanini et al. 2016a). Through laboratory adaptive evolution (using methyl red as pH indicator for screening on solid glycerol medium) and medium optimization, the titer of MA reached to 196 g/L, the yield was 0.82 g/g glycerol and the overall productivity was 0.39 g/L/h. The strain was further studied in a fed-batch bioreactor, and found that the titer was 195 ± 15 g/L with an overall production rate of 0.74 ± 0.06 g/L/h when the pH was controlled at 6.5 (automatic NaOH addition) and CaCO3 was added to the fermentation system (Zambanini et al. 2016b). It is noteworthy that the energy required for cooling is significantly reduced, because the fermentation is insensitive to high temperatures of 37 °C. Using industrial wastewater as substrate to produce chemical products related to industry has economic and ecological production process, which is of great significance for sustainable development of bio-economy.

However, the decrease of pH may seriously affect the yield of MA. When the pH decreased to 4.5, the yield of MA was only 9 + 1 g/L. It has been reported that an antiport with protons, leading to additional H^+ ions pumped against the proton motive force and increases ATP consumption, is the most possible mechanism has been reported for exporting dicarboxylic acids at low pH (Jamalzadeh et al. 2012).

3.6 Other Organic Acids

3.6.1 Lactic Acid

Lactic acid (2-hydroxypropionic acid) is a natural organic acid that has received extensive attention due to its important applications in the food, cosmetics, pharmaceutical, and chemical industries (Moon et al. 2012). Lactic acid has a broad application space as a synthetic raw material for polylactic acid; polylactic acid is the most commercially valuable degradable industrial plastic raw material. At the same time, lactic acid as a chemical will be used to synthesize biodegradable polymers, oxygenates, green solvents and plant growth promoters (Datta and Henry 2006).

The synthesis of lactic acid includes both chemical synthesis and fermentation. The vast majority of lactic acid in the world is synthesized by fermentation, which is due to the faster acid production rate and higher yield. The microorganisms for lactic acid production by fermentation mainly includes Lactobacillus, *Bacillus subtilis*, *Escherichia coli* and *Corynebacterium glutamicum*. *Lactobacillus* are harmless, they have a long history of industrial production and have no adverse effect on the health of consumers and producers, thus have important commercial value.

It has been reported that some *Bacillus* can also be utilized in the production of lactic acid, including *B. coagulans*, *B. thermophilus*, *B. licheniformis*, *B. subtilis*, and the like. Compared to *Lactobacillus*, *Bacillus* produces lactic acid with certain advantages, which will likely reduce the production cost of lactic acid. It has been reported that Bacillus in open fermentation applications include *B. licheniformis* (Wang et al. 2011), *Bacillus 36D1* and *2–6* (Zhao et al. 2010). Ye et al. studied a *B. coagulans C106* that fermented lactic acid with xylose at 50 °C (Ye et al. 2013a). The medium was not sterilized. The productivity was 7.5 g/L/h by fed-batch, and the yield of lactic acid was 215.7 g/L. *Bacillus* can use hexose and pentose which from hydrolysis of lignocellulosic biomass to produce lactic acid, and *B. coagulans 36D1* passes the pentose phosphate pathway, and the maximum yield of lactic acid is 1 g/g (Patel et al. 2006).

C. glutamicum is a Gram-positive bacterium that grows rapidly, and is a saprophytic organism that can produce a variety of organic acids using limited oxygen. Studies have shown that after fermentation with a higher concentration of bacteria on inorganic salt medium for 360 h, the mass concentration of the cells is 60 g/L, and the titer of L-lactic acid is 42.9 g/L (Okino et al. 2005). Similarly, *C. glutamicum ΔldhA/pCRB20* expresses the D-lactate dehydrogenase gene from *L. delbrueckii* by genetic engineering, and is fermented for 30 h in batches. The high-concentration cells were fermented on the inorganic salt medium with glucose, lactic acid can reach 120 g/L (Okino et al. 2008). Therefore, *C. glutamicum* does not require a nutrient-complex medium to produce high-yield lactic acid. However, the production of acetic acid and formic acid during fermentation results in a low yield of lactic acid, which remains a problem to be solved.

3.6.2 Butyric Acid

Butyric acid is an important source of energy for human colonic cecal epithelial cells and plays an important role in maintaining intestinal stability and preventing colorectal cancer. Butyric acid also has the function of inhibiting the proliferation and differentiation of tumor cells, enhancing the immune function of the body, and preventing colitis. Butyric acid is not only an important raw material for synthetic fragrances and other fine chemical products, but also widely used in food and pharmaceuticals. Butyric acid is the main end product in the fermentation of several *Anaerobic Clostridium* (*C. acetobutylicum*, *Clostridium butyricum*, *Clostridium beijerii* et al.) (Jang and Sang 2014). Among them, *C. butyricum* has the characteristics of simple nutrient requirements, high yield of butyric acid and better tolerance to toxic metabolites, so it has proved to be one of the most promising *Clostridia* in the industrial production of butyric acid (Fu et al. 2017). The specific metabolic pathway for butyric acid synthesis in *Clostridium*, by using glucose and xylose as carbon sources, which eventually produces the main product butyric acid and accompanied by a small amount of lactic acid, acetic acid and butanol (Liu et al. 2010b). A mutant *C. butyricum* was constructed by deleting acetate kinase gene, via cell-immobilized in a fiber bed bioreactor, the final butyric acid concentration and yield reached 50.1 g/L and 0.45 g/g, respectively.

3.6.3 Gluconic Acid

Gluconic acid is an important intermediate in chemical, pharmaceutical and food products. As a sour agent, it is also used to prepare household and factory cleaning agents, auxiliaries for fabric processing, metal rust removers, plasticizers for concrete in the construction industry et al. (Ramachandran et al. 2006). For gluconic acid production, the biological fermentation methods including bacterial or fungal fermentation, immobilized cells and immobilized enzyme. Under most fermentation conditions, calcium gluconate were produced, and then via ion exchange, evaporation, concentration and crystallization to obtain gluconic acid (Parimal et al. 2019).

The biological fermentation method requires many processes such as cultivating strains, screening strains and sterilization, and has strict temperature requirements, many by-products, long cycle, and the addition of bacteria and other impurities during the production of gluconic acid, affecting gluconic acid. The purity of the product, so its development is urgently needed to solve many technical problems.

3.7 Conclusions

In the past decades, the high yield of organic acids has been improved by microbial fermentation. In recent years, with the development of metabolic engineering, synthetic biology and systems biology, the efficient synthesis of organic acids has been guaranteed, and the design and construction of microbial cell factories have made a qualitative leap forward. More importantly, the pursuit of high production and productivity as well as production to meet the economic needs of industry. Sustainable economic production depends on the optimal combination of metabolic pathways, substrate absorption and assimilation, intermediates and product transport.

References

Adeoye AO, Lateef A, Gueguim-Kana EB. Optimization of citric acid production using a mutant strain of *Aspergillus niger* on cassava peel substrate. Biocatal Agric Biotechnol. 2015;4:568–74.

Ahn JH, Jang YS, Lee SY. Production of succinic acid by metabolically engineered microorganisms. Curr Opin Biotechnol. 2016;42:54–66.

Andersen MR, Lehmann L, Nielsen J. Systemic analysis of the response of *Aspergillus niger* to ambient pH. Genome Biol. 2009;10:R47.

Andersen MR, Salazar MP, Schaap PJ, van de Vondervoort PJ, Culley D, Thykaer J, et al. Comparative genomics of citric-acid-producing *Aspergillus niger* ATCC 1015 versus enzyme-producing CBS 513.88. Genome Res. 2011;21:885–97.

Aramaki K, Ogawa A, Tsukahara M, Kunieda H. Formation of microemulsions in aqueous NaCl/sodium (3-dodecanoyloxy-2-hydroxy-propyl) succinate/glycerol mono (2-ethylhexyl) ether/oil systems. J Dispers Sci Technol. 2002;23:29–36.

Arisan-Atac I, Wolschek MF, Kubicek CP. Trehalose-6-phosphate synthase A affects citrate accumulation by *Aspergillus niger* under conditions of high glycolytic flux. FEMS Microbiol Lett. 1996;140:77–83.

Arslan D, Steinbusch KJJ, Diels L, De Wever H, Buisman CJN, Hamelers HVM. Effect of hydrogen and carbon dioxide on carboxylic acids patterns in mixed culture fermentation. Bioresour Technol. 2012;118:227–34.

Asai T, Aida K, Sugisaki Z, Yakeishi N. On α-ketoglutaric acid fermentation. J Gen Appl Microbiol. 1955;1(4):308–46.

Bari MN, Alam MZ, Muyibi SA, Jamal P, Abdullah AM. Improvement of production of citric acid from oil palm empty fruit bunches: optimization of media by statistical experimental designs. Bioresour Technol. 2009;100:3113–20.

Barrett DG, Yousaf MN. Poly (triol alpha-ketoglutarate) as biodegradable, chemoselective, and mechanically tunable elastomers. Macromolecules. 2008;41(17):6347–52.

Battat E, Peleg Y, Bercovitz A, Rokem JS, Goldberg I. Optimization of L-malic acid production by *Aspergillus flavus* in a stirred fermentor. Biotechnol Bioeng. 1991;37:1108–16.

Becker J, Wittmann C. Advanced biotechnology: metabolically engineered cells for the bio-based production of chemicals and fuels, materials, and health-care products. Angew Chem Int Ed. 2015;54(11):3328–50.

Becker J, Lange A, Fabarius J, Wittmann C. Top value platform chemicals: bio-based production of organic acids. Curr Opin Biotechnol. 2015;36:168–75.

Betiku E, Adesina OA. Statistical approach to the optimization of citric acid production using filamentous fungus *Aspergillus niger* grown on sweet potato starch hydrolyzate. Biomass Bioenergy. 2013;55:350–4.

Bradfield MF, Nicol W. Continuous succinic acid production from xylose by *Actinobacillus succinogenes*. Bioprocess Biosyst Eng. 2016;39(2):233–44.

Bradfield MFA, Mohagheghi A, Salvachúa D, Smith H, Black BA, Dowe N, Beckham GT, Nicol W. Continuous succinic acid production by *Actinobacillus succinogenes* on xylose-enriched hydrolysate. Biotechnol Biofuels. 2015;8(1):181.

Brown SH, Bashkirova L, Berka R, Chandler T, Doty T, McCall K, McCulloch M, McFarland S, Thompson S, Yaver D, Berry A. Metabolic engineering of *Aspergillus oryzae* NRRL 3488 for increased production of L-malic acid. Appl Microbiol Biotechnol. 2013;97(20):8903–12.

Brunhuber NMW, Thoden JB, Blanchard JS, Vanhooke JL. *Rhodococcus* L-phenylalanine dehydrogenase: kinetics, mechanism, and structural basis for catalytic specifity. Biochemistry. 2000;39(31):9174–87.

Cao W, Wang Y, Luo J, Yin J, Xing J, Wan Y. Succinic acid biosynthesis from cane molasses under low pH by *Actinobacillus succinogenes* immobilized in luffa sponge matrices. Bioresour Technol. 2018;268:45–51.

Capuder M, Solar T, Bencina M, Legisa M. Highly active, citrate inhibition resistant form of *Aspergillus nige*r 6-phosphofructo-1-kinase encoded by a modified pfkA gene. J Biotechnol. 2009;144:51–7.

Carsanba E, Papanikolaou S, Fickers P, Erten H. Screening various *Yarrowia lipolytica* strains for citric acid production. Yeast. 2019; https://doi.org/10.1002/yea.3389.

Carvalho M, Roca C, Reis MAM. Improving succinic acid production by *Actinobacillus succinogenes* from raw industrial carob pods. Bioresour Technol. 2016;218:491–7.

Cavallo E, Charreau H, Cerrutti P, Foresti ML. *Yarrowia lipolytica*: a model yeast for citric acid production. FEMS Yeast Res. 2017;17(8)

Chen X, Xu G, Xua N, Zou W, Zhu P, Liu L, Chen J. Metabolic engineering of *Torulopsis glabrata* for malateproduction. Metab Eng. 2013;19:10–6.

Chen H, He X, Geng H, Liu H. Physiological characterization of ATP-citrate lyase in *Aspergillus niger*. J Ind Microbiol Biotechnol. 2014a;41:721–31.

Chen C, Ding S, Wang D, Li Z, Ye Q. Simultaneous saccharification and fermentation of cassava to succinic acid by *Escherichia coli* NZN111. Bioresour Technol. 2014b;163:100–5.

Chen X, Wu J, Song W, Zhang L, Wang H, Liu L. Fumaric acid production by *Torulopsis glabrata*: engineering the urea cycle and the purine nucleotide cycle. Biotechnol Bioeng. 2015;112(1):156–67.

Chen PC, Zheng P, Ye XY, Ji F. Preparation of a. succinogenes immobilized microfiber membrane for repeated production of succinic acid. Enzym Microb Technol. 2017a;98:34–42.

Chen X, Wang Y, Dong X, Hu G, Liu L. Engineering rTCA pathway and C4-dicarboxylate transporter for l-malic acid production. Appl Environ Microbiol. 2017b;101:4041–52.

Chernyavskaya OG, Shishkanova NV, Finogenova TV. Biosynthesis of α-ketoglutaric acid from ethanol by yeasts. Appl Microbiol Biotechnol. 1997;33(2):261–5.

Chernyavskaya O, Shishkanova N, Il'chenko A, Finogenova T. Synthesis of α-ketoglutaric acid by *Yarrowia lipolytica* yeast grown on ethanol. Appl Microbiol Biotechnol. 2000;53(2):152–8.

Chi Z, Wang ZP, Wang GY, Khan I, Chi ZM. Microbial biosynthesis and secretion of l-malic acid and its applications. Crit Rev Biotechnol. 2016a;36(1):99–107.

Chi Z, Liu GL, Liu CG, Chi ZM. Poly (β-l-malic acid) (PMLA) from *Aureobasidium spp.* and its current proceedings. Appl Environ Microbiol. 2016b;100:3841–51.

Dai Z, Mao X, Magnuson JK, Lasure LL. Identification of genes associated with morphology in *Aspergillus niger* by using suppression subtractive hybridization. Appl Environ Microbiol. 2004;70:2474–85.

Datta R, Henry M. Lactic acid: recent advances in products, processes and technologies—a review. J Chem Technol Biotechnol. 2006;81(7):1119–29.

de Jongh WA, Nielsen J. Enhanced citrate production through gene insertion in *Aspergillus niger*. Metab Eng. 2008;10:87–96.

Demirel G, Yaykaşlı KO, Yaşar A. The production of citric acid by using immobilized *Aspergillus niger* A-9 and investigation of its various effects. Food Chem. 2005;89:393–6.

Dessie W, Xin F, Zhang W, Jiang Y, Wu H, Ma J, Jiang M. Opportunities, challenges, and future perspectives of succinic acid production by *Actinobacillus succinogenes*. Appl Microbiol Biotechnol. 2018;102(23):9893–910.

Dhillon GS, Brar SK, Verma M, Tyagi RD. Apple pomace ultrafiltration sludge – a novel substrate for fungal bioproduction of citric acid: optimisation studies. Food Chem. 2011;128:864–71.

Dong X, Chen X, Qian Y, Wang Y, Wang L, Qiao W, Liu L. Metabolic engineering of *Escherichia coli* W3110 to produce L-malate. Biotechnol Bioeng. 2017;114:656–64.

Feng J, Yang J, Li X, Guo M, Wang B, Yang ST, Zou X. Reconstruction of a genome-scale metabolic model and in silico analysis of the polymalic acid producer *Aureobasidium pullulans* CCTCC M2012223. Gene. 2017;607:1–8.

Finogenova TV, Morgunov IG, Kamzolova SV, Chernyavskaya OG. Organic acid production by the yeast *Yarrowia lipolytica*: a review of prospects. Appl Biochem Microbiol. 2005;41(5):418–25.

Förster A, Jacobs K, Juretzek T, Mauersberger S, Barth G. Overexpression of the ICL1 gene changes the product ratio of citric acid production by *Yarrowia lipolytica*. Appl Microbiol Biotechnol. 2017;77:861–9.

Fu GY, Lu Y, Chi Z, Liu GL, Zhao SF, Jiang H, et al. Cloning and characterization of a pyruvate carboxylase gene from *Penicillium rubens* and overexpression of the genein the yeast *Yarrowia lipolytica* for enhanced citric acid production. Mar Biotechnol (NY). 2016;18:1–14.

Fu H, Yang ST, Wang M, Wang J, Tang IC. Butyric acid production from lignocellulosic biomass hydrolysates by engineered *Clostridium tyrobutyricum* overexpressing xylose catabolism genes for glucose and xylose co-utilization. Bioresour Technol. 2017;234:389–96.

Gao C, Yang X, Wang H, Rivero CP, Li C, Cui Z, et al. Robust succinic acid production from crude glycerol using engineered *yarrowia lipolytica*. Biotechnol Biofuels. 2016;9(1):179.

Grobler J, Bauer F, Subden RE, Van Vuuren HJ. The *mae1* gene of *Schizosaccharomyces pombe* encodes apermease for malate and other C4 dicarboxylicacids. Yeast. 1995;11:1485–91.

Guarnieri MT, Chou YC, Salvachúa D, Mohagheghi A, St. John PC, Peterson DJ, Bomble YJ, Beckham GT. Metabolic engineering of *Actinobacillus succinogenes* provides insights into succinic acid biosynthesis. Appl Environ Microbiol. 2017;83(17):e00996–17.

Guettler MV, Rumler D, Jain MK. *Actinobacillus succinogenes sp. nov.*, a novel succinic-acid-producing strain from the bovine rumen. Int J Syst Evol Microbiol. 1999;49:207–16.

Hattori T, Kino K, Kirimura K. Regulation of alternative oxidase at the transcription stage in *Aspergillus niger* under the conditions of citric acid production. Curr Microbiol. 2009;58:321–5.

Holz M, André F, Mauersberger S, Barth G. Aconitase overexpression changes the product ratio of citric acid production by *Yarrowia lipolytica*. Appl Microbiol Biotechnol. 2009;81:1087–96.

Hossain GS, Li J, Shin HD, Liu L, Wang M, Du G, Chen J. Improved production of α-ketoglutaric acid (α-KG) by a *Bacillus subtilis* whole-cell biocatalyst via engineering of l-amino acid deaminase and deletion of the α-KG utilization pathway. J Biotechnol. 2014;187:71–7.

Hossain GS, Shin HD, Li J, Wang M, Du G, Liu L, Chen J. Integrating error-prone PCR and DNA shuffling as an effective molecular evolution strategy for the production of α-ketoglutaric acid by l-amino acid deaminase. RSC Adv. 2016;6(52):46149–58.

Hou L, Liu L, Zhang H, Zhang L, Zhang L, Zhang J, et al. Functional analysis of the mitochondrial alternative oxidase gene (aox1) from *Aspergillus niger* CGMCC 10142 and its effects on citric acid production. Appl Microbiol Biotechnol. 2018;102:7981–95.

Il'Chenko AP, Chernyavskaya OG, Shishkanova NV, Finogenova TV. Metabolism of *Yarrowia lipolytica* grown on ethanol under conditions promoting the production of α-ketoglutaric and citric acids: a comparative study of the central metabolism enzymes. Microbiology. 2002;71(3):269–74.

Imandi SB, Bandaru VV, Somalanka SR, Bandaru SR, Garapati HR. Application of statistical experimental designs for the optimization of medium constituents for the production of citric acid from pineapple waste. Bioresour Technol. 2008;99:4445–50.

Isar J, Agarwal L, Saran S, Saxena RK. A statistical method for enhancing the production of succinic acid from *Escherichia coli* under anaerobic conditions. Bioresour Technol. 2006;97(13):1443–8.

Jamalzadeh E, Verheijen PJ, Heijnen JJ, Van Gulik WM. pH-dependent uptake of fumaric acid in *Saccharomyces cerevisiae* under anaerobic conditions. Appl Environ Microbiol. 2012;78:705–16.

Jang YS, Sang YL. Metabolic engineering of *Clostridium acetobutylicum*, for highly selective butyric acid production. New Biotechnol. 2014;31(11):S161.

Jantama K, Haupt MJ, Svoronos SA, Zhang X, Moore JC, Shanmugam KT, Ingram LO. Combining metabolic engineering and metabolic evolution to develop nonrecombinant strains of *Escherichia coli* C that produce succinate and malate. Biotechnol Bioeng. 2008;99:1140–53.

Kadooka C, Izumitsu K, Onoue M, Okutsu K, Yoshizaki Y, Takamine K, et al. Mitochondrial citrate transporters CtpA and YhmA are required for extracellular citric acid accumulation and contribute to cytosolic acetyl coenzyme a generation in *Aspergillus luchuensis* mut. kawachii. Appl Environ Microbiol. 2019;85(8)

Karthikeyan A, Sivakumar N. Citric acid production by Koji fermentation using banana peel as a novel substrate. Bioresour Technol. 2010;101:5552–6.

Kirimura K, Kobayashi K, Ueda Y, Hattori T. Phenotypes of gene disruptants in relation to a putative mitochondrial malate-citrate shuttle protein in citric acid-producing *Aspergillus niger*. Biosci Biotechnol Biochem. 2016;80:1737–46.

Knuf C, Nookaew I, Remmers I, Khoomrung S, Brown S, Berry A, Nielsen J. Physiological characterization of the high malic acid-producing *Aspergillus oryzae* strain 2103a-68. Appl Microbiol Biotechnol. 2014;98(8):3517–27.

Koepsell HJ, Stodola FH, Sharpe ES. Production of α-Ketoglutarate in glucose oxidation by Pseudomonas fluorescens. J Am Chem Soc. 1952;74(20):5142–4.

Kumar Gupta G, De S, Franco A, Balu AM, Luque R. Sustainable biomaterials: current trends, challenges and applications. Molecules. 2015;21:E48.

Li X, Liu Y, Yang Y, Zhang H, Wang H, Wu Y, Zhang M, Sun T, Cheng J, Wu X, Pan L, Jiang S, Wu H. High levels of malic acid production by the bioconversion of corn straw hydrolyte using an isolated *Rhizopus delemar* strain. Biotechnol Bioprocess Eng. 2014;19:478–92.

Liang L, Liu R, Li F, Wu M, Chen K, Ma J, Jiang M, Wei P, Ouyang P. Repetitive succinic acid production from lignocellulose hydrolysates by enhancement of ATP supply in metabolically engineered *Escherichia coli*. Bioresour Technol. 2013;143:405–12.

Liu L, Li Y, Zhu Y, Du G, Chen J. Redistribution of carbon flux in *Torulopsis glabrata* by altering vitamin and calcium level. Metab Eng. 2007;9(1):21–9.

Liu YP, Zheng P, Sun ZH, Ni Y, Dong JJ, Zhu LL. Economical succinic acid production from cane molasses by *Actinobacillus succinogenes*. Bioresour Technol. 2008;99(6):1736–42.

Liu XY, Chi Z, Liu GL, Wang F, Madzak C, Chi ZM. Inulin hydrolysis and citric acid production from inulin using the surface-engineered *Yarrowia lipolytica* displaying inulinase. Metab Eng. 2010a;12:469–76.

Liu X, Zhu Y, Yang ST. Construction and characterization of ack, deleted mutant of *Clostridium tyrobutyricum*, for enhanced butyric acid and hydrogen production. Biotechnol Prog. 2010b;22(5):1265–75.

Liu XY, Chi Z, Liu GL, Madzak C, Chi ZM. Both decrease in ACL1 gene expression and increase in ICL1 gene expression in marine-derived yeast *Yarrowia lipolytica* expressing INU1 gene enhance citric acid production from inulin. Mar Biotechnol (NY). 2013a;15:26–36.

Liu L, Hossain GS, Shin HD, Li J, Du G, Chen J. One-step production of α-ketoglutaric acid from glutamic acid with an engineered l-amino acid deaminase from *Proteus mirabilis*. J Biotechnol. 2013b;164(1):97–104.

Liu R, Liang L, Li F, Wu M, Chen K, Ma J, Jiang M, Wei P, Ouyang P. Efficient succinic acid production from lignocellulosic biomass by simultaneous utilization of glucose and xylose in engineered *Escherichia coli*. Bioresour Technol. 2013c;149:84–91.

Liu J, Xie Z, Shin HD, Li J, Du G, Chen J, Liu L. Rewiring the reductive tricarboxylic acid pathway and L-malate transport pathway of *Aspergillus oryzae* for overproduction of L-malate. J Biotechnol. 2017a;253:1–9.

Liu J, Li J, Shin HD, Liu L, Du G, Chen J. Protein and metabolic engineering for the production of organic acids. Bioresour Technol. 2017b;239:412–21.

Liu J, Li J, Liu Y, Shin H-d, Ledesma-Amaro R, Du G, Chen J, Liu L. Synergistic rewiring of carbon metabolism and redox metabolism in cytoplasm and mitochondria of *Aspergillus oryzae* forincreased L-malate production. ACS Synth Biol. 2018;7(9):2139–47.

Lockwood LB, Stodola FH. Preliminary studies on the production of alpha-ketoglutaric acid by Pseudomonas fluorescens. J Biol Chem. 1946;164:81–3.

Maharaj K, Bradfield MF, Nicol W. Succinic acid-producing biofilms of *Actinobacillus succinogenes*: reproducibility, stability and productivity. Appl Microbiol Biotechnol. 2014;98(17):7379–86.

Meijer S, Nielsen ML, Olsson L, Nielsen J. Gene deletion of cytosolic ATP: citrate lyase leads to altered organic acid production in *Aspergillus niger*. J Ind Microbiol Biotechnol. 2009;36:1275–80.

Mirbagheri M, Nahvi I, Emtiazi G, Darvishi F. Enhanced production of citric acid in *Yarrowia lipolytica* by Triton X-100. Appl Biochem Biotechnol. 2011;165:1068–74.

Molla G, Melis R, Pollegioni L. Breaking the mirror: l-amino acid deaminase, a novel stereoselective biocatalyst. Biotechnol Adv. 2017;35(6):657–68.

Moon SY, Hong SH, Kim TY, Lee SY. Metabolic engineering of *Escherichia coli* for the production of malic acid. Biochem Eng J. 2008;40(2):312–20.

Moon S, Wee Y, Choi G. A novel lactic acid bacterium for the production of high purity L-lactic acid, *Lactobacillus paracasei subsp. paracasei* CHB2121. J Biosci Bioeng. 2012;114(2):155–9.

Morimoto Y, Honda K, Ye X, Okano K, Ohtake H. Directed evolution of thermotolerant malic enzyme for improved malate production. J Biosci Bioeng. 2014;117(2):147–52.

Niu P, Dong X, Wang Y, Liu L. Enzymatic production of α-ketoglutaric acid from l-glutamic acid via l-glutamate oxidase. J Biotechnol. 2014;179:56–62.

Niu J, Arentshorst M, Nair PD, Dai Z, Baker SE, Frisvad JC, et al. Identification of a classical mutant in the industrial host *Aspergillus niger* by systems genetics: laea is required for citric acid production and regulates the formation of some secondary metabolites. Genes Genomes Genet. 2015;6:193–204.

Ochoa-Estopier A, Guillouet SE. D-stat culture for studying the metabolic shifts from oxidative metabolism to lipid accumulation and citric acid production in *Yarrowia lipolytica*. J Biotechnol. 2014;170:35–41.

Ohno Y, Nakamori T, Zheng H, Suye S. Reverse reaction of malic enzyme for $HCO3^-$ fixation into pyruvic acid to synthesize L-malic acid with enzymatic coenzyme regeneration. Biosci Biotechnol Biochem. 2008;72:1278–82.

Okino S, Inui M, Yukawa H. Production of organic acids by *Corynebacterium glutamicum* under oxygen deprivation. Appl Microbiol Biotechnol. 2005;68(4):475–80.

Okino S, Suda M, Fujikura K, Inui M, Yukawa H. Production of d-lactic acid by *Corynebacterium glutamicum* under oxygen deprivation. Appl Microbiol Biotechnol. 2008;78(3):449–54.

Papadaki E, Mantzouridou FT. Citric acid production from the integration of Spanish-style green olive processing wastewaters with white grape pomace by *Aspergillus niger*. Bioresour Technol. 2019;280:59–69.

Papagianni M. Advances in citric acid fermentation by *Aspergillus niger*: biochemical aspects, membrane transport and modeling. Biotechnol Adv. 2007;25:244–63.

Papanikolaou S, Galiotou-Panayotou M, Fakas S, Komaitis M, Aggelis G. Citric acid production by *Yarrowia lipolytica* cultivated on olive-mill wastewater-based media. Bioresour Technol. 2008;99:2419–28.

Parimal P, Ramesh K, Subhamay B. Purification and concentration of gluconic acid from an integrated fermentation and membrane process using response surface optimized conditions. Front Chem Sci Eng. 2019;13(1):152–63.

Patel MA, Ou MS, Harbrucker R, Aldrich HC, Buszko ML, Ingram LO, Shanmugam KT. Isolation and characterization of acid-tolerant, thermophilic bacteria for effective fermentation of biomass-derived sugars to lactic acid. Appl Environ Microbiol. 2006;72(5):3228–35.

Peleg Y, Rokem JS, Goldberg I, Pines O. Inducible overexpression of the FUM1 gene in *saccharomyces cerevisiae*: localization of fumarase and efficient fumaric acid bioconversion to L-malic acid. Appl Environ Microbiol. 1990;56:2777–83.

Poorahong S, Santhosh P, Ramírez GV, Tseng TF, Wong JI, Kanatharana P, Thavarungkul P, Wang J. Development of amperometric α-ketoglutarate biosensor based on ruthenium-rhodium modified carbon fiber enzyme microelectrode. Biosens Bioelectron. 2011;26(8):3670–3.

Qin J, Zhou YJ, Krivoruchko A, Huang M, Liu L, Khoomrung S, Siewers V, Jiang B, Nielsen J. Modular pathway rewiring of *Saccharomyces cerevisiae* enables high-level production of L-ornithine. Nat Commun. 2015;6:8224.

Ramachandran S, Fontanille P, Pandey A, Larroche C. Gluconic acid: properties, applications and microbial production. Food Technol Biotechnol. 2006;44(2):185–95.

Ruijter GJ, Panneman H, Visser J. Overexpression of phosphofructokinase and pyruvate kinase in citric acid-producing *Aspergillus niger*. Biochim Biophys Acta. 1997;1334:317–26.

Ruijter GJ, van de Vondervoort PJ, Visser J. Oxalic acid production by *Aspergillus niger*: an oxalate-non-producing mutant produces citric acid at pH 5 and in the presence of manganese. Microbiology. 1999;145:2569–76.

Ruijter GJ, Panneman H, Xu D, Visser J. Properties of *Aspergillus niger* citrate synthase and effects of citA overexpression on citric acid production. FEMS Microbiol Lett. 2000;184:35–40.

Rymowicz W, Fatykhova AR, Kamzolova SV, Rywinska A, Morgunov IG. Citric acid production from glycerol-containing waste of biodiesel industry by *Yarrowia lipolytica* in batch, repeated batch, and cell recycle regimes. Appl Microbiol Biotechnol. 2010;87:971–9.

Salvachúa D, Mohagheghi A, Smith H, Bradfield MFA, Nicol W, Black BA, Biddy MJ, Dowe N, Beckham GT. Succinic acid production on xylose-enriched biorefinery streams by *Actinobacillus succinogenes* in batch fermentation. Biotechnol Biofuels. 2016;9:28.

Sawant O. Fungal citric acid production using waste materials: a mini-review. J Microbiol Biotechnol Food Sci. 2018;8:821–8.

Shi X, Chen Y, Ren H, Liu D, Zhao T, Zhao N, Ying H. Economically enhanced succinic acid fermentation from cassava bagasse hydrolysate using *Corynebacterium glutamicum* immobilized in porous polyurethane filler. Bioresour Technol. 2014;174:190–7.

Show PL, Oladele KO, Siew QY, Aziz Zakry FA, Lan JCW, Ling TC. Overview of citric acid production from *Aspergillus niger*. Front Life Sci. 2015;8:271–83.

Sitepu IR, Garay LA, Sestric R, Levin D, Block DE, German JB, Boundy-Mills KL. Oleaginous yeasts for biodiesel: current and future trends in biology and production. Biotechnol Adv. 2014;32(7):1336–60.

Song CW, Lee SY. Combining rational metabolic engineering and flux optimization strategies for efficient production of fumaric acid. Appl Microbiol Biotechnol. 2015;99(20):8455–64.

Steiger MG, Rassinger A, Mattanovich D, Sauer M. Engineering of the citrate exporter protein enables high citric acid production in *Aspergillus niger*. Metab Eng. 2019;52:224–31.

Stols L, Donnelly MI. Production of succinic acid through overexpression of NAD(+)-dependent malic enzyme in an *Escherichia coli* mutant. Appl Environ Microbiol. 1997;63:2695–701.

Stottmeister U, Aurich A, Wilde H, Andersch J, Schmidt S, Sicker D. White biotechnology for green chemistry: fermentative 2-oxocarboxylic acids as novel building blocks for subsequent chemical syntheses. J Ind Microbiol Biotechnol. 2005;32(11–12):651–64.

Sun X, Wu H, Zhao G, Li Z, Wu X, Liu H, et al. Morphological regulation of *Aspergillus niger* to improve citric acid production by chsC gene silencing. Bioprocess Biosyst Eng. 2018;41:1029–38.

Taing O, Taing K. Production of malic and succinic acids by sugar-tolerant yeast *Zygosaccharomyces rouxii*. Eur Food Res Technol. 2006;224:343–7.

Thakker C, Martínez I, Li W, San K-Y, Bennett G. Metabolic engineering of carbon and redox flow in the production of small organic acids. J Ind Microbiol Biotechnol. 2015;42(3):403–22.

Urbance SE, Pometto AL, DiSpirito AA, Denli Y. Evaluation of succinic acid continuous and repeat-batch biofilm fermentation by *Actinobacillus succinogenes* using plastic composite support bioreactors. Appl Microbiol Biotechnol. 2004;65(6):664–70.

Van der Werf MJ, Guettler MV, Jain MK, Zeikus JG. Environmental and physiological factors affecting the succinate product ratio during carbohydrate fermentation by *Actinobacillus sp.* 130Z. Arch Microbiol. 1997;167(6):332–42.

Vasco-Cardenas MF, Banos S, Ramos A, Martin JF, Barreiro C. Proteome response of *Corynebacterium glutamicum* to high concentration of industrially relevant C(4) and C(5) dicarboxylic acids. J Proteome. 2013;85:65–88.

Vieille BDSVJ. Respiratory glycerol metabolism of *Actinobacillus succinogenes* 130Z for succinate production. J Ind Microbiol Biotechnol. 2014;41:1339–52.

Vuoristo KS, Mars AE, Sanders JPM, Eggink G, Weusthuis RA. Metabolic engineering of TCA cycle for production of chemicals. Trends Biotechnol. 2016;34(3):191–7.

Wang Q, Zhao X, Chamu J, Shanmugam KT. Isolation, characterization and evolution of a new thermophilic *Bacillus licheniformis* for lactic acid production in mineral salts medium. Bioresour Technol. 2011;102(17):8152–8.

Wang J, Zhang B, Zhang J, Wang H, Zhao M, Wang N, Dong L, Zhou X, Wang D. Enhanced succinic acid production and magnesium utilization by overexpression of magnesium transporter mgtA in *Escherichia coli* mutant. Bioresour Technol. 2014;170:125–31.

Wang L, Zhang J, Cao Z, Wang Y, Gao Q, Zhang J, et al. Inhibition of oxidative phosphorylation for enhancing citric acid production by *Aspergillus niger*. Microb Cell Factories. 2015;14:7.

Wang L, Cao Z, Hou L, Yin L, Wang D, Gao Q, et al. The opposite roles of agdA and glaA on citric acid production in *Aspergillus niger*. Appl Environ Microbiol. 2016;100:5791–803.

Werpy TA, Holladay JE, White JF. Top value added chemicals from biomass: I. results of screening for potential candidates from sugars and synthesis gas. Synth Fuels. 2004; https://doi.org/10.2172/926125.

West TP. Malic acid production from thin stillage by *Aspergillus* species. Biotechnol Lett. 2011;33:2463–7.

Wu H, Li Q, Li ZM, Ye Q. Succinic acid production and CO_2 fixation using a metabolically engineered *Escherichia coli* in a bioreactor equipped with a self-inducing agitator. Bioresour Technol. 2012;107:376–84.

Xie G, West TP. Citric acid production by *Aspergillus niger* ATCC 9142 from a treated ethanol fermentation co-product using solid-state fermentation. Lett Appl Microbiol. 2009;48:639–44.

Xu J, Su X-F, Bao J-W, Zhang H-J, Zeng X, Tang L, Wang K, Zhang J-H, Chen X-S, Mao Z-G. A novel cleaner production process of citric acid by recycling its treated wastewater. Bioresour Technol. 2016;211:645–53.

Yan D, Wang C, Zhou J, Liu Y, Yang M, Xing J. Construction of reductive pathway in *Saccharomyces cerevisiae* for effective succinic acid fermentation at low pH value. Bioresour Technol. 2014a;156:232–9.

Yan Q, Zheng P, Dong J-J, Sun Z-H. A fibrous bed bioreactor to improve the productivity of succinic acid by *Actinobacillus succinogenes*. J Chem Technol Biotechnol. 2014b;89(11):1760–6.

Yang G, Jahan MS, Ahsan L, Zheng L, Ni Y. Recovery of acetic acid from pre-hydrolysis liquor of hardwood kraft-based dissolving pulp production process by reactive extraction with triisooctylamine. Bioresour Technol. 2013;138:253–8.

Ye X, Honda K, Morimoto Y, Okano K, Ohtake H. Direct conversion of glucose to malate by synthetic metabolic engineering. J Biotechnol. 2013a;164(1):34–40.

Ye L, Zhou X, Hudari MS, Li Z, Wu JC. Highly efficient production of l-lactic acid from xylose by newly isolated *Bacillus coagulans* C106. Bioresour Technol. 2013b;132:38–44.

Yin X, Madzak C, Du G, Zhou J, Chen J. Enhanced alpha-ketoglutaric acid production in *Yarrowia lipolytica* WSH-Z06 by regulation of the pyruvate carboxylation pathway. Appl Microbiol Biotechnol. 2012;96(6):1527–37.

Yin X, Li JH, Shin HD, Du GC, Liu L, Chen J. Metabolic engineering in the biotechnological production of organic acids in the tricarboxylic acid cycle of microorganisms. Advances and prospects. Biotechnol Adv. 2015;33(6):830–41.

Yin X, Shin HD, Li J, Du G, Liu L, Chen J. Comparative genomics and transcriptome analysis of *Aspergillus niger* and metabolic engineering for citrate production. Sci Rep. 2017;7:41040.

Yovkova V, Otto C, Aurich A, Mauersberger S, Barth G. Engineering the α-ketoglutarate overproduction from raw glycerol by overexpression of the genes encoding $NADP^+$-dependent isocitrate dehydrogenase and pyruvate carboxylase in *Yarrowia lipolytica*. Appl Microbiol Biotechnol. 2014;98(5):2003–13.

Yuzbashev TV, Yuzbasheva EY, Sobolevskaya TI, Laptev IA, Vybornaya TV, Larina AS, Matsui K, Fukui K, Sineoky SP. Production of succinic acid at low pH by a recombinant strain of the aerobic yeast *Yarrowia lipolytica*. Biotechnol Bioeng. 2010;107(4):673–82.

Yuzbashev TV, Bondarenko PY, Sobolevskaya TI, Yuzbasheva EY, Laptev IA, Kachala VV, Fedorov AS, Vybornaya TV, Larina AS, Sineoky SP. Metabolic evolution and (13) C flux analysis of a succinate dehydrogenase deficient strain of *Yarrowia lipolytica*. Biotechnol Bioeng. 2016;113(11):2425–32.

Zambanini T, Sarikaya E, Kleineberg W, Buescher JM, Meurer G, Wierckx N, Blank LM. Efficient malic acid production from glycerol with *Ustilago trichophora* TZ1. Biotechnol Biofuels. 2016a;9:67.

Zambanini T, Kleineberg W, Sarikaya E, Buescher JM, Meurer G, Wierckx N, Blank LM. Enhanced malic acid production from glycerol with high-cell density *Ustilago trichophora* TZ1 cultivations. Biotechnol Biofuels. 2016b;9:135.

Zelle RM, de Hulster E, Van Winden WA, De Waard P, Dijkema C, Winkler AA, Geertman JMA, Van Dijken JP, Pronk JT, Van Maris AJA. Malic acid production by *Saccharomyces cerevisiae*: engineering of pyruvate carboxylation, oxaloacetate reduction, and malate export. Appl Environ Microbiol. 2008;74:2766–77.

Zelle RM, de Huister E, Kloezen W, Pronk JT, van Maris AJA. Key process conditions for production of C4 dicarboxylic acids in bioreactor batch cultures of an engineered *Saccharomyces cerevisiae* strain. Appl Environ Microbiol. 2010;76:744–50.

Zhang X, Wang X, Shanmugam KT, Ingram LO. L-malate production by metabolically engineered *Escherichia coli*. Appl Environ Microbiol. 2011;77:427–34.

Zhang N, Jiang JC, Yang J, Wei M, Zhao J, Xu H, et al. Citric acid production from acorn starch by tannin tolerance mutant *Aspergillus niger* AA120. Appl Biochem Biotechnol. 2018;188(1):1–11.

Zhao B, Wang L, Ma C, Yang C, Xu P, Ma Y. Repeated open fermentative production of optically pure L-lactic acid using a thermophilic *Bacillus sp.* strain. Bioresour Technol. 2010;101(16):6494–8.

Zheng P, Dong J-J, Sun Z-H, Ni Y, Fang L. Fermentative production of succinic acid from straw hydrolysate by *Actinobacillus succinogenes*. Bioresour Technol. 2009a;100(8):2425–9.

Zheng H, Ohno Y, Nakamori T, Suye S. Production of l-malic acid with fixation of HCO_3^- by malic enzyme-catalyzed reaction based on regeneration of coenzyme on electrode modified by layer-by-layer self-assembly method. J Biosci Bioeng. 2009b;107:16–20.

Zheng P, Zhang K, Yan Q, Xu Y, Sun Z. Enhanced succinic acid production by *Actinobacillus succinogenes* after genome shuffling. J Ind Microbiol Biotechnol. 2013;40(8):831–40.

Zhou J, Yin X, Madzak C, Du G, Chen J. Enhanced α-ketoglutarate production in *Yarrowia lipolytica* WSH-Z06 by alteration of the acetyl-CoA metabolism. J Biotechnol. 2012;161(3):257–64.

Zou X, Zhou Y, Yang ST. Production of polymalic acid and malic acid by *Aureobasidium pullulans* fermentation and acid hydrolysis. Biotechnol Bioeng. 2013;110:2105–13.

Chapter 4
Microbial Production of Oligosaccharides and Polysaccharides

Rongzhen Tian, Yanfeng Liu, and Long Liu

4.1 Glucosamine and N-Acetylglucosamine

Glucosamine (GlcN) is a derivative obtained by substituting one hydroxyl group of glucose with an amino group. Glucosamine and its acetylated derivative, N-acetylglucosamine (GlcNAc), have many physiological functions in the treatment of human osteoarthritis, repair of cartilage, maintenance of joint function, delaying the development of knee osteoarthritis, and improving skin moisture (Kubomura et al. 2017; Pavelká et al. 2002). Therefore, it is widely used in health care products, cosmetics and clinical (Chen et al. 2010; Dostrovsky et al. 2011; Nakamura 2011). Currently, GlcN and GlcNAc have become one of the most common non-vitamin, non-mineral dietary supplements in many countries including China, the United States, and many European countries (Dostrovsky et al. 2011; Kennedy 2005). As the aging of population and the scope of application continues to expand, the demand for GlcN and GlcNAc will continue to grow.

GlcN and GlcNAc are essential components of chitin in the epidermis of crustaceans and are currently produced primarily from crabs and shrimp shells (Sitanggang et al. 2010). However, the limitations of variable raw material supply and the presence of allergens make it not only unable to meet the growing demand, but also limit the range of applications of GlcN and GlcNAc. The production of GlcN and GlcNAc using microorganisms has been extensively studied and proved to be an effective solution. Filamentous fungi, including three wild-type fungi, *Rhizopus oligosorus*, *Monascus pilosus* and *Aspergillus sp.*, have been developed for the biological production of GlcN due to their widespread use in the production of organic acids,

R. Tian · Y. Liu · L. Liu (✉)
Key Laboratory of Carbohydrate Chemistry and Biotechnology, Ministry of Education, Jiangnan University, Wuxi, China

Key Laboratory of Industrial Biotechnology, Ministry of Education, Jiangnan University, Wuxi, China
e-mail: longliu@jiangnan.edu.cn

L. Liu, J. Chen (eds.), *Systems and Synthetic Biotechnology for Production of Nutraceuticals*, https://doi.org/10.1007/978-981-15-0446-4_4

enzymes, antibiotics and other fine chemicals (Liu et al. 2013a). However, the highest GlcN yield reported is only 14.37 g/l (Zhang et al. 2012a). The disadvantages of low yield and low productivity weaken the economic competitiveness of the fungal biosynthesis of GlcN and GlcNAc. In addition, by genetic engineering of the synthesis and transport pathways of *E. coli*, optimizing the culture conditions, the yield of GlcNAc reached 110 g/l (Liu et al. 2013a).

However, due to the susceptibility of *E. coli* to phage contamination and the presence of endotoxin, the use of *B. subtilis* expression systems for the production of food and pharmaceutical grades of GlcN and GlcNAc is receiving increasing attention (Liu et al. 2013a). *B. subtilis* can produce a precursor compound of GlcNAc, Glucosamin-6-P (GlcN-6-P), by its own phosphoglucose isomerase (PGI) and GlcN synthase (GlmS) (Liu et al. 2013a). Therefore, biosynthesis of GlcNAc by *B. subtilis* by introducing an exogenous GlcN-6-P N-acetyltransferase gene (Gna1) is theoretically feasible. However, biosynthesis of GlcNAc in *B. subtilis* has three major problems. First, *B. subtilis* can utilize extracellular GlcN and GlcNAc as carbon sources, which may lead to degradation of the product. The second problem is that GlmS is strongly inhibited by GlcN-6-P, and accumulation of GlcN-6-P inhibits cell growth. Third, the biosynthetic pathways of GlcN and GlcNAc share a common precursor with glycolysis and peptidoglycan synthesis (Liu et al. 2013a). How to properly distribute metabolic flux is a huge challenge. In order to solve these problems, various metabolic engineering strategies have been applied in the production of GlcNAc using *B. subtilis*. First, the genes for the uptake and degradation of GlcNAc by *B. subtilis*, including *nagA* (encoding GlcNAc-6-phosphate deacetylase), *gamA* (encoding GlcN-6-phosphate deaminase), and *nagB* (encoding GlcN-6-P deaminase), were knocked out to prevent the consumption of products (Liu et al. 2013b). Then, in order to increase metabolic flux, *glmS*, *gna1* (from Saccharomyces cerevisiae S288C), *pgi* and other genes have been overexpressed by various metabolic engineering strategies, including the use of high-copy plasmids, combining different promoters, 5′-terminus fusion engineering, etc. (Liu et al. 2013b; Ling et al. 2017; Ma et al. 2019) DNA-guided scaffold system was also used to modulate the activities of key enzymes GlmS and GNA1 (Liu et al. 2014a). In addition, more work is devoted to regulating the anabolic pathway of the product and its competitive metabolic pathways. Liu et al. divided the GlcNAc synthesis-related metabolic network into three modules, including GlcNAc synthesis, glycolysis, and peptidoglycan synthesis (Liu et al. 2014b). The flux of different modules was optimized by combining different promoter types and strengths, synthetic small regulatory RNAs and Hfq protein. Gu et al. inhibit the competitive pathways of GlcNAc synthesis by a initiation codon optimization strategy, including glycolysis, peptidoglycan synthesis pathway, pentose phosphate pathway and tricarboxylic acid cycle (Gu et al. 2017). Niu et al. dynamically control the flux of peptidoglycan synthesis and glycolytic pathways in *B. subtilis* by designing a GlcN-6-P responsive *glmS* ribozyme switch (Niu et al. 2018). Wu et al. downregulated the expression of three key genes of pentose phosphate pathway (HMP), glycolysis, and peptidoglycan synthesis pathway (PSP) by optimizing the xylose-induced CRISPR interference (CRISPRi) system to achieve inhibition of competitive pathways (Wu et al. 2018). In addition to solving the three problems that have been raised, there is also some work devoted to saving unnecessary energy and carbon or reducing the overflow of metabolites in non-GlcNAc synthetic pathways. By knocking out the sporulation gene spo0A, blocking the overflow of acetoin, overcoming the overflow of pyruvic acid, the waste of energy and substrate is greatly reduced (Liu et al. 2014a; Ma et al. 2018, 2019). There is also works to reduce the bottleneck of the GlcNAc biosynthetic pathway. Liu et al. proposed a new strategy for detecting bottlenecks in metabolic pathways. Through the use of kinetic models and the determination of metabolic kinetics, it is predicted that the ineffective loop between GlcNAc-6-P and GlcNAc is a major problem in the GlcNAc biosynthetic pathway (Liu et al. 2016). By redesigning the central carbon and redox metabolism, Gu et al. not only not only alleviated the pathway bottleneck, but also reduced the overflow of acetyl-CoA and eliminated the formation of by-products (Gu et al. 2019). At present, the reported productivity of *B. subtilis* GlcNAc has exceeded 100 g/L, which has great potential for the commercial supply of large-scale GlcNAc.

4.2 Heparin

Glycosaminoglycans (GAGs) are a class of linear polysaccharides with varying lengths, backbone sugars, and modifications consisting of repeating disaccharide units. Due to its extensive biological and physiological functions, a series of GAGs with great potential for application have been intensively studied as health care products and therapeutic drugs (Chavaroche et al. 2013; Williams et al. 2018; Deangelis 2002; Linhardt 2003; Suflita et al. 2015). Heparin (HP) is a very important member of the GAG family. It is a polysaccharide which is alternately linked by Glucuronic acid (GlcUA) and N-acetylglucosamine (GlcNAc) via β-1,4 and α-1,4 glycosidic bonds and modified by a certain degree of sulfation (Kang et al. 2018; Rabenstein 2002). It is widely distributed on the cell surface and plays an important role in cell recognition, signal transmission, and tissue development. HP is used clinically as an anticoagulant and antithrombotic drug (Cziraky and Spinler 1993; Onishi et al. 2016). In addition, HP and derivatives also have great potential in the treatment of tumor and inflammation (Fareed et al. 2018). As early as 1939, bovine bovine-derived was recognized by the FDA (Hemker 2016). And recently, due to the aging of the world population, the demand for HP has increased dramatically.

For the production of HP, it is currently mainly extracted from animal tissues (mainly animal small intestine mucosa). However, extracting HP from animal tissues often faces multiple problems (Bode et al. 2016). First, the amount of HP in the intestinal tissue tends to vary depending on the animal's feed and living environment. Second, animal-derived HP often has problems with allergen contamination (Liu et al. 2009a). Oversulfated chondroitin sulfate is confirmed to be a contaminant in animal-derived heparin (Guerrini et al. 2008). In addition, the low degree of sulfation of HP extracted from animals results in lower activity units and is prone to complications such as thrombocytopenia (Warkentin et al. 1995; Stallforth et al. 2009;

Sandercock and Leong 2017). The use of chemical methods for the synthesis of HP has also been explored, but due to complex sulfation steps and long half-life of the product, efficient chemical synthesis has not been developed for large-scale production of HP (Zhang et al. 2008; Boltje et al. 2009). Therefore, there is a need to produce HP using a lower cost, more stable, and safer alternative. Since HP is naturally biosynthesized, the use of microbial cell factories for efficient biosynthesis of HP seems to be a viable approach. Ideally, co-express all the enzymes of the heparin synthesis in a common strain could synthesize heparin products in vivo. Unfortunately, 3′-phosphoadenosine-5′-phosphosulfate (PAPS), the cofactor for sulfotransferase reactions is exclusively produced intracellularly while precursor heparosan remains on the cell surface, forming phosphatidic acid molecules in the outer membrane (Barreteau et al. 2012). Heparosan is an unsulfated polysaccharide that is important in the manufacture of cosmetics and pharmaceuticals, especially as a precursor to HP. And it has been reported that precursor heparosan can chemically synthesize HP in high yield (Zhang et al. 2008; Laremore et al. 2009). Thus, using the microbial synthetic heparosan to produce bioactive HP by semi-chemical synthesis and chemoenzymatic modification is a promising approach.

Heparosan is naturally found in *E. coli* K5 and *Pasteurella multocida* (Vann et al. 1981; Deangelis and White 2002). Since the size of *E. coli* K5 heparosan is closer to that of heparin (average molecular weight is 10–20 KDa), and the genetic manipulation method of *E. coli* is more mature, it is generally preferred to use *E. coli* K5 to biosynthesize heparosan. The yield of heparosan achieved 15 g/L in a fed-batch fermentor culture grown on defined medium containing glucose (Wang et al. 2010). UDP-glucuronic acid (UDP-GlcA) and UDP-N-acetyl-D-glucosamine (UDP-GlcNAc) are two precursor compounds of heparosan. In the heparin biosynthetic pathway, the synthesis of UDP-GlcA is considered to be a rate-limiting step, and therefore, the initial metabolic engineering efforts focused on increasing the activity of UDP-glucose dehydrogenase (UDPGDH). However, an increase in the concentration of UDP-GlcA reduces the heparosan titer, which shifts the direction of metabolic engineering to the balance of the two precursor materials (Roman et al. 2003). Experiments in a cell-free reaction system have shown that the relative ratio of UDP-GlcNAc and UDP-GlcA affects the rate of heparin polymerization and its chain length (Lidholt et al. 1988). In addition, there are three major problems in the biosynthesis of heparosan using *E. coli* K5. The first is that heparin biosynthesis pathway has a common step with *E. coli* cell wall biosynthesis, which may affect cell growth. The second is how to promote heparosan detachment from the surface of the cell into the medium. The third problem is that *E. coli* K is a pathogenic bacterium that may cause urinary tract infections, so there is a potential danger of using it for the production of heparosan (Wang et al. 2011; Hänfling et al. 1996).

In order to solve the above problems, many efforts have been made, in particular, various strains have been developed for the production of heparosan (Barreteau et al. 2012; Wang et al. 2010; Deangelis 2012). First, a safer strain, *E. coli* BL21, was developed for biosynthesis of heparosan. The heparosan titer was 1.88 g/L by inducing expression of four genes *pKfiA*, *pKfiB*, *pKfiAC*, and *pKfiD* from *E. coli* K5 and optimization of fermentation (Zhang et al. 2012b). Although the titer is much lower than *E. coli* K5, this work not only indicates the importance of the balance of the two precursors, but also shows that controlling the growth of cells to a certain extent is conducive to the accumulation of heparosan. At the same time, it was found that there was no heparosan in the cells, indicating that the transport system BL21 of *E. coli* was sufficient to transport all intracellular heparin sugars to the outside of the cells. *B. subtilis*, a GRAS (generally recognized as safe) strain, can also be used to produce heparosan (van Dijl and Hecker 2013). Through the construction and optimization of the metabolic pathway, and after fermentation optimization, the heparosan titer reached 5.82 g/L, with great potential for biosynthesis of heparosan (Jin et al. 2016a). In addition, cyanobacteria, also known as the GARS strain, has also been shown to be useful in the production of heparosan. By expressing heparosan synthase derived from pathogenic *P. multocida*, *Synechococcus elongatus* PCC 7942 produced 2.8 μg/L of heparin under photoautotrophic conditions (Sarnaik et al. 2019). In general, heparosan's microbial synthesis has great potential due to its own advantages. But large-scale production of heparosan still requires more in-depth metabolic engineering and fermentation optimization of engineered strains.

4.3 Chondroitin Sulfate

Chondroitin sulfate (CS) is a widely distributed GAG in the form of CS-proteopolysaccharide in cartilage connective tissue, which usually plays a role in maintaining cell structure and facilitating nutrient and oxygen diffusion. It consists of repeating disaccharide units (-β-1, 4-GlcA-β-1, 3-GalNAc-; GlcA, glucuronic acid; GalNAc, N-acetylgalactosamine) with varying degrees of sulfation (Dickendesher et al. 2012). CS is mainly classified into CS-A (chondroitin sulfate A,[GlcA-β-1, 3-GalNAc(4S)]), CS-C (chondroitin sulfate C, [GlcA-β-1, 3-GalNAc(6S)]), CS-D (chondroitin sulfate D,[GlcA(2S)-β-1, 3-GalNAc(6S)]), CS-E (chondroitin sulfate E, [GlcA-β-1, 3-GalNAc(4S, 6S)]); CS-O (chondroitin,[GlcA-β-1, 3-GalNAc]) according to the difference in its sulfation pattern (Mende et al. 2016). The main application of CS is to treat osteoarthritis and promote cartilage regeneration. The European League Against Rheumatism (EULAR) recommends the use of CS for the treatment of knee and hand osteoarthritis (McAlindon et al. 2000). At the same time, CS exhibits its anti-inflammatory activity by systemically releasing interleukins in the intestine. In addition, several studies have shown that CS also has a variety of functions such as promoting tissue regeneration, helping to diagnose and treat cancer, as an antiviral drug (Cimini et al. 2010a). Therefore, it is a very important chemical in medical and health care.

Currently, CS is mainly produced by chemical extraction of animal cartilage tissue (eg, bovine trachea, pig nasal septum, chicken keel, etc.) (He et al. 2015). However, the extraction process is not only laborious, but also requires consumption of a large amount of sodium hydroxide, urea, cysteine (or guanidine hydrochloride) and trichloroacetic acid for tissue digestion and deproteinization, which is liable to cause environmental pollution (Shi et al. 2014). At the same time, the contamination brought about by the production of raw materials is also an important issue. For example, keratin sulfate and oversulfated chondroitin sulfate can cause serious consequences (Gucrrini et al. 2008; Rainsford 2009). In addition, the extraction of CS from animal tissues may present a potential risk of interspecies virus and prion transmission. Therefore, there is a need to develop more secure and low cost alternatives to produce a particular form of CS. Several chemical methods and enzymatic methods have been established for synthesis of CS and its oligosaccharides (Kobayashi et al. 2003). For example, using a glycosyltransferase PMCS derived from Pasteurella multocida to synthesize a chondroitin chain containing a specific number of residues; synthesizing CS having a well-defined structure by hyaluronidase-catalyzed polymerization; synthesizing CS by immobilized chondroitin polymerase mutants (Shi et al. 2014; Fujikawa et al. 2005). However, the use of enzymatic methods has several disadvantages, including expensive saccharide precursors, dedicated reactors, which limits its large-scale application (Shi et al. 2014). In addition to enzymatic methods, microorganisms have also been developed for the production of CS due to the naturally occurring CS or CS-like compounds and easily modified properties.

Since the capsular polysaccharide of *E. coli* K4 is almost identical to CS, the first study of the biotechnological production process for the production of CS by fermentation was carried out using *E. coli* K4 (Cimini et al. 2010a). Through fermentation optimization, the titer of CS was only 300 mg/L (Manzoni et al. 1996). Subsequent work was mainly carried out around metabolic engineering of *E. coli* K4. The two premise substances of CS are the same as HP, which are UDP-GlcNAc and UDP-GlcA. Therefore, the metabolic engineering strategy is mainly to direct the metabolic flow to the synthesis of the two precursors. To achieve this, genes and proteins directly related to CS biosynthesis can be modified, such as overexpressing chondroitin polymerase to increase the level of UDP-GlcNAc (Cimini et al. 2010b). This can also be achieved by modulating transcription factors of CPS biosynthesis, such as expression of the transcriptional activator RfaH to positively regulate polysaccharide biosynthesis, overexpression of the transcriptional regulator SlyA to enhance the expression of CS biosynthesis-related gene cluster (Cimini et al. 2013; Wu et al. 2013). In addition, by profiling the nucleotide carbohydrater precursors in *E. coli* K4 to guide fermentation optimization, the titer of CS can also be increased (Restaino et al. 2017). Although the production level of CS of *E. coli* K4 is quite high, *E. coli* K4 is a pathogenic bacteria as described previously. Other safer non-pathogenic hosts have also been developed for the production of CS, including *E. coli* BL21 and *B. subtilis* 168, which are widely used to express non-native genes to produce a variety of natural products and proteins. By using a high-copy plasmid heterologously expressing three genes *kfoA*, *kfoC* and *kfoF* from chondroitin biosynthesis of *E. coli* K4 in *E. coli* BL21 and performing fermentation optimization, the highest fermentation yield of CS was 2.4 g/L, which is almost equivalent to *E. coli* K4 (He et al. 2015). However, the supply of the two precursors still has great potential for improvement, and a large amount of chondroitin is also found to accumulate in the cells. This suggests that further metabolic engineering strategies and optimized media composition are still necessary to obtain better industrially competitive strains. In addition, the use of *B. subtilis* biosynthesis CS has also been reported. By heterologous expression of the key genes of CS biosynthesis *kfoC*, *kfoA*, optimization of metabolic pathways and fermentation conditions, the final CS titer was 5.22 g/L (Jin et al. 2016a). Furthermore, a new biological method was developed for de novo biosynthesis of CS-A and CS-C (Zhou et al. 2018). The strategy consists of two steps: first, CS-O is produced from sucrose by a constructed *B. subtilis* cell factory, and then CS is converted to CSA and CSC using a specific sulfation modification system. Through deep pathway engineering and fed-batch fermentation optimization, CS production reached 7.15 g/L. The emergence of various microbial synthesis platforms of CS-O has laid the foundation for large-scale production of homogeneous CS and their oligosaccharides with different sulfation patterns by combining with in vitro enzymatic sulfation and hydrolysis.

4.4 Hyaluronic Acid

Hyaluronic acid (HA) is a natural linear polymer composed of a disaccharide repeating unit of β-1,3-N-acetylglucosamine and β-1,4-glucuronic acid, with a molecular weight from 10^4 to 10^7 daltons. HA is widely present in animals and is generally present in the form of hyaluronate. Due to its excellent viscoelasticity, high moisturizing ability, high biocompatibility, and no immunogenicity and toxicity, it is widely used in nutrition, cosmetics and medicine (Liu et al. 2011).

Traditionally HA has been extracted from cock combs, but in order to avoid potential toxins, it is now mainly produced by microbial fermentation with low cost and less polluted environment, in particular *Streptococcus sp.* and *B. subtilis*. Although HA has now successfully used the industrial strain of *Streptococcus zooepidemicus* to achieve mass production (Liu et al. 2011). There are several key issues in the production of HA by large-scale fermentation. First, since HA has a high viscosity, the fermenter usually faces problems such as poor mixing and low oxygen transmission rate, which severely limits the yield of HA. Second, there is a strong competition between HA synthesis and cell growth metabolic pathways, which may inhibit cell growth and reduce HA production. Finally, the accumulation of lactic acid, the main by-product of HA fermentation, strongly inhibits cell growth and HA biosynthesis. Therefore, optimization of the culture medium and fermentation process is a huge challenge (Liu et al. 2011). These problems have been solved to some extent by optimization of pH, temperature, agitation speed, aeration rate, shear stress, dissolved oxygen, bioreactor type and fermentation mode during fermentation (Zhang et al. 2006; Liu et al. 2009b; Chong et al. 2005). At the same time, molecular weight is an important quality parameter for commercial HA products. HA with high molecular weight (greater than 10 kDa) is widely used in ophthalmology, orthopedics, wound healing and cosmetics, while HA with relatively low molecular weight (2–3.5 kDa) can promote angiogenesis, help anti-tumor and anti-inflammatory (Stern et al. 2006; Toole et al. 2008). HA of different molecular weights can be produced to some extent by changes in fermentation conditions. Although *Streptococcus sp.* has achieved great success in the industrial production of HA, since *S. zooepidemicus* is pathogenic, the development of other alternative strains for HA production is receiving increasing attention. In particular, *B. subtilis*, as a GRAS strain, has been extensively studied in the production of HA. The production of HA of a specific molecular weight by *B. subtilis* 168 has been achieved by metabolic engineering. Even the production of HA using *B. subtilis* achieved the highest reported HA titer of HA produced by the microbial strain (Jin et al. 2016b). The microbial production of HA is one of the most successful examples of the current use of microbial production, and its microbial production strategy can be used for microbial production of other polysaccharides.

4.5 Human Milk Oligosaccharides

Human milk oligosaccharides (HMOs) are the third most abundant solid component in human milk and consist of five monosaccharide building blocks, including: glucose (Glc), galactose (Gal), N-acetylglucosamine (GlcNAc), fucose (Fuc) and sialic acid (Bode 2012, 2015; Bych et al. 2019). Due to different chain lengths and varying degrees of fucosylation, sialylation, the number of HMOs currently identified has exceeded 150 (Bode et al. 2016). HMOs can be used as prebiotics to help shape the microbiota composition and reduce harmful bacteria in the gut, playing a very important role in infant health (Kunz 2012). In addition, it can promote the maturation of the immune system and prevent pathogens from attaching to host cells (Elison et al. 2016; Puccio et al. 2017). As a result, the market for HMOs is getting bigger, especially as a health supplement to improve infant health.

For the production of HMOs, it is apparent that large-scale production cannot be achieved from the extraction of breast milk (Bych et al. 2019). At the same time, the chemical synthesis of HMOs has a high production cost due to the complicated production process, the many reaction steps and the limited availability of raw materials, which limits the large-scale production and application of HMOs (Bych et al. 2019). In addition, methods for chemically enzymatically synthesizing structurally diverse HMOs have also been developed. However, the titer reported so far is only in the milligram range. The large-scale production of HMO by microorganisms has proven to be a potentially viable method (Prudden et al. 2017; Xiao et al. 2016; Zhao et al. 2016). 2′-fucosyllactose (2′-FL) is one of the most abundant HMOs in human milk and has a relatively simple structure. Currently, 2′-FL has been produced by *E. coli*, *Saccharomyces cerevisiae*, *Lactobacillus* and *Bacillus* (Bych et al. 2019). *E. coli* has become a major strain in research and commercial production 2′-FL because of its potent lactose uptake and HMOs output system, high growth rate, and mature genetic manipulation tools (Dumon et al. 2001; Samain et al. 1997). The biosynthesis of 2′-FL in engineered *E. coli* requires three elements including two precursors and one enzyme, which are also essential for the biosynthesis of other HMOs (eg, difucosyllactose (DFL), 3-fucosyllactose (3-FL) and lacto-N-neotetraose (LNnT)) (Bode et al. 2016). The first precursor is lactose, which is the receptor for the exogenous sugars of most HMOs based on fermentation. Wild-type *E. coli* can effectively absorb and degrade lactose by the enzyme lactose permease (LacY) and β-galactosidase (LacZ), so the *lacZ* gene is generally knocked out to prevent degradation of lactose (Chin et al. 2015, 2016). Second, the guanosine diphosphate fucose (GDP-L-fucose) needs to overproduced intracellularly (Chin et al. 2015). GDP-L-fucose is a precursor of capsular isoflavonic acid and is also a fucosyl donor of 2′-FL. There are currently three strategies to increase the level of GDP-L-fucose, including increasing the expression level of the GDP-L-fucose biosynthesis pathway, knocking out or down-regulating the biotransformation gene of GDP-L-fucose, targeting the potentially limiting co-factors GTP and NADPH, adding L-fucose during the fermentation process (Lee et al. 2009, 2011, 2012; Huang et al. 2017; Becker and Lowe 2003). However, GDP-L-fucose consistently stays at the mg/L level, indicating that the focus needs to be to increase the flux of the 2′-FL biosynthetic pathway rather than increasing the absolute concentration of the substrate (Lee et al. 2011, 2012). Third, fucosyltransferase that transfer the sugar moieties from the sugar nucleotides to lactose are needed. Helicobacter pylori-derived α1,2-fucosyltransferase (FucT2) and *E. coli*-derived α1,2-fucosyltransferases (wbgL and wbsJ) have been used for biosynthesis of 2′-FL (Engels and Elling 2013; Shao et al. 2003). In order to meet the growing market demand, the amplification of the fermentation process is necessary to produce HMOs on a large scale. However, factors such as longer mixing times, genetic instability, and defects in the carbon source itself are huge challenges. Despite this, 2′-FL and LNnT have been successfully industrialized using microorganisms through optimization of fermentation conditions, highly innovative strategies (such as synthetic product addiction), and metabolic pathway remodeling (Bych et al. 2019). The fermentor volume reported so far exceeds 200 m^3, and the highest reported 2′-FL titer currently reported is 180 g/L (Bych et al. 2019). Through microbial metabolic engineering strategies and fermentation optimization, HMOs are being produced at a lower cost for larger scale, and have appeared as nutritional supplements in infant formula in 2016 (Bych et al. 2019). There will be more rapid development and wider application in the future. Finally, the biosynthesis of more diverse kinds of HMOs using *E. coli* and other GARS strains has great potential.

4.6 Chitin Oligosaccharides

Chitin oligosaccharides (Cos) is a series of linear oligosaccharides composed of N-acetyl-β-d-glucosamine (GlcNAc) monosaccharide repeat units linked by β-1-4 glycosidic bonds. It has been widely used in the food, pharmaceutical, cosmetic, agricultural and water treatment industries due to its biological functions such as anti-oxidation, anti-inflammatory, anti-tumor and anti-hypertension (Xu et al. 2010; Fernandes et al. 2008, 2010; Benhabiles et al. 2012; Huang et al. 2006). Currently, the main commercial source of chitin is crab and shrimp shells. In industrial processing, chitin is extracted from crustaceans by two separation steps: demineralization using HCl and deproteinization using aqueous NaOH (Jung and Park 2014). However, due to the potential allergen and environmental friendliness of chemical extraction, metabolic engineering of microbial cells represents the most promising strategy, and engineered *E. coli* that can produce a mixture of COs has been reported. In addition, *B. subtilis* can also produce a mixture of COs as an expression host, and it has a unique advantage as a GRAS strain (Ling et al. 2018). The main difficulties in using microorganisms to produce COs are: the intracellular concentration of UDP-GlcNAc, the sugar donor of COs, may limit the synthesis of CO; the competition between carbon and energy consumption between the COs synthesis pathway and the glycolysis pathway increases the difficulty of the fermentation process, because the balance between cell growth and COs synthesis is very important (Ling et al. 2018). The use of microbial production of well-defined COs has shown new potential by overexpressing metabolic pathways to achieve metabolic flux expansion, increased precursor supply, dynamic pathway regulation, and modular pathway engineering (Blazeck and Alper 2010; Schumann 2007; Guiziou et al. 2016).

4.7 Xanthan Gum

Xanthan gum (XG) is a linear polysaccharide composed of a β-d-glucose skeleton, and is linked to a trisaccharide side chain having a certain degree of acetone and acetylation in every other glucose (Palaniraj and Jayaraman 2011; Rosalam and England 2006). Due to its ideal water solubility, emulsifying properties, and thickening properties, it is widely used in many industries such as food, oil recovery, pharmaceuticals, cosmetics, waterborne coatings and textiles (Palaniraj and Jayaraman 2011; Rosalam and England 2006; Mittal et al. 2016). At the same time, XG is a microbial high molecular weight exopolysaccharide produced by *Xanthomonas*, so industrial XG is mainly produced by fermentation (Kumar et al. 2018). Through fermentation optimization, such as feed technology, temperature, pH, stirring and adding defoamer during the fermentation process, the production technology of XG is relatively mature (Palaniraj and Jayaraman 2011). There are several huge challenges in the fermentation production phase. First of all, similar to HA, since the fermentation broth produced in the production stage is very viscous, the agitation in the fermenter requires a considerable balance between cell destruction and oxygen delivery (Rosalam and England 2006). The preferred strategy is to freely move the liquid media and air passing through the porous fibrous matrix, which not only improves oxygen transfer, but also enhances reaction rate and cell growth. Second, the choice of carbon and nitrogen sources is also very important. Through exploration, glucose or sucrose is generally used as a carbon source, and glutamic acid is used as a nitrogen source. Starch, glycerin, barley flour, corn flour and sugar cane molasses have also been tested to reduce production costs (Wang et al. 2016). Further, many different complexity kinetic models have been developed for the design and expansion of XG fermentation bioreactors. The kinetic model can describe the consumption of carbon sources and the production of XG at different times. Some models also describe changes in nitrogen source concentration and oxygen consumption. These models can help to better understand growth restriction factors and production-restricted nutrients in the production of XG (Faria et al. 2010; Casas et al. 2000; Quinlan 1986). The fermentation optimization of the XG production process is very successful, which fully satisfies the large-scale production and commercial application of XG, making XG the most commercially industrial glue obtained by fermentation.

4.8 Concluding Remarks

Due to the intensive research on various oligosaccharides and polysaccharides in recent years, their important functions have been paid more and more attention (Elshahawi et al. 2015; Shi 2016). The biosynthesis of oligosaccharides and polysaccharides using microbial fermentation has been recognized as the most promising alternative to traditional animal tissue extraction or chemical production method synthesis. In order to produce oligosaccharides and polysaccharides more safely and efficiently with microorganisms, more efficient strategies for metabolic engineering and fermentation optimization are being developed and applied, such as multi-omics, genomics-based metabolic models and genome-scale engineering tools. More cheap and safe oligosaccharides and polysaccharides will be widely used as nutraceuticals in the future.

References

Barreteau H, Richard E, Drouillard S, Samain E, Priem B. Production of intracellular heparosan and derived oligosaccharides by lyase expression in metabolically engineered E. coli K-12. Carbohydr Res. 2012;360:19–24. https://doi.org/10.1016/j.carres.2012.07.013.

Becker DJ, Lowe JB. Biosynthesis and biological function in mammals. Glycobiology. 2003;13:41R–53R. https://doi.org/10.1093/glycob/cwg054.

Benhabiles MS, et al. Antibacterial activity of chitin, chitosan and its oligomers prepared from shrimp shell waste. Food Hydrocoll. 2012;29:48–56. https://doi.org/10.1016/j.foodhyd.2012.02.013.

Blazeck J, Alper H. Systems metabolic engineering. Genome-scale models and beyond. Biotechnol J. 2010;5:647–59. https://doi.org/10.1002/biot.200900247.

Bode L. Human milk oligosaccharides. Every baby needs a sugar mama. Glycobiology. 2012;22:1147–62. https://doi.org/10.1093/glycob/cws074.

Bode L. The functional biology of human milk oligosaccharides. Early Hum Dev. 2015;91:619–22. https://doi.org/10.1016/j.earlhumdev.2015.09.001.

Bode L, et al. Overcoming the limited availability of human milk oligosaccharides. Challenges and opportunities for research and application. Nutr Rev. 2016;74:635–44. https://doi.org/10.1093/nutrit/nuw025.

Boltje TJ, Buskas T, Boons G-J. Opportunities and challenges in synthetic oligosaccharide and glycoconjugate research. Nat Chem. 2009;1:611–22. https://doi.org/10.1038/nchem.399.

Bych K, et al. Production of HMOs using microbial hosts — from cell engineering to large scale production. Curr Opin Biotechnol. 2019;56:130–7. https://doi.org/10.1016/j.copbio.2018.11.003.

Casas J, Santos V, García-Ochoa F. Xanthan gum production under several operational conditions. Molecular structure and rheological properties☆. Enzyme Microb Technol. 2000;26:282–91. https://doi.org/10.1016/S0141-0229(99)00160-X.
Chavaroche AA, van den Broek LAM, Eggink G. Production methods for heparosan, a precursor of heparin and heparan sulfate. Carbohydr Polym. 2013;93:38–47. https://doi.org/10.1016/j.carbpol.2012.04.046.
Chen J-K, Shen C-R, Liu C-L. N-acetylglucosamine. Production and applications. Mar Drugs. 2010;8:2493–516. https://doi.org/10.3390/md8092493.
Chin Y-W, Kim J-Y, Lee W-H, Seo J-H. Enhanced production of 2′-fucosyllactose in engineered Escherichia coli BL21star(DE3) by modulation of lactose metabolism and fucosyltransferase. J Biotechnol. 2015;210:107–15. https://doi.org/10.1016/j.jbiotec.2015.06.431.
Chin Y-W, Seo N, Kim J-H, Seo J-H. Metabolic engineering of Escherichia coli to produce 2′-fucosyllactose via salvage pathway of guanosine 5′-diphosphate (GDP)-l-fucose. Biotechnol Bioeng. 2016;113:2443–52. https://doi.org/10.1002/bit.26015.
Chong BF, Blank LM, Mclaughlin R, Nielsen LK. Microbial hyaluronic acid production. Appl Microbiol Biotechnol. 2005;66:341–51. https://doi.org/10.1007/s00253-004-1774-4.
Cimini D, Restaino OF, Catapano A, De Rosa M, Schiraldi C. Production of capsular polysaccharide from Escherichia coli K4 for biotechnological applications. Appl Microbiol Biotechnol. 2010a;85:1779–87. https://doi.org/10.1007/s00253-009-2261-8.
Cimini D, et al. Improved fructosylated chondroitin production by kfoC overexpression in E. coli K4. J Biotechnol. 2010b;150:324–31. https://doi.org/10.1016/j.jbiotec.2010.09.954.
Cimini D, De Rosa M, Carlino E, Ruggiero A, Schiraldi C. Homologous overexpression of rfaH in E. coli K4 improves the production of chondroitin-like capsular polysaccharide. Microb Cell Fact. 2013;12:46. https://doi.org/10.1186/1475-2859-12-46.
Cziraky MJ, Spinler SA. Low-molecular-weight heparins for the treatment of deep-vein thrombosis. Clin Pharm. 1993;12:892–9.
Deangelis PL. Evolution of glycosaminoglycans and their glycosyltransferases. Implications for the extracellular matrices of animals and the capsules of pathogenic bacteria. Anat Rec. 2002;268:317–26. https://doi.org/10.1002/ar.10163.
Deangelis PL. Glycosaminoglycan polysaccharide biosynthesis and production. Today and tomorrow. Appl Microbiol Biotechnol. 2012;94:295–305. https://doi.org/10.1007/s00253-011-3801-6.
Deangelis PL, White CL. Identification and molecular cloning of a heparosan synthase from Pasteurella multocida type D. J Biol Chem. 2002;277:7209–13. https://doi.org/10.1074/jbc.M112130200.
Dickendesher TL, et al. NgR1 and NgR3 are receptors for chondroitin sulfate proteoglycans. Nat Neurosci. 2012;15:703 EP. https://doi.org/10.1038/nn.3070.
Dostrovsky NR, Towheed TE, Hudson RW, Anastassiades TP. The effect of glucosamine on glucose metabolism in humans. A systematic review of the literature. Osteoarthr Cartil. 2011;19:375–80. https://doi.org/10.1016/j.joca.2011.01.007.
Dumon C, et al. In vivo fucosylation of lacto-N-neotetraose and lacto-N-neohexaose by heterologous expression of Helicobacter pylori α-1,3 fucosyltransferase in engineered Escherichia coli. Glycoconj J. 2001;18:465–74. https://doi.org/10.1023/A:1016086118274.
Elison E, et al. Oral supplementation of healthy adults with 2′-O-fucosyllactose and lacto-N-neotetraose is well tolerated and shifts the intestinal microbiota. Br J Nutr. 2016;116:1356–68. https://doi.org/10.1017/S0007114516003354.
Elshahawi SI, Shaaban KA, Kharel MK, Thorson JS. A comprehensive review of glycosylated bacterial natural products. Chem Soc Rev. 2015;44:7591–697. https://doi.org/10.1039/c4cs00426d.
Engels L, Elling L. WbgL. A novel bacterial α1,2-fucosyltransferase for the synthesis of 2′-fucosyllactose. Glycobiology. 2013;24:170–8. https://doi.org/10.1093/glycob/cwt096.
Fareed J, Bacher P, Jeske W. Advances in heparins and related research. An epilogue. Molecules. 2018;23:pii: E390. https://doi.org/10.3390/molecules23020390.

Faria S, Vieira PA, Resende MM, Ribeiro EJ, Cardoso VL. Application of a model using the phenomenological approach for prediction of growth and xanthan gum production with sugar cane broth in a batch process. LWT- Food Sci Technol. 2010;43:498–506. https://doi.org/10.1016/j.lwt.2009.09.018.

Fernandes JC, et al. Antimicrobial effects of chitosans and chitooligosaccharides, upon Staphylococcus aureus and Escherichia coli, in food model systems. Food Microbiol. 2008;25:922–8. https://doi.org/10.1016/j.fm.2008.05.003.

Fernandes JC, et al. Anti-inflammatory activity of chitooligosaccharides in vivo. Mar Drugs. 2010;8:1763–8. https://doi.org/10.3390/md8061763.

Fujikawa S-i, Ohmae M, Kobayashi S. Enzymatic synthesis of chondroitin 4-sulfate with well-defined structure. Biomacromolecules. 2005;6:2935–42. https://doi.org/10.1021/bm050364p.

Gu Y, et al. Rewiring the glucose transportation and central metabolic pathways for overproduction of N-acetylglucosamine in *Bacillus subtilis*. Biotechnol J. 2017;12:1700020. https://doi.org/10.1002/biot.201700020.

Gu Y, et al. Synthetic redesign of central carbon and redox metabolism for high yield production of N-acetylglucosamine in *Bacillus subtilis*. Metab Eng. 2019;51:59–69. https://doi.org/10.1016/j.ymben.2018.10.002.

Guerrini M, et al. Oversulfated chondroitin sulfate is a contaminant in heparin associated with adverse clinical events. Nat Biotechnol. 2008;26:669–75. https://doi.org/10.1038/nbt1407.

Guiziou S, et al. A part toolbox to tune genetic expression in Bacillus subtilis. Nucleic Acids Res. 2016;44:7495–508. https://doi.org/10.1093/nar/gkw624.

Hänfling P, Shashkov AS, Jann B, Jann K. Analysis of the enzymatic cleavage (beta elimination) of the capsular K5 polysaccharide of Escherichia coli by the K5-specific coliphage. Reexamination. J Bacteriol. 1996;178:4747–50. https://doi.org/10.1128/jb.178.15.4747-4750.1996.

He W, et al. Production of chondroitin in metabolically engineered E. coli. Metab Eng. 2015;27:92–100. https://doi.org/10.1016/j.ymben.2014.11.003.

Hemker HC. A century of heparin. Past, present and future. J Thromb Haemost. 2016;14:2329–38. https://doi.org/10.1111/jth.13555.

Huang R, Mendis E, Rajapakse N, Kim S-K. Strong electronic charge as an important factor for anticancer activity of chitooligosaccharides (COS). Life Sci. 2006;78:2399–408. https://doi.org/10.1016/j.lfs.2005.09.039.

Huang D, et al. Metabolic engineering of Escherichia coli for the production of 2′-fucosyllactose and 3-fucosyllactose through modular pathway enhancement. Metab Eng. 2017;41:23–38. https://doi.org/10.1016/j.ymben.2017.03.001.

Jin P, et al. Efficient biosynthesis of polysaccharides chondroitin and heparosan by metabolically engineered Bacillus subtilis. Carbohydr Polym. 2016a;140:424–32. https://doi.org/10.1016/j.carbpol.2015.12.065.

Jin P, Kang Z, Yuan P, Du G, Chen J. Production of specific-molecular-weight hyaluronan by metabolically engineered Bacillus subtilis 168. Metab Eng. 2016b;35:21–30. https://doi.org/10.1016/j.ymben.2016.01.008.

Jung W-J, Park R-D. Bioproduction of chitooligosaccharides. Present and perspectives. Mar Drugs. 2014;12:5328–56. https://doi.org/10.3390/md12115328.

Kang Z, et al. Bio-based strategies for producing glycosaminoglycans and their oligosaccharides. Trends Biotechnol. 2018;36:806–18. https://doi.org/10.1016/j.tibtech.2018.03.010.

Kennedy J. Herb and supplement use in the US adult population. Clin Ther. 2005;27:1847–58. https://doi.org/10.1016/j.clinthera.2005.11.004.

Kobayashi S, Fujikawa S-i, Ohmae M. Enzymatic synthesis of chondroitin and its derivatives catalyzed by hyaluronidase. J Am Chem Soc. 2003;125:14357–69. https://doi.org/10.1021/ja036584x.

Kubomura D, Ueno T, Yamada M, Nagaoka I. Evaluation of the chondroprotective action of N-acetylglucosamine in a rat experimental osteoarthritis model. Exp Ther Med. 2017;14:3137–44. https://doi.org/10.3892/etm.2017.4849.

Kumar A, Rao KM, Han SS. Application of xanthan gum as polysaccharide in tissue engineering. A review. Carbohydr Polym. 2018;180:128–44. https://doi.org/10.1016/j.carbpol.2017.10.009.

Kunz C. Historical aspects of human milk oligosaccharides. Adv Nutr (Bethesda, Md). 2012;3:430S–9S. https://doi.org/10.3945/an.111.001776.

Laremore TN, Zhang F, Dordick JS, Liu J, Linhardt RJ. Recent progress and applications in glycosaminoglycan and heparin research. Curr Opin Chem Biol. 2009;13:633–40. https://doi.org/10.1016/j.cbpa.2009.08.017.

Lee W-H, Han N-S, Park Y-C, Seo J-H. Modulation of guanosine 5′-diphosphate-d-mannose metabolism in recombinant Escherichia coli for production of guanosine 5′-diphosphate-l-fucose. Bioresour Technol. 2009;100:6143–8. https://doi.org/10.1016/j.biortech.2009.07.035.

Lee W-H, Chin Y-W, Han NS, Kim M-D, Seo J-H. Enhanced production of GDP-L-fucose by overexpression of NADPH regenerator in recombinant Escherichia coli. Appl Microbiol Biotechnol. 2011;91:967–76. https://doi.org/10.1007/s00253-011-3271-x.

Lee W-H, Shin S-Y, Kim M-D, Han NS, Seo J-H. Modulation of guanosine nucleotides biosynthetic pathways enhanced GDP-l-fucose production in recombinant Escherichia coli. Appl Microbiol Biotechnol. 2012;93:2327–34. https://doi.org/10.1007/s00253-011-3776-3.

Lidholt K, Riesenfeld J, Jacobsson KG, Feingold DS, Lindahl U. Biosynthesis of heparin. Modulation of polysaccharide chain length in a cell-free system. Biochem J. 1988;254:571–8. https://doi.org/10.1042/bj2540571.

Ling M, et al. Combinatorial promoter engineering of glucokinase and phosphoglucoisomerase for improved N-acetylglucosamine production in *Bacillus subtilis*. Bioresour Technol. 2017;245:1093–102. https://doi.org/10.1016/j.biortech.2017.09.063.

Ling M, Li J, Du G, Liu L. Metabolic engineering for the production of chitooligosaccharides. Advances and perspectives. Emerg Top Life Sci. 2018;2:377. https://doi.org/10.1042/ETLS20180009.

Linhardt RJ. 2003 Claude S. Hudson Award address in carbohydrate chemistry. Heparin: structure and activity. J Med Chem. 2003;46:2551–64. https://doi.org/10.1021/jm030176m.

Liu H, Zhang Z, Linhardt RJ. Lessons learned from the contamination of heparin. Nat Prod Rep. 2009a;26:313–21. https://doi.org/10.1039/b819896a.

Liu L, Sun J, Xu W, Du G, Chen J. Modeling and optimization of microbial hyaluronic acid production by Streptococcus zooepidemicus using radial basis function neural network coupling quantum-behaved particle swarm optimization algorithm. Biotechnol Prog. 2009b;25:1819–25. https://doi.org/10.1002/btpr.278.

Liu L, Liu Y, Li J, Du G, Chen J. Microbial production of hyaluronic acid. Current state, challenges, and perspectives. Microb Cell Fact. 2011;10:99. https://doi.org/10.1186/1475-2859-10-99.

Liu L, et al. Microbial production of glucosamine and N-acetylglucosamine. Advances and perspectives. Appl Microbiol Biotechnol. 2013a;97:6149–58. https://doi.org/10.1007/s00253-013-4995-6.

Liu Y, et al. Pathway engineering of *Bacillus subtilis* for microbial production of N-acetylglucosamine. Metab Eng. 2013b;19:107–15. https://doi.org/10.1016/j.ymben.2013.07.002.

Liu Y, et al. Spatial modulation of key pathway enzymes by DNA-guided scaffold system and respiration chain engineering for improved N-acetylglucosamine production by *Bacillus subtilis*. Metab Eng. 2014a;24:61–9. https://doi.org/10.1016/j.ymben.2014.04.004.

Liu Y, et al. Modular pathway engineering of *Bacillus subtilis* for improved N-acetylglucosamine production. Metab Eng. 2014b;23:42–52. https://doi.org/10.1016/j.ymben.2014.02.005.

Liu Y, et al. A dynamic pathway analysis approach reveals a limiting futile cycle in N-acetylglucosamine overproducing Bacillus subtilis. Nat Commun. 2016;7:11933. https://doi.org/10.1038/ncomms11933.

Ma W, et al. Metabolic engineering of carbon overflow metabolism of *Bacillus subtilis* for improved N-acetyl-glucosamine production. Bioresour Technol. 2018;250:642–9. https://doi.org/10.1016/j.biortech.2017.10.007.

Ma W, et al. Combinatorial fine-tuning of GNA1 and GlmS expression by 5′-terminus fusion engineering leads to overproduction of N-Acetylglucosamine in *Bacillus subtilis*. Biotechnol J. 2019;14:1800264. https://doi.org/10.1002/biot.201800264.

Ma W, et al. Combinatorial pathway enzyme engineering and host engineering overcomes pyruvate overflow and enhances overproduction of N-acetylglucosamine in *Bacillus subtilis*. Microb Cell Fact. 2019;18(1). https://doi.org/10.1186/s12934-018-1049-x.

Manzoni M, Bergomi S, Molinari F, Cavazzoni V. Production and purification of an extracellularly produced K4 polysaccharide from Escherichia coli. Biotechnol Lett. 1996;18:383–6. https://doi.org/10.1007/BF00143456.

McAlindon TE, LaValley MP, Gulin JP, Felson DT. Glucosamine and chondroitin for treatment of osteoarthritis a systematic quality assessment and meta-analysis. JAMA. 2000;283:1469–75. https://doi.org/10.1001/jama.283.11.1469.

Mende M, et al. Chemical synthesis of glycosaminoglycans. Chem Rev. 2016;116:8193–255. https://doi.org/10.1021/acs.chemrev.6b00010.

Mittal H, Kumar V, Saruchi, Ray SS. Adsorption of methyl violet from aqueous solution using gum xanthan/Fe3O4 based nanocomposite hydrogel. Int J Biol Macromol. 2016;89:1–11. https://doi.org/10.1016/j.ijbiomac.2016.04.050.

Nakamura H. Application of glucosamine on human disease—osteoarthritis. Carbohydr Polym. 2011;84:835–9. https://doi.org/10.1016/j.carbpol.2010.08.078.

Niu T, et al. Engineering a glucosamine-6-phosphate responsive glmS ribozyme switch enables dynamic control of metabolic flux in *Bacillus subtilis* for overproduction of N-Acetylglucosamine. ACS Synth Biol. 2018;7:2423–35. https://doi.org/10.1021/acssynbio.8b00196.

Onishi A, St Ange K, Dordick JS, Linhardt RJ. Heparin and anticoagulation. Front Biosci (Landmark Ed). 2016;21:1372–92.

Palaniraj A, Jayaraman V. Production, recovery and applications of xanthan gum by *Xanthomonas campestris*. J Food Eng. 2011;106:1–12. https://doi.org/10.1016/j.jfoodeng.2011.03.035.

Pavelká K, et al. Glucosamine sulfate use and delay of progression of knee osteoarthritis. A 3-year, randomized, placebo-controlled, double-blind study. Arch Intern Med. 2002;162:2113–23. https://doi.org/10.1001/archinte.162.18.2113.

Prudden AR, et al. Synthesis of asymmetrical multiantennary human milk oligosaccharides. Proc Natl Acad Sci. 2017;114:6954. https://doi.org/10.1073/pnas.1701785114.

Puccio G, et al. Effects of infant formula with human milk oligosaccharides on growth and morbidity. A randomized multicenter trial. J Pediatr Gastroenterol Nutr. 2017;64:624–31. https://doi.org/10.1097/MPG.0000000000001520.

Quinlan AV. Kinetics of secondary metabolite synthesis in batch culture when two different substrates limit cell growth and metabolite production. Xanthan synthesis by Xanthomonas campestrisa. Ann N Y Acad Sci. 1986;469:259–69. https://doi.org/10.1111/j.1749-6632.1986.tb26503.x.

Rabenstein DL. Heparin and heparan sulfate. Structure and function. Nat Prod Rep. 2002;19:312–31.

Rainsford KD. Importance of pharmaceutical composition and evidence from clinical trials and pharmacological studies in determining effectiveness of chondroitin sulphate and other glycosaminoglycans. A critique. J Pharm Pharmacol. 2009;61:1263–70. https://doi.org/10.1211/jpp.61.10.0001.

Restaino OF, Di Lauro I, Di Nuzzo R, De Rosa M, Schiraldi C. New insight into chondroitin and heparosan-like capsular polysaccharide synthesis by profiling of the nucleotide sugar precursors. Biosci Rep. 2017;37:BSR20160548. https://doi.org/10.1042/BSR20160548.

Roman E, Roberts I, Lidholt K, Kusche-Gullberg M. Overexpression of UDP-glucose dehydrogenase in Escherichia coli results in decreased biosynthesis of K5 polysaccharide. Biochem J. 2003;374:767–72. https://doi.org/10.1042/BJ20030365.

Rosalam S, England R. Review of xanthan gum production from unmodified starches by Xanthomonas comprestris sp. Enzyme Microb Technol. 2006;39:197–207. https://doi.org/10.1016/j.enzmictec.2005.10.019.

Samain E, Drouillard S, Heyraud A, Driguez H, Geremia RA. Gram-scale synthesis of recombinant chitooligosaccharides in Escherichia coli. Carbohydr Res. 1997;302:35–42. https://doi.org/10.1016/S0008-6215(97)00107-9.

Sandercock PA, Leong TS. Low-molecular-weight heparins or heparinoids versus standard unfractionated heparin for acute ischaemic stroke. Cochrane Database Syst Rev. 2017;4:CD000119. https://doi.org/10.1002/14651858.CD000119.pub4.

Sarnaik A, et al. Metabolic engineering of cyanobacteria for photoautotrophic production of heparosan, a pharmaceutical precursor of heparin. Algal Res. 2019;37:57–63. https://doi.org/10.1016/j.algal.2018.11.010.

Schumann W. Production of recombinant proteins in Bacillus subtilis. Adv Appl Microbiol. 2007;62:137–89.

Shao J, Li M, Jia Q, Lu Y, Wang PG. Sequence of Escherichia coli O128 antigen biosynthesis cluster and functional identification of an α-1,2-fucosyltransferase. FEBS Lett. 2003;553:99–103. https://doi.org/10.1016/S0014-5793(03)00980-3.

Shi L. Bioactivities, isolation and purification methods of polysaccharides from natural products. A review. Int J Biol Macromol. 2016;92:37–48. https://doi.org/10.1016/j.ijbiomac.2016.06.100.

Shi Y-g, et al. Chondroitin sulfate. Extraction, purification, microbial and chemical synthesis. J Chem Technol Biotechnol. 2014;89:1445–65. https://doi.org/10.1002/jctb.4454.

Sitanggang AB, Wu H-S, Wang SS, Ho Y-C. Effect of pellet size and stimulating factor on the glucosamine production using *Aspergillus* sp. BCRC 31742. Bioresour Technol. 2010;101:3595–601. https://doi.org/10.1016/j.biortech.2009.12.084.

Stallforth P, Lepenies B, Adibekian A, Seeberger PH. Carbohydrates. A frontier in medicinal chemistry. J Med Chem. 2009;52:5561–77. https://doi.org/10.1021/jm900819p.

Stern R, Asari AA, Sugahara KN. Hyaluronan fragments. An information-rich system. Eur J Cell Biol. 2006;85:699–715. https://doi.org/10.1016/j.ejcb.2006.05.009.

Suflita M, Fu L, He W, Koffas M, Linhardt RJ. Heparin and related polysaccharides. Synthesis using recombinant enzymes and metabolic engineering. Appl Microbiol Biotechnol. 2015;99:7465–79. https://doi.org/10.1007/s00253-015-6821-9.

Toole B, Ghatak S, Misra S. Hyaluronan oligosaccharides as a potential anticancer therapeutic. Curr Pharm Biotechnol. 2008;9:249–52. https://doi.org/10.2174/138920108785161569.

van Dijl J, Hecker M. Bacillus subtilis. From soil bacterium to super-secreting cell factory. Microb Cell Fact. 2013;12:3. https://doi.org/10.1186/1475-2859-12-3.

Vann WF, Schmidt MA, Jann B, Jann K. The structure of the capsular polysaccharide (K5 antigen) of urinary-tract-infective Escherichia coli 010:K5:H4. Eur J Biochem. 1981;116:359–64. https://doi.org/10.1111/j.1432-1033.1981.tb05343.x.

Wang Z, et al. E. coli K5 fermentation and the preparation of heparosan, a bioengineered heparin precursor. Biotechnol Bioeng. 2010;107:964–73. https://doi.org/10.1002/bit.22898.

Wang Z, Dordick JS, Linhardt RJ. Escherichia coli K5 heparosan fermentation and improvement by genetic engineering. Bioeng Bugs. 2011;2:63–7. https://doi.org/10.4161/bbug.2.1.14201.

Wang Z, Wu J, Zhu L, Zhan X. Activation of glycerol metabolism in Xanthomonas campestris by adaptive evolution to produce a high-transparency and low-viscosity xanthan gum from glycerol. Bioresour Technol. 2016;211:390–7. https://doi.org/10.1016/j.biortech.2016.03.096.

Warkentin TE, et al. Heparin-induced thrombocytopenia in patients treated with low-molecular-weight heparin or unfractionated heparin. N Engl J Med. 1995;332:1330–6. https://doi.org/10.1056/NEJM199505183322003.

Williams A, Linhardt RJ, Koffas MA. Metabolic engineering of capsular polysaccharides. Emerg Top Life Sci. 2018;2:337. https://doi.org/10.1042/ETLS20180003.

Wu Q, et al. Transcriptional engineering of Escherichia coli K4 for fructosylated chondroitin production. Biotechnol Prog. 2013;29:1140–9. https://doi.org/10.1002/btpr.1777.

Wu Y, et al. CRISPRi allows optimal temporal control of N-acetylglucosamine bioproduction by a dynamic coordination of glucose and xylose metabolism in *Bacillus subtilis*. Metab Eng. 2018;49:232–41. https://doi.org/10.1016/j.ymben.2018.08.012.

Xiao Z, et al. Chemoenzymatic synthesis of a library of human milk oligosaccharides. J Org Chem. 2016;81:5851–65. https://doi.org/10.1021/acs.joc.6b00478.

Xu Q, et al. Chitooligosaccharides protect human embryonic hepatocytes against oxidative stress induced by hydrogen peroxide. Marine Biotechnol. 2010;12:292–8. https://doi.org/10.1007/s10126-009-9222-1.

Zhang J, Ding X, Yang L, Kong Z. A serum-free medium for colony growth and hyaluronic acid production by *Streptococcus zooepidemicus* NJUST01. Appl Microbiol Biotechnol. 2006;72:168–72. https://doi.org/10.1007/s00253-005-0253-x.

Zhang Z, et al. Solution structures of chemoenzymatically synthesized heparin and its precursors. J Am Chem Soc. 2008;130:12998–3007. https://doi.org/10.1021/ja8026345.

Zhang J, Liu L, Li J, Du G, Chen J. Enhanced glucosamine production by *Aspergillus* sp. BCRC 31742 based on the time-variant kinetics analysis of dissolved oxygen level. Bioresour Technol. 2012a;111:507–11. https://doi.org/10.1016/j.biortech.2012.02.063.

Zhang C, et al. Metabolic engineering of Escherichia coli BL21 for biosynthesis of heparosan, a bioengineered heparin precursor. Metab Eng. 2012b;14:521–7. https://doi.org/10.1016/j.ymben.2012.06.005.

Zhao C, et al. The one-pot multienzyme (OPME) synthesis of human blood group H antigens and a human milk oligosaccharide (HMOS) with highly active Thermosynechococcus elongatus α1–2-fucosyltransferase. Chem Commun. 2016;52:3899–902. https://doi.org/10.1039/C5CC10646J.

Zhou Z, et al. A microbial–enzymatic strategy for producing chondroitin sulfate glycosaminoglycans. Biotechnol Bioeng. 2018;115:1561–70. https://doi.org/10.1002/bit.26577.

Chapter 5
Microbial Production of Flavonoids

Sonam Chouhan, Kanika Sharma, Sanjay Guleria, and Mattheos A. G. Koffas

5.1 Introduction

Plants have always been important to mankind. Apart from their role as food, plants have been utilized in folklore medicine as drugs in the treatment of diseases (Petrovska 2012). Secondary metabolites have been attributed for such medicinal properties of plants. Food as well as pharmaceutical industries are highly interested in the health-promoting impacts of secondary metabolites of plants and a continuous search to develop novel, safe as well as effective drugs or food additives is in progress. Food preservatives are the stuffs that are incorporated into food in order to steady or inhibit food decay due to micro-organisms or oxidation. Present-day food preservatives are generally synthetic chemicals like benzoates nitrates, sorbates and nitrites (Silva and Lidon 2016) that are often times associated with perceived or real health risks. Such health risks comprise gastrointestinal disorders, allergic reactions and cancer (Etemadi et al. 2017). Thus, there is large current attention in additional "natural" sources of food preservatives that can be volatile oils, plant extracts or purified secondary metabolites (Gassara et al. 2016). Plant-derived secondary metabolites can be categorized into three classes, on the basis of their biosynthetic genesis: (1) flavonoids and allied phenolic and polyphenolic compounds (2) terpenoids (3) nitrogen-containing alkaloids and sulphur containing compounds (Kabera

S. Chouhan · K. Sharma · S. Guleria
Natural Product Laboratory, Division of Biochemistry, Faculty of basic Sciences, Sher-e Kashmir University of Agricultural Sciences and Technology, Jammu, Jammu and Kashmir, India

M. A. G. Koffas (✉)
Department of Chemical and Biological Engineering, Rensselaer Polytechnic Institute, Troy, NY, USA

Department of Biological Sciences, Center for Biotechnology and Interdisciplinary Studies, Rensselaer Polytechnic Institute, Troy, NY, USA
e-mail: Koffam@rpi.edu

L. Liu, J. Chen (eds.), *Systems and Synthetic Biotechnology for Production of Nutraceuticals*, https://doi.org/10.1007/978-981-15-0446-4_5

et al. 2014). Flavonoids represent a big class of plant-extracted natural products that can be additionally categorized into six subclasses relying on the alterations in the heterocyclic C-ring: isoflavones, flavones, flavonols, flavanones, the catechins or flavanols and anthocyanidins. Such compounds exhibit a huge span of biological actions, resulting in a number of researches on such secondary metabolites (Zhang et al. 2018). Flavonoids extracted from plants have been greatly utilized in nutraceuticals, functional foods as well as cosmetics in the international trade (Panche et al. 2016); few of them are possible candidates for the treatment of diseases, like Alzheimer's, type II diabetes and cancer (Baptista et al. 2014; Ibrahim et al. 2014). In recent times, flavonoids are produced by extraction from plants. This can be a time consuming procedure that needs the utilization of non-environmentally secure solvents and often times fails to isolate complex compounds produced in the plants in small amounts (Trantas et al. 2015). In addition to this, extraction from plants is depended upon seasonal as well as environmental alterations like soil composition and nutrition. Chemical production may be a possibility, but this is restricted due to the complicated chemical constitution of the desirable products (Delmulle et al. 2017). Therefore, synthesis of flavonoids from microbes has achieved much interest because of dearth of accessible plant sources for the synthesis of flavonoids and a great amount of metabolic engineering researches have been revealed for the synthesis of flavonoids in microbes (Xiu et al. 2017).

Although it has been ably established that microbes possess the capacity to synthesize flavonoids, difficulties still exist, chiefly in converting such microbes into well-organized microbial cell factories. The present book chapter deals with the general strategies as well as tools that are being utilized to create industrially viable microbe-based systems for the synthesis of flavonoids. Therefore, this review gives a comprehensive outline of the distinct aspects that are required in the construction of microbial synthesis of flavonoids (Delmulle et al. 2017).

5.2 Overview of Flavonoids

Flavonoids belong to the general family of phenolics, having C6–C3–C6 as their general structural formula. The two C6 units (Ring A and Ring B) are joined by a heterocyclic ring resulting in the formation of a phenylpropanoid core having 15-carbon atoms (Chouhan et al. 2017). Flavonoids include a large number of phytochemicals that can be further categorized into six major subgroups on the basis of the differences in the heterocyclic carbon ring: flavones, flavonols, flavanones, isoflavones, the catechins or flavanols and anthocyanidins (Malla et al. 2013). A large number of biological activities have been demonstrated by these compounds (Vila-Real et al. 2011). Mainly through the modification of rings B and C in several ways like oxidation, methylation, alkylation, methoxylation, rearrangement, C- and O-glycosylation and hydroxylation, more than 9000 derivatives of flavonoids are formed. On the basis of position of attachment ring B to the ring C, flavonoids can be categorized into three major classes: common flavonoids, neoflavonoids and isoflavonoids (Pandey et al. 2016). The common flavonoids class comprises various subclasses like flavan, flavone, flavanone, flavanonol, flavanol, flavonol, and anthocyanin on the basis of the alterations to the ring C (dehydrogenation of C2, hydroxylation at C3 or C4, and oxidation at C4). Isoflavones like genistein, represent a large group of compounds wherein Ring B is joined at the C3 position of Ring C. Isoflavones are mostly found in the leguminous plants. Flavanones, flavones, flavanonols and flavonols comprise a huge sub-group of flavonoids being the most common as well as almost ubiquitous, throughout the plant kingdom. Flavonols, also known by the name flavan-3-ols or catechins, differ from most of the flavonoids as they lack double bond between C2 and C3, moreover, there is no C4 carbonyl in the Ring C. The monomeric form of flavonols i.e. catechins and epicatechins as well as their derivatives are among the vital flavonoids found in cacao beans and tea-leaves (Prior et al. 2001; Si et al. 2006). Polymers can be formed from catechins and epicatechin and these polymers are commonly known as proanthocyanidins. Anthocyanidins are produced from the acid catalyzed cleavage of this last group of compounds or are directly synthesized from flavanone precursors using a four-step pathway. Anthocyanins commonly represent the glycosidic types of such compounds in the plant kingdom. On the basis of the hydroxylation as well as the pattern of methoxylation on the ring B, and the glycosylation with distinct sugar units, more than 500 anthocyanins have been identified in nature (Tsao 2010). Anthocyanins have been primarily attributed for imparting bright colors to the flowering plants. Apart from being serving as color pigments, synthesis of flavonoids in plants also occurs in response to the changes associated with environment, usually as a defense mechanism like subjection to ultraviolet rays, infringement by pathogens (Winkel-Shirley 2001). Structures of the common flavonoids' subclasses are depicted below in Fig. 5.1.

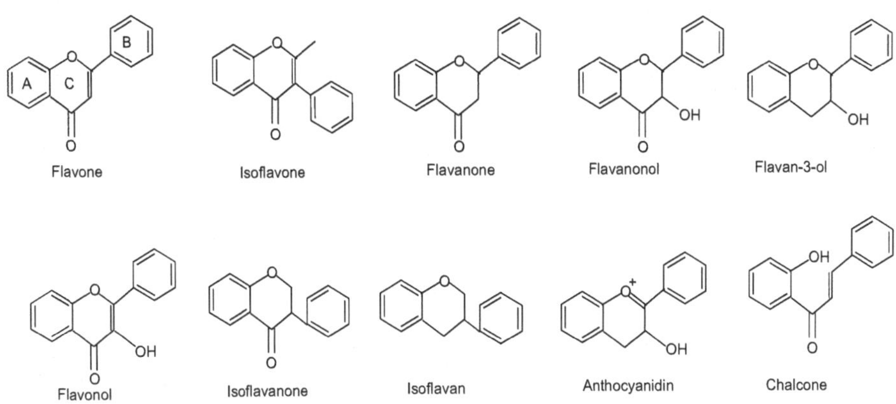

Fig. 5.1 Structures of the common flavonoids subclasses

5.3 Health Promoting Characteristics of Flavonoids

Flavonoids are synthesized in plants when exposed to different kinds of biotic and abiotic stresses such as microbial invasions, physical injury, environmental stresses, etc. (Treutter 2006). Flavonoids have important health benefits on humans (Williams et al. 2004). Absorption of flavonoids can occur via the gastrointestinal (GI) tract, but amounts of flavonoids in plasma are usually quite low i.e.1 μmol/l because of their rapid metabolism (Tsao 2010). Flavonoids and other phenolic compounds are among the plant-derived secondary metabolites that have been reported to possess several pharmacological activities (Costa et al. 2016; Rasines-Perea and Teissedre 2017). Different classes of flavonoid compounds possess several valuable attributes like antibacterial, antiviral, antioxidant and anti-cancer properties (Djouossi et al. 2015; Lani et al. 2016). Apart from these, flavonoids also possess activity against degenerative diseases like Alzheimer's disease (Bakhtiari et al. 2017), cardiovascular diseases (Lovegrove et al. 2017) and cancer (Youns and Hegazy 2017). A great amount of such diseases have been associated with oxidative stress resulting in the presence of reactive oxygen as well as reactive nitrogen species in tissues. By acting as antioxidants, flavonoids aid from this oxidative stress. Flavonoids have the ability to suppress the production of free radicals through the inhibition of the synthesis of or deactivation of the active species as well as precursors of free radicals. Apart from this, flavonoids can also play the role of radical scavengers associated with the lipid peroxidation chain reactions. Various catechins, that are members of another huge group of flavonoids, may cause activation of AMP-activated protein kinase (AMPK) which in turn has a key role in regulating the metabolism of lipids and glucose. Upon activation, AMPK elevates cellular energy levels through the inhibition of pathways related to anabolism like the production of lipids and glucose as well as stimulation of catabolic pathways like the oxidation of fat and glucose (Leiherer et al. 2013). It has also been revealed by animal studies that some catechins may possess anti-diabetic properties by enhancing the function of beta cells of pancreas (Ortsater et al. 2012). Increased use of flavonoids in skin care and cosmetic products as UV rays protectant has also been reported (Saewan and Jimtaisong 2015). Moreover, plasma rich in geneistein (isoflavone) has been associated with reduced risk of type 2 diabetes in women. Isoflavone has been associated with the anti-diabetic effect of isoflavone for their physiological alterations (Ko et al. 2015). Regarding human health, attention in anthocyanins originated from their antioxidant properties (Tsuda 2012). Citrus flavonoids, various subgroups of flavonoids comprising flavanones and O-polymethoxylated flavonoids, exhibit remarkable lipid and lipoprotein-reduction capacity, decrease gathering of hepatic lipid as well as inhibit excess synthesis of lipoprotein, regularize the sensitivity of insulin, blunt tissue inflammation as well as decrease the development of atherosclerosis. Such advantageous metabolic results are conciliated, partially by regularization of hepatic fatty acid metabolism, increasing insulin signaling as well as decrease in the reaction of inflammation (Assini et al. 2013). One of the most plentiful as well as crucial flavonoids that is present in almost all citrus fruits is naringenin. It has been revealed that naringenin has an essential part in the instigation of apoptotic cell death in the cancer cells (Park et al. 2017). Intriguingly, naringenin has also been found to possess anti-dengue virus potential through damaging the infection caused by four dengue virus serotypes in human cells (Frabasile et al. 2017). When diabetic mice were utilized as model systems, it was found that the glucosylated flavanone hesperidin as well as the flavanone naringenin were extremely efficient in increasing the metabolism of lipid through the modification of hepatic enzyme activities. Apart from this, simultaneous reduction of blood sugar levels also occurred by down-regulation of the hepatic glucose-6-phosphatase as well as GLUT2 while at the same time up-regulation of the adipocyte GLUT4 and hepatic glucokinase was also found (Ae Park et al. 2006).

5.4 Significance of Microbial Production

The universal existence of flavonoids in plants makes them a part of the everyday human diet. Everyday overall intake of flavonoids may surpass 1 g, with prime sources being vegetables and fruits as well as beverages like cocoa, tea, beer and wine (Zamora-Ros et al. 2016). Food additives, pharmaceuticals, nutraceuticals, cosmetics and others are some of the commercial products that use flavonoids. International trade as well as demand for flavonoids was valued more than 840 million USD during the year 2015 and is expected to pass 1 trillion USD after the year 2020 (https://www.zionmarketresearch.com/news/global-flavonoids-market). In the present time, the elevating demands for flavonoids as well as other plant-derived secondary metabolites cannot be met by extracting them from plants as plants synthesize such compounds in restricted quantities, under certain environmental conditions or some type of abiotic or biotic stresses. As a result, production of flavonoids from plants requires complicated extraction processes including toxic chemicals as well as complex downstream handling (Paterson and Anderson 2005; Keasling 2010; Ng et al. 2019). Extraction of flavonoids from plants is the default practice, but it has remarkable drawbacks including high cost and remarkable carbon footprint because of unnecessary energy as well as solvent needs. One of the alternative techniques for the production of flavonoids is chemical production but this is not easy to be scaled up and cannot easily do certain chemical modifications, such as selected hydroxylations as well as glycosylations (De Luca et al. 2012).

Metabolic engineering is another approach to improve the concentration of flavonoids in plants, but the complex nature of plant cells, their multicellular constitution as well as the complicated and stern biosynthetic modulation causes hurdles. Utilization of plant cell cultures is another methodology for flavonoid production, and anthocyanins have been synthesized using such methods. However, this approach too has several drawbacks like culture heterogeneity, variation in the product titers, unstable cultures, less growth rates, accumulation as well as vulnerability to stresses (Wilson and Roberts 2012). It has been shown by several studies that elicitation can be utilized for elevating the synthesis of secondary metabolites via several elicitors like salicylic acid, chitosan methyl jasmonate, and metal ions (Karla et al. 2016).

Microbial bio-fermentation of metabolically engineered microbes is being praised repeatedly as a predominant substitute and is receiving increasing attention for various reasons. This is because microbial synthesis has no seasonal and regional dependence, ease of scale-up and ability to convert non-complex feedstock like glucose and oxygen and/or carbon dioxide, to the desired product. In addition, common microbial workhorses like the bacterium *Escherichia coli* and the yeast *Saccharomyces cerevisiae*, are genetically tractable with abundant bioengineering tools available for genetic and metabolic engineering purposes. Contrarily, plants accommodate the distinct secondary metabolites to greatly varying levels, generally producing a range of flavonoids as well as their derivatives (Ng et al. 2019). Furthermore, engineered microbial cell factories allow waste recycling that further makes the synthesis process cost-effective while encouraging an ideal economic system aimed at minimizing waste and making the most of resources (circular economy) (Ong et al. 2017). In addition, several natural and new flavonoid by-products can be produced in microbial cell factories through metabolic engineering, synthetic biology as well as protein engineering possessing more pharmaceutical and nutraceutical value (Kolewe et al. 2008). Using microorganisms for production of flavonoids can be more economical because it requires less time, is ecofriendly, also reduces the loss of pathway intermediates to challenging pathways frequently found in the natural host and has easy downstream management (Leonard and Koffas 2007; Wu et al. 2013).

5.5 Biosynthesis of Flavonoids

Flavonoids occur in the plants in response to environmental changes, usually as a defense mechanism like subjection to the UV rays, violation by pathogen (Winkel-Shirley 2001). Cytosol is the main site for the biosynthesis of flavonoids, with some downstream supplementary enzymes grouped inside the plastids. (Ng et al. 2019). Flavonoids are synthesized from the phenylpropanoid route, wherein deamination of L-phenylalanine is done by the enzyme phenylalanine ammonia-lyase (PAL) to produce cinnamic acid which acts as a substrate for the hydroxylation reaction to *p*-coumaric acid via the enzyme cinnamic-4-hydroxylase (C4H) a P450 hydroxylase that requires a cytochrome P450 reductase (CPR) (Bahaudin et al. 2018). The enzyme 4-coumarate: CoA ligase (4CL) catalyzes the production of 4-coumaroyl-CoA in the subsequent step. Following this, condensation of three molecules of malonyl-CoA occurs with one molecule of CoA ester through the enzyme chalcone synthase (CHS) resulting in the production of chalcones. It has been estimated that more than 9000 flavonoids are derived from chalcones through the action of several enzymes such as isomerases, oxido-reductases, hydroxylases and post alteration enzymes such as glycosyltransferases, acyltransferases and methyltransferases,

(Veitch and Grayer 2011; Iwashina 2015). Downstream flavonoids are derived from (2*S*)-flavanones via the stereospecific isomerization of chalcones in a reaction catalyzed by the enzyme chalcone isomerase (CHI). The enzyme flavanone 3β-hydroxylase (FHT) catalyzes the hydroxylation of (2*S*)- flavanones at the 3-carbon position yielding dihydroflavanols which are then reduced by the enzyme dihydroflavonol 4-reductase (DFR) at 4-carbon position to yield leucoanthocyanidins. Leucoanthocyanidins are relatively unstable molecules and are reduced by the enzyme leucoanthocyanidin reductase (LAR) to produce flavan-3-ols or catechins. Such compounds, or their reduced forms like flavan-3-ols by the enzyme leucoanthocyanidin reductase (LAR), can get further oxidized by the enzyme anthocyanidin synthase (ANS) to generate the unstable flavylium cation anthocyanins, that gets further attached with a glucosyl residue at the C3 position in the ring C to yield anthocyanin-3-*O*-glucosides likecyanidin-3-*O*-glucoside (C3G). Other alterations like methylation, hydroxylation, acylation on the ring skeleton and glycosylation at other hydroxyl groups produce various other anthocyanin molecules (Zha and Koffas 2018). Moreover, the structural variety and related structures like condensed tannins, isoflavonoids, stilbenes and aurones are produced by enzymes that catalyze the addition of different functional groups. The different biologically active potentials of flavonoids are attributed to this functionalization. (Chouhan et al. 2017). Pathway representing biosynthesis of flavonoids is shown in Fig. 5.2.

5.6 Current and Emerging Techniques in Microbial Production of Flavonoids

Metabolic engineering is modification of metabolic pathways through recombinant DNA technology to produce excess amounts of industrially-important chemicals, fuels and pharmaceutical products (Bailey 1991). The process of metabolic engineering has been largely applied in micro-organisms in order to develop sustainable and green approaches utilizing renewable feedstocks having less energy needs, and less waste release than the traditional extraction processes which depend on plant amounts and utilization of arable land (Marienhagen and Bott 2013). Production of flavonoids by metabolic engineering is more suitable than chemical synthesis and plant extraction. This approach employs overexpressing target genes, alteration of corresponding metabolic pathways as well as stoichiometric evaluation which primarily focuses on elevating titers and productivity (Bahaudin et al. 2018). Regarding the synthesis of flavonoids through the approach of metabolic engineering from micro-organisms, it needs selecting and optimizing the host strain, verification of targets for gene alterations as well as knowledge of the enzymes that are involved in the biosynthetic pathways. Usually, biosynthesis of natural products through metabolic engineering of micro-organisms involves the following steps: bioprospecting as well as construction of recombinant route (recombineering); selection and cloning or creation of heterologous genes; selection of the host for synthesis, selection

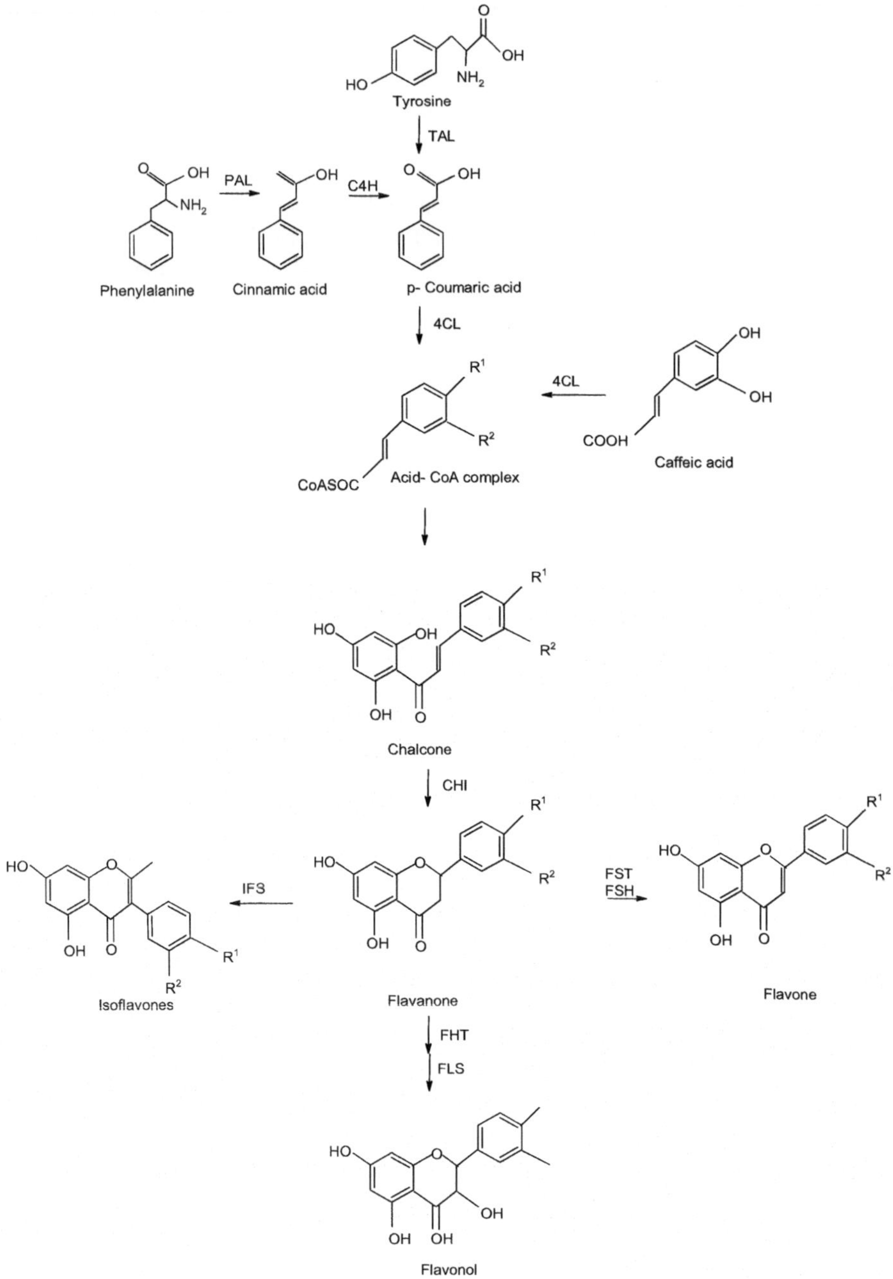

Fig. 5.2 Pathway of flavonoid biosynthesis in plants. *PAL* phenyl ammonia lyase, *TAL* tyrosine ammonia lyase, *C4H* cinnamate 4-hydroxylase, *4CL* 4-coumarate: CoA ligase, *CHI* stilbene synthase, *IFS* isoflavone synthase, *FSI* soluble flavones synthase, *FSH* membrane- bound flavones synthase, *FHT* flavanone 3β- hydroxylase, *FLS* flavonol synthase. (Adapted from Cress et al. 2013)

of vector, and modification of heterologous genes into host; optimization of expression, folding, and action of plant proteins in the microbial hosts (often through protein engineering); strain enhancement through carbon flux redistribution, toxicity depletion, transporter engineering, expulsion of regulatory limitations, enzyme colocalization or compartmentalization as well as route stabilization (Cress et al. 2015; He et al. 2017; Vemuri and Aristidou 2005).

5.7 Selecting a Production Platform

A prime essential feature while targeting the synthesis of flavonoid from micro-organisms is selecting the synthesis host. Despite the evolutionary distance separates plants and bacteria, several biosynthetic pathways of plant have been recreated in bacterial systems (Kotopka et al. 2018). On a large scale, *E. coli* and *S. cerevisiae* are the prokaryotic and eukaryotic organisms that have been used extensively for metabolic engineering of natural products. In the past few years, several new experimental tools for metabolic engineering have been developed as well as employed which allow recreation of complicated pathways to synthesize flavonoids in these two microbes. Prokaryotic micro-organisms like *E. coli*, are advantageous as they grow rapidly, they are easy to handle and there is a large toolbox available for their engineering. The prokaryotic micro-organisms are devoid of the particular eukaryotic plant organelles which are necessary for the biosynthesis of flavonoids, such as the endoplasmic reticulum (ER), on which cytochrome P450 enzymes are attached. Such enzymes are necessary for the functionalization of flavonoids (Ayabe and Akashi 2006).

S. cerevisiae is a Generally Regarded As Safe (GRAS) organism and is genetically tractable. This permits the expression of eukaryotic proteins that require attachment to cell organelles (Pandey et al. 2016). However, in comparison to the prokaryotes, *S. cerevisiae* exhibits slower growth and is not as genetically tractable as *E.coli*. Despite extensive work on microbial flavonoid production, the current production titers achieved are still far from the desired manufacturing levels (Delmulle et al. 2017).

5.7.1 Escherichia coli

Hwang et al. first demonstrated the synthesis of plant-derived flavonoids namely naringenin chalcone and pinocembrin chalcone from the recombinant *E. coli cells* by utilizing three enzymes from distinct sources: coumarate: coenzyme A ligase (4CL) from *Streptomyces coelicolor*, phenylalanine ammonia lyase (PAL) from *Rhodotorula rubra* and chalcone synthase (CHS) from *Glycyrrhiza echinata.* The recombinant *E. coli* strain was fed with tyrosine and phenylalanine respectively (Hwang et al. 2003). Likewise, Leonard et al. utilized two isoforms of the enzyme flavones synthase (FS) [FSI is soluble and FS II is membrane bound] for engineering the yeast strains and synthesized several flavones (chrysin, apigenin) as well as intermediary flavanones (naringenin, eriodictyol and pinocembrin) by utilizing phenylpropanoid precursors. *E. coli* strains, that expressed five plant derived genes for the synthesis of flavones were also engineered with the enzyme flavone synthase (FSI) obtained from parsley that produced luteolin, genkwanin, and apigenin in significant quantities after 24 h culture (Leonard et al. 2005).

Miyahisa and coworkers used CHS, PAL, as well as 4CL with the enzyme chalcone isomerase (CHI) in a vector for optimizing gene expression. This recombinant *E. coli construct* elevated the yield of naringenin to 60 mg/L (Miyahisa et al. 2005). Further in the year 2006, Miyahisa et al. synthesized 9 mg/L of the flavone chrysin in recombinant *E. coli* cells (Miyahisa et al. 2006).

In 2005, Yan et al. reported the production of eriodictyol with a production titer of 6.5 mg/L (Yan et al. 2005a). Further in the same year, Yan et al. synthesized pinocembrin with a yield of 16.3 mg/L from recombinant *E. coli* cells (Yan et al. 2005a). The foremost trial for the recombinant biosynthesis of anthocyanin was done by Yan et al. in the year 2005. The genes of enzyme flavanone 3β-hydroxylase (F3H) and ANS were cloned from *Malus domestica*, DFR from *Anthurium andraeanum*, and the genes of the enzyme flavonoid 3-O-glucosyltransferase (F3GT) were cloned from *Petunia hybrida*. The recombinant *E. coli* strain synthesized 6.0 mg/L of cyanidin 3-*O*-glucoside as well as 5.6 mg/L pelargonidin 3-*O*-glucoside, wherein naringenin and eriodictyol served as the precursors (Yan et al. 2005a). In 2007, Katsuyama et al. reported the synthesis of 33 mg/L of flavonols and 102 mg/L flavanones from engineered *E.coli* cells using 4CL (*Lithospermum erythrohizon*), CHS, CHI (*Glyccyrrhiza echinata*), STS (*Arachis hypogaea*), FNS (*Petroselinum crismum*), F3H, FLS (Citrus), ACC (*Cornybacterium glutamicum*) as the biosynthetic components (Katsuyama et al. 2007). Likewise, Leonard et al. reported the synthesis of 52 mg/L eriodictyol from recombinant *E.coli* cells (Leonard et al. 2007).

Adequate availability of UDP-glucose is important to synthesize anthocyanins in an effective method. It has been attained via overexpressing UDP-glucose biosynthetic genes (*pyrE, pyrR, cmk, ndk, pgm, galU*) accompanied by restricted retardation of the UDP-glucose utilization pathways. Such alterations resulted in 20-times increased synthesis of cyanidin 3-*O*-glucoside (Leonard et al. 2008). Further, 710 mg/L pinocembrin [(2*S*)-Flavanones] production was also reported in engineered *E. coli* cells (Leonard et al. 2008). Another research revealed a 57.8% elevated synthesis of cyanidin 3-*O*-glucoside via overexpressing *pgm* and *galU* with simultaneous expression of ANS and 3GT (Yan et al. 2008).

Further in 2011, Santos et al. reported the synthesis of 84 mg/L of (2*S*)-flavanones naringenin using recombinant *E. coli* cells (Santos et al. 2011). Similarly, Xu et al. synthesized 474 mg/L of naringenin from recombinant *E. coli* cells that had been engineered based on predictions derived from stoichiometric-based modeling for improved availability of malonyl-CoA (Xu et al. 2011a). In 2012, Malla et al. synthesized 30 mg/L of 7-O-methyl aromadendrin from engineered *E. coli* cells carrying genes for the enzymes 4CL (*Petroselinum crispum*), CHS (*Petunia hybrida*), CHI (*M. sativa*) as the biosynthetic components (Malla et al. 2012). Also, 40.02 mg/L of pinocembrin was synthesized in 2013 using recombinant *E. coli* cells (Wu et al. 2013). a titer that was further increased to 100.64 mg/L in 2014, again by improving the intracellular availability of malonyl-CoA (Wu et al. 2014). Likewise in the same year, Zhu et al. synthesized 107 mg/L of the (2*S*)-Flavanones eriodictyol from engineered *E. coli* cells (Zhu et al. 2014).

Effective synthesis of anthocyanins requires optimization of several other factors like induction time-point, pH, temperature, level of dissolved oxygen and substrate feeding. Inducing the anthocyanin pathway at the stationary phase was found to be appropriate for the synthesis of cyanidin 3-*O*-glucoside. Moreover, overexpression of YadH, a cyanidin 3-*O*-glucoside-related efflux pump, resulted in a 15% increase in anthocyanin production. Further, yield of cyanidin 3-*O*-glucoside was increased through the removal of another efflux pump TolC that is likely accountable for the uptake of the substrate catechin (Lim et al. 2015). Further in the same year, Zhao et al. reported the synthesis of flavan-3-ol in recombinant *E. coli* cells (Zhao et al. 2015). A study conducted by Lee et al. by reported 30 mg/L of apigenin production from the recombinant cells (Lee et al. 2015).

Kaempferol (KMF) is a natural occurring flavonol that is also known by the name robigenin or 3, 4, 5, 7-tetrahydroxyflavone. It has several pharmacological as well as biological potentials like antimicrobial, antioxidant, anticancer, anti-inflammatory, neuroprotective, antidiabetic, cardioprotective, anti-osteoporotic, anti-allergic activities estrogenic/anti-estrogenic and anxiolytic, analgesic (Chen and Chen 2013). Kaempferol has been chiefly obtained through conventional plant extraction. But, the price of Kaempferol synthesis is large because of its very less content in plants (Agar et al. 2015).

A study conducted by Pei et al. designed a recombinant *Escherichia coli* strain for the effective production of kaempferol. Through optimization of fed-batch fermentation conditions, maximal synthesis of kaempferol was achieved with a production titer of 1184.2 ± 16.5 mg/L, representing the greatest amount of kaempferol production from naringenin that has been reported (Pei et al. 2018). Moreover, 37.55 ± 1.62 mg/L of kaempferol has also been synthesized in recombinant *E. coli cells* at a temperature of 40 °C for 40–50 min in a system comprised of 10% glycerol in 100 mM Tris-HCl (pH 7.2), 0.01 mM ferrous ion and 25 μg/mL of the recombinant enzymes flavonol synthase as well as flavanone 3-hydroxylase (Zhang et al. 2018).

Tokuyama et al. reported that magnesium starvation in *Escherichia coli* cells leads to the synthesis of naringenin having the largest synthesis yield of 144 ± 15 μM and specific productivity of 127 ± 21 μmol $gCDW^{-1}$. This was attributed to the fact that starvation of vital nutrients like magnesium, nitrogen, phosphorus and sulfur makes cells enter into stationary phase, which in turn triggers the synthesis of target metabolites as cells do not utilize carbon for the production of biomass (Tokuyama et al. 2018).

Baicalein and scutellarein are plant-based bioactive flavones but their supply is solely based on plant extraction. Lia et al. constructed a biosynthetic pathway in *E. coli* for the efficient production of these two flavones. For this purpose they utilized flavonoid biosynthetic pathway genes from five different species i.e.

4-coumarate-coenzyme A ligase from *Petroselinum crispum* (4CL), phenylalanine ammonia lyase from *Rhodotorula toruloides* (PAL), chalcone synthase from *Petunia hybrida* (CHS) an oxidoreductase flavone synthase I from *P. crispum* (FNSI) and chalcone isomerase from *Medicago sativa* (CHI); all these resulted in the synthesis of the intermediates chrysin and apigenin through the feeding of phenylalanine and tyrosine as precursors. When several versions were assessed with different P450 derivatives, it was observed that the construction expressing 2B1 having a 22-amino acid N-terminal truncated flavone C-6 hydroxylase from *S. baicalensis* (F6H) as well as partner P450 reductase from *Arabidopsis thaliana* (AtCPR) was most effective to synthesize scutellarein with a production titer of 47.1 mg/L as well as baicalein having a yield of 8.5 mg/L, when supplemented with 0.5 g/L phenylalanine and tyrosine for 48 h of the fermentation process. Further, when the availability of malonyl-CoA was enhanced, synthesis of baicalein reached 23.6 mg/L while the titer of scutellarein reached 106.5 mg/L in a flask culture (Lia et al. 2019).

Likewise in the same year, Pandey et al. reported the synthesis of 12 mg/L (28 mol/L) of scutellarin A from apigenin by the recombinant *E. coli* BL21 (DE3) strain (Pandey et al. 2019). The synthesis of some flavonoids in recombinant *E. coli* cells is summarized in Table 5.1.

5.7.2 Saccharomyces cerevisiae

In 2005, Yan et al. utilized *S. cerevisiae* for the construction and introduction of a four-step flavanone biosynthetic pathway. It was observed that the recombinant yeast strain carrying the biosynthetic components C4H (*Arabidopsis thaliana*), 4CL (*Petroselinum crismum*), CHS, CHI (*Petunia x hybrida*) synthesized 62 times more naringenin and 22 times more pinocembrin when fed with phenylpropanoid acids as compared to the earlier utilized prokaryotic strains (Yan et al. 2005b). Likewise in the same year, Jiang et al. synthesized naringenin with a production titer of 7 mg/L from *S. cerevisiae* having the biosynthetic components PAL (*Rhodosporidium toruloides*), 4CL (*Arabidopsis thaliana*), CHS (*Hypericum androsaemum*) (Jiang et al. 2005). In 2009, Trantas et al. synthesized genistein with a production titer of 8 mg/L from *S. cerevisiae* (Trantas et al. 2009). Koopman et al. reported the synthesis of 400 μM naringenin in the culture medium by the strains of the yeast *S. cerevisiae* carrying PAL, C4H, CPR, 4CL, CHS, CHI (*Arabidopsis thaliana*), TAL (*Rhodobacter capsulatus*), ARO4^{G2265} (*S. cerevisiae*) as the biosynthetic components (Koopman et al. 2012).

Further, synthesis of kaempferol with a production titer of 66 mg/L has also been reported from the yeast *S. cerevisiae* (Duan et al. 2017). In the year 2018, Levisson et al., reported the de novo synthesis of the anthocyanin pelargonidin 3-*O*-glucoside from yeast strain *S. cerevisiae.* For this purpose specific genes from *A. thaliana* and *Gerbera hybrida* were introduced into *S. cerevisiae* (Levisson et al. 2018). Optimization of the parameters for flavonoid production resulted into pelargonidin titers of 0.01 μmol/gCDW, whereas kaempferol as well as dihydrokaempferol were

Table 5.1 Table representing synthesis of some flavonoids in the recombinant *Escherichia coli* cells

S.No.	Microorganism	Biosynthetic components	Product	Yield	References
1.	*Escherichia coli*	F3H (*Malus domestica*), ANS (*Malus domestica*), DFR (*Anthurium andraeanum*), F3GT (*Petunia hybrida*)	Cyanidin 3-O-glucoside	6 mg/L	Yan et al. (2005a)
2.	*Escherichia coli*	F3H (*Malus domestica*), ANS (*Malus domestica*), DFR (*Anthurium andraeanum*), F3GT (*Petunia hybrida*)	Pelargonidin 3-O-glucoside	5.6 mg/L	Yan et al. (2005a)
3.	*Escherichia coli*	PAL (*Rhodotorula rubra*), 4CL (*Streptomyces coelicolor*), CHS (*Glycyrrhiza echinata*), CHI (*Pureria lobata*), ACC (*Cornybacterium glutamicum*)	Naringenin	57 mg/L	Miyahisa et al. (2005)
4.	*Escherichia coli*	L-Phenylalanine	Chrysin (Flavones)	9 mg/L	Miyahisa et al. (2006)
5.	*Escherichia coli*	4CL (*Lithospermum erythrohizon*), CHS, CHI (*Glyccyrrhiza echinata*), STS (*Arachis hypogaea*), FNS (*Petroselinum crismum*), F3H, FLS (Citrus), ACC (*Cornybacterium glutamicum*).	Flavonols	33 mg/L	Katsuyama et al. (2007)

(continued)

Table 5.1 (continued)

S.No.	Microorganism	Biosynthetic components	Product	Yield	References
6.	*Escherichia coli*	4CL (*Lithospermum erythrohizon*), CHS, CHI (*Glycyrrhiza echinata*), STS (*Arachis hypogaea*), FNS (*Petroselinum crismum*), F3H, FLS (Citrus), ACC (*Cornybacterium glutamicum*).	Flavanones	102 mg/L	Katsuyama et al. (2007)
7.	*Escherichia coli*	–	Eriodictyol	52 mg/L	Leonard et al. (2007)
8.	*Escherichia coli*	pyrE, pyrR, cmk, ndk, pgm, galU	Cyanidin 3-O-glucoside	Increased by 20-fold	Leonard et al. (2008)
9.	*Escherichia coli*	–	Pinocembrin [(2S)-Flavanone]	710 mg/L	Leonard et al. (2008)
10.	*Escherichia coli*	pgm and galU along with the expression of ANS and 3GT	Cyanidin 3-O-glucoside	57.8% increase in the production of cyanidin 3-O-glucoside	Yan et al. (2008)
11.	*Escherichia coli*	4CL (P*etroselinum* crismum), CHS (*Petunia x hybrida*), CHI (*Medicago sativa*), ACC (*Photorhabdus luminescens*), PGK, PDH (*Escherichia coli*)	Naringenin [(2S) Flavanone]	474 mg/L	Xu et al. (2011a)
12.	*Escherichia coli*	–	Naringenin ((2S)-Flavanone)	84 mg/L	Santos et al. (2011)
13.	*Escherichia coli*	4CL (P. crispum), CHS (*Petunia hybrid*), CHI (*Medicago sativa*)	7-O-methyl aromadendrin	30 mg/L	Malla et al. (2012)
14.	*Escherichia coli*	–	Pinocembrin	40.02 mg/L	Wu et al. (2013)

(continued)

Table 5.1 (continued)

S.No.	Microorganism	Biosynthetic components	Product	Yield	References
15.	*Escherichia coli*	matB, matC	Naringenin ((2S)-Flavanone)	100.64 mg/L	Wu et al. (2014)
16.	*Escherichia coli*	–	Eriodictyol ((2S)-Flavanone)	107 mg/L	Zhu et al. (2014)
17.	*Escherichia coli*	–	Anthocyanin	15% more production	Lim et al. (2015)
18.	*Escherichia coli*	–	Apigenin (Flavones)	30 mg/L	Lee et al. (2015)
19.	*Escherichia coli*	–	Flavan-3-ol	–	Zhao et al. (2015)
20.	*Escherichia coli*	PAL (*Rhodotorula glutinis*), 4CL (*Petroselinum crispum*), CHS (*Petunia hybrid*), CHI (*Medicago sativa*)	Pinocembrin	525.8 mg/L	Wu et al. (2016)
21.	*Escherichia coli*	–	Flavan-3-ols	58-fold improvement	Jones et al. (2016)
22.	*Escherichia coli*	15 exogenous or modified pathway enzymes from diverse plants as well as other microbes	Pelargonidin 3-*O*-glucoside (anthocyanin)	9.5 mg/L	Jones et al. (2017)
23.	*Escherichia coli*	–	Naringenin	127 ± 21 μmol gCDW-1	Tokuyama et al. (2018)
24.	*Escherichia coli*	–	Kaempferol	37.55 ± 1.62 mg/L	Zhang et al. (2018)
25.	*Escherichia coli*	–	Kaempferol	1184.2 ± 16.5 mg/L	Pei et al. (2018)
26.	*Escherichia coli*	–	Baicalein and Scutellarein	23.6 mg/L and 106.5 mg/L	Lia et al. (2019)
27.	*Escherichia coli*	–	Scutellarein A	12 mg/L	Pandey et al. (2019)

Table 5.2 Table representing synthesis of some flavonoids in the recombinant *Saccharomyces cerevisae* cells

S.No.	Microorganism	Biosynthetic components	Product	Yield	References
1.	*Saccharomyces cerevisiae*	PAL (*Rhodosporidium toruloides*), 4CL (*Arabidopsis thaliana*), CHS (*Hypericum androsaemum*)	Naringenin	7 mg/L	Jiang et al. (2005)
2.	*Saccharomyces cerevisae*	C4H (*Arabidopsis thaliana*), 4CL (*Petroselinum. crismum*), CHS, CHI (*Petunia x hybrida*)	Naringenin	28.3 mg/L	Yan et al. (2005b)
3.	*Saccharomyces cerevisae*	p-Coumaric acid	Genistein	8 mg/L	Trantas et al. (2009)
4.	*Saccharomyces cerevisiae*	PAL, C4H, CPR, 4CL, CHS, CHI (Arabidopsis thaliana), TAL (*Rhodobacter capsulatus*), ARO4^{G226S} (*Saccharomyces cerevisiae*)	Naringenin	109 mg/L	Koopman et al. (2012)
5.	*Saccharomyces cerevisiae*	Naringenin	Kaempferol	66 mg/L	Duan et al. (2017)
6.	*Saccharomyces cerevisiae*	Specific genes from*Arabidopsis thaliana*and *Gerbera hybrida*	Kaempferol	5 mg/L	Levisson et al. (2018)
7.	*Saccharomyces cerevisiae*	Specific genes from*Arabidopsis thaliana*and *Gerbera hybrida*	Dihydrokaempferol	44 mg/L	Levisson et al. (2018)

produced with a titer of 20 μM [5 mg/L] and 150 μM [44 mg/L], respectively. Table 5.2 summarizes the synthesis of flavonoids using recombinant *S. cerevisae.*

5.8 Carbon Flux Manipulation Towards Heterologous Production Pathways

Carbon flux manipulation for improving the availability of cofactors and precursors involved in flavonoid biosynthesis is an important target of metabolic engineering. Malonyl CoA is the central metabolite in the synthesis of various primary and secondary metabolites including flavonoids but its intracellular availability for the production of important secondary metabolites is often limited due to the competition with essential cellular metabolism. Microbes usually maintain low intracellular concentration of malonyl-CoA by tightly regulating its biosynthesis which is a major bottleneck in the polyphenol production using *E. coli* and *S. cerevisiae* as production platforms (Lim et al. 2011; Kim and Ahn 2014; Yang et al. 2015). Some chemical inhibitors such as cerulenin have been used to redirect malonyl CoA towards the polyphenol production. These inhibitors that target the fatty acid metabolic pathways, are expensive and also slow down the cell growth and biomass accumulation at higher concentrations. So, metabolic engineering can be used as potential alternative to increase the availability of this important molecule for flavonoid synthesis without compromising cell viability. The intracellular pool of malonyl CoA and hence synthesis of flavonoids in heterologous hosts has been optimized by the application of various rational and computational engineering approaches as depicted in Fig. 5.3 (Leonard et al. 2007; Santos et al. 2011; Xu et al. 2011a, b, 2014; Bhan et al. 2012).

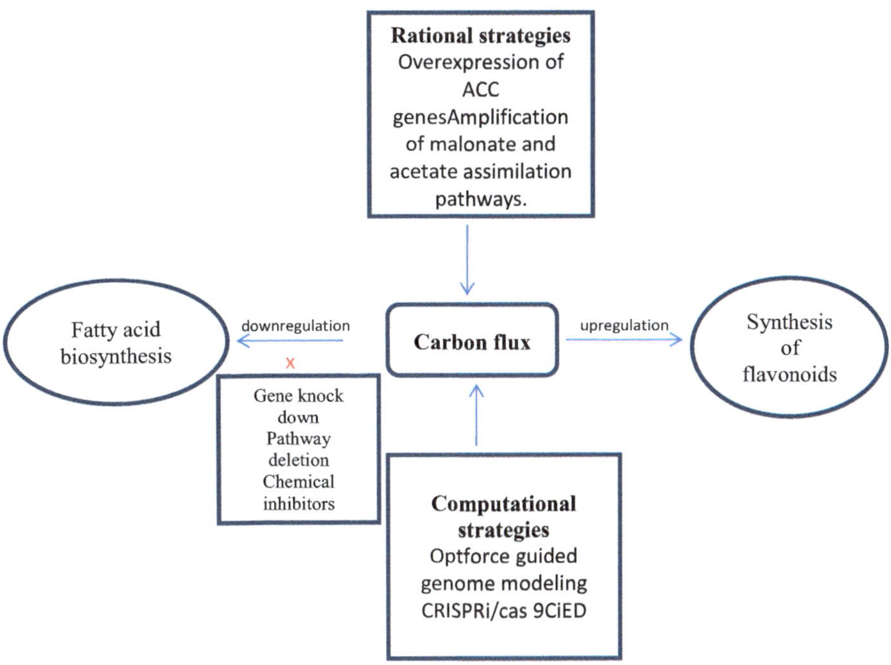

Fig. 5.3 Rational and computational strategies for increasing the availability of malonyl CoA for flavonoid biosynthesis in microbial cell factories

5.8.1 Rational Design

The rational engineering strategy for increasing the carbon flux involves deletion of the competing pathways, overexpression of the rate-limiting enzymes, or gene deregulation. The correct manipulation of the metabolic flux requires detailed knowledge of the metabolic pathways. However, this strategy has some limitations as it neglects the complexity of the metabolic pathways and focuses on only a part of the pathway without considering the interdependency within and between different pathways due to which optimization of one parameter may negatively influence the other (Juminaga et al. 2012). Using rational design, optimization of strain performances requires multiple rounds of strain construction, selection, and optimization which is often labor intensive, time consuming, and expensive. Early metabolic engineering efforts typically relied on rational engineering approaches which does not take into account the broad spectrum of metabolism. Due to the interdependency between different metabolic pathways, combinatorial engineering has been developed for optimizing the pathway flux which involves global fine tuning of the metabolic pathways by targeting multiple genetic targets concomitantly in a random manner (Boock et al. 2015; McNerney et al. 2015). The carbon flux in microbial cell factories is diverted towards the pathway of interest by manipulating the target enzymes, downregulation of the competing pathways and upregulation of the desired pathways (Leonard et al. 2007). The intracellular pool of malonyl-CoA and its flux towards the flavonoid biosynthetic pathway has been optimized by overexpression of the acetyl CoA carboxylase (ACC) genes, acetate assimilation genes, and malonate assimilation genes in most of the studies thus improving the synthesis of malonyl CoA whereas the genes involved in the consumption of malonyl-CoA in non-target pathways are deleted (Leonard et al. 2007; Zha et al. 2009).

Over expression of the acetyl-CoA carboxylase, catalyzing the conversion of acetyl CoA to malonyl CoA can improve the intracellular pool of malonyl CoA (Zha et al. 2009; Xu et al. 2011a; Yang et al. 2018). A 15-fold increase in intracellular level of malonyl-CoA was achieved by the combination of several strategies involving overexpression of genes encoding for acetyl-CoA synthetases, knockout of competing pathways and elimination of malonyl-CoA degrading pathways (Zha et al. 2009). The availability of malonyl-CoA precursor in flavanone-producing recombinant *E. coli* strains has been improved by engineering of central metabolic pathways (Leonard et al. 2007). The intracellular malonyl-CoA pool was increased by coordinated overexpression of four acetyl-CoA carboxylase (ACC) subunits from *Photorhabdus luminescens* (PlACC) combined with biotin ligase (BirA) and genes of the acetate assimilation pathways including acetate kinase A (*ackA*), phosphate acetyltransferase (*pta*) and acetyl-CoA synthase (*acs*) resulting in an increase of upto 1379%, 183%, and 373%, in the production of pinocembrin, naringenin, and eriodictyol respectively (Leonard et al. 2007). Similarly off-target consumption of malonyl-CoA in the primary metabolic pathways (fatty acid biosynthesis) has been reduced by knock out of fatty acid biosynthesis genes in order to improve the titer of other valuable malonyl-CoA derived products (Zha et al. 2009; Chen et al. 2017;

Yang et al. 2018). Malonyl CoA pool has also been enhanced by the introduction of genes involved in malonate assimilation pathway (encoded by genes *matB* and *matC*) in *E. coli* from *R. trifolii* which leads to direct conversion of malonate to malonyl-CoA instead of the native conversion from glucose. Expression of these genes along with plant biosynthetic genes (Pc4CL2, PhCHS, and MsCHI) resulted in over 250% increase in flavanone production. In addition, the fatty acid biosynthesis pathway was inhibited by the addition of cerulenin (inhibitor of fatty acid biosynthesis genes) in the medium which resulted in 900% increase in the production of flavanones as compared to the *E. coli* strain expressing only flavonoid biosynthetic genes (Leonard et al. 2008). Cerulenin acts as the inhibitor of β-ketoacyl-acyl carrier protein synthase enzymes (FabB and FabF), which catalyse the condensation of malonyl ACP with acyl-ACP for the extension of fatty acid chain (Schujman et al. 2006, 2008). The high cost of cerulenin and its negative impact on cell viability prohibits its use for commercial scale fermentation production (Davis et al. 2000). Synthetic antisense RNA technology can also be applied to redirect the carbon flux towards desired pathway by down regulating some genes to optimize the flux towards desired pathway. In one approach, synthetic antisense RNAs (asRNAs) were used to downregulate the expression of pathway specific genes for optimization of the flux of interest. High intracellular concentration of malonyl CoA in *E. coli* was attained by knocking down of fatty acid biosynthetic genes (*fabD*) involved in the pathways that diverge carbon from the flavonoid pathway to fatty acid biosynthesis. 1.70 and 1.53 fold higher resveratrol (268.2 mg/L) and naringenin (91.31 mg/L) respectively were produced by the recombinant *E. coli* harboring plant derived naringenin and resveratrol biosynthesis genes in contrast to the control strain lacking antisense RNA coding plasmids (Yang et al. 2015). Chen et al. used a synthetic malonyl CoA biosensor for screening the phosphorylation site mutations of acetyl-CoA carboxylase (Acc1p) in *S. cerevisiae.* Out of 13 mutated phosphorylation sites, a combination of three site mutations in Acc1p, S686A, S659A, and S1157A, resulted in increased malonyl-CoA availability (Chen et al. 2018).

5.8.2 Computational Tools

Traditional metabolic engineering relied on conventional analytical techniques such as HPLC and LC for the quantification of malonyl-CoA in order to know the effectiveness of various engineering strategies. Moreover, it has been found that the manipulation of local metabolic networks leads to only small increases in the carbon flux. Simple investigation of the metabolic pathway is not sufficient for determination of the potential genetic targets for manipulation in order to redirect pathway flux as the cellular metabolic pathways are very complex and interconnected. Recently, prediction of genetic perturbation targets, for channeling the carbon flux towards target compounds, is carried out by computational algorithms based on flux balance analysis (FBA) due to rapid advancements in systems and synthetic biology (Stephanopoulos et al. 2004; Medema et al. 2012; Lanza et al. 2012).

Cellular malonyl-CoA concentration can be enhanced by using an integrated computational and experimental approach (Fowler et al. 2009; Xu et al. 2011b; Bhan et al. 2013). Computational models aid in identification of the genes that are directly as well as indirectly involved in malonyl-CoA metabolism, prediction of the target genes for knock out or overexpression to improve the availability of malonyl-CoA for the synthesis of desired compounds, and also predict the potential effect of the genetic manipulation on cell growth and metabolism. Genome scale metabolic modeling is an effective strategy that has been successfully applied to choose the best combination of approaches in order to improve the malonyl-CoA yield. Optimum biomass production and naringenin synthesis in *E. coli* was carried out by balancing the distribution of precursors between different pathways using a genome-scale metabolic network model in *E. coli* (Xu et al. 2011a). Cipher of Evolutionary Design (CiED), a computational model, was used to design a new E. coli strain with enhanced carbon flux towards malonyl CoA and other cofactors (Fowler et al. 2009). OptForce is another computational tool, based on constraint based genome modeling, which has been used to redirect the malonyl CoA pool towards the pathway of interest by predicting the target genetic perturbations (Ranganathan et al. 2010). Optforce involves the combination of computational and experimental approaches for the optimization of carbon flux in heterologous production pathways. Xu et al. attempted to improve the intracellular availability of malonyl-CoA in *E. coli* by the synergistic effect of various Optforce guided genetic perturbations. Genes encoding fumarase (*fumC*) and succinyl-CoA synthetase (*sucC*) were identified as knock out targets (Δ*fumC* and Δ*sucC*) whereas the genes encoding ACC, phosphoglycerate kinase (PGK) and pyruvate dehydrogenase (PDH) were identified as over-expression targets in order to redirect carbon flux towards malonyl-CoA for the production of flavonoids (Xu et al. 2011b). The engineered strain displayed 660% and 420% increase in naringenin and eriodictyol production, respectively (Xu et al. 2011b). In a similar approach, a mutant strain produced 60% higher resveratrol than the control when the OptForce predictions were applied in the case of resveratrol production in *E.coli* (Bhan et al. 2013).

Carbon flux and hence flavonoid production has also been optimized by using combination of CRISPR/dCas9 and OptForce (Cress et al. 2015). Fine tuning of the malonyl-CoA flux towards the heterologous pathway and biomass accumulation was attained by the application of CRISPRi. The potential target genes were identified by CRISPRi mediated repression of multiple genes involved in central metabolic pathways which resulted in an increase of over 223% in the intracellular malonyl-CoA level and thus high yield of (2*S*)-naringenin (421.6 mg/L) (Wu et al. 2015a, b). Chen et al. (2018) designed a synthetic malonyl-CoA biosensor and used it to screen phosphorylation site mutations of Acetyl CoA carboxylase (Acc1p) in *S. cerevisiae*. Thirteen phosphorylation sites were mutated, and a combination of three site mutations in Acc1p, S686A, S659A, and S1157A, was found to increase malonyl-CoA availability. Li et al. (2015) reported the design of a malonyl-CoA sensor in *S. cerevisiae* using an adapted bacterial transcription factor FapR and its corresponding operator *fapO*, which is capable of reporting intracellular malonyl-CoA levels. By combining this sensor with a genome-wide overexpression library, they identified two novel gene targets that improved intracellular malonyl-CoA concentration. They further utilized the resulting recombinant yeast strain to produce a valuable compound, 3-hydroxypropionic acid, from malonyl-CoA and enhanced its titer by 120%. Yang et al. 2018, reported development of a colorimetric malonyl-CoA biosensor applicable in three industrially important bacteria: *E. coli*, *Pseudomonas putida*, and *Corynebacterium glutamicum*. For target screening, a 1858 synthetic small regulatory RNA library was constructed and applied to find 14 knockdown gene targets that generally enhanced malonyl-CoA levels in *E. coli*. Knocking down these genes alone or in combination, and also in multiple different *E. coli* strains for two polyketide production cases, allowed rapid development of engineered strains capable of enhanced production of 6-methylsalicylic acid, aloesone, resveratrol, and naringenin to 440.3, 30.9, 51.8, and 103.8 mg/L, respectively (Yang et al. 2015).

5.8.3 Protein Engineering

The optimum production of secondary metabolites can be achieved by using combination of metabolic engineering and protein engineering. Heterologous metabolic pathways are optimized by rational and combinatorial genetic engineering as well as molecular level control mechanisms (Boyle and Silver 2012). But these approaches do not play any significant role in improving the properties of the enzymes. Protein engineering improves the kinetics of enzyme-catalyzed biochemical reactions by creating improved enzymes, capable of catalyzing both existing and novel reactions (Sawayama et al. 2009; Lee et al. 2010). Recently, computational design and combinatorial strategies have enhanced the efficiency of rational, semi-rational, and random mutagenesis approaches of enzyme engineering leading to increased/decreased/altered specificity, increased activity and solubility/stability of enzymes (Stefan 2010; Bottcher and Bornscheuer 2010). Rational design of enzymes depends on the ability to derive structure–function relationships. Overexpression of an enzyme is not the solution for increasing the product titer in case of enzymes with low turnover or poor expression. In such situations, properties of enzymes can be improved through evolutionary or rational engineering methods in order to increase the pathway efficiency. Increased production of desired compounds (natural and unnatural) can be achieved by creating novel enzymes with desired catalytic and kinetic efficiency. The enzymes can be tailored by protein engineering so as to enhance the efficiency of the constructed pathway (Wang et al. 2011a; Lee et al. 2012). Various metabolic engineering methods like site directed mutagenesis, mutasynthesis, synthesis of novel enzymes, codon optimization, synthesis of fusion proteins, screening and selection of efficient enzymes from different genetic sources, enzyme engineering or directed evolution can be used to enhance the catalytic power of the enzymes involved in the biosynthesis of flavonoids in microbial cell factories (Tee and Wong 2017; Wang et al. 2011a, b). Available knowledge of enzyme structure, information about activity/inhibitory domains of protein, and functional or evolutionary relationships are used to make decisions on the type of protein engineering strategies to be used. Rational engineering approaches are used when the information about protein structure is available whereas directed evolution is the method of choice when no information is available regarding protein/enzymes. Directed evolution involves the selection of enzyme with improved attributes by screening of a library of mutant forms of the target enzyme created by random mutagenesis. The application of combinatorial and computational approaches have increased the effectiveness of directed evolution. Also, other protein engineering tools such as site-directed mutagenesis and mutasynthesis have been applied to improve production of both natural and non-natural flavonoids, isoflavonoids, and other plant natural product derivatives (Chemler et al. 2007; Fowler et al. 2011; Bhan et al. 2015). The tailored enzymes possess enhanced kinetic and catalytic activity, improved heterologous expression, stability, elimination of undesirable side reactions, have the ability to accept structurally related substrates, and have altered substrate and product regiospecificity (He et al. 2006; Katsuyama et al. 2007; Chemler et al. 2007; Bhan et al. 2014). Protein engineering approach has been summarized in Fig. 5.4 shown below.

Protein engineering combined with metabolic engineering is a useful strategy for the creation of desired enzymes which possess the catalytic power for the synthesis of both natural and unnatural compounds in heterologous production hosts thus diversifying the range of flavonoid compounds (Chemler et al. 2007, 2010). For example, the production of resveratrol in *S. cerevisiae* was increased from 650 to 5250 μg/L by using translational fusion protein composed of At4CL (chalcone synthase from *Arabidopsis thaliana*) and VvSTS (stilbene synthase from *Vitis vinifera*) generated by replacing stop codon of At4CL by a three amino acid linker followed by VvSTS open reading frame. Further improvement in the production titer of resveratrol as well as *p*-coumaric acid was achieved by using yeast preferred codon optimized RsTAL (TAL from *R. sphaeroides*) in combination with the fusion protein (At4CL-VvSTS). The combination of metabolic engineering and protein engi-

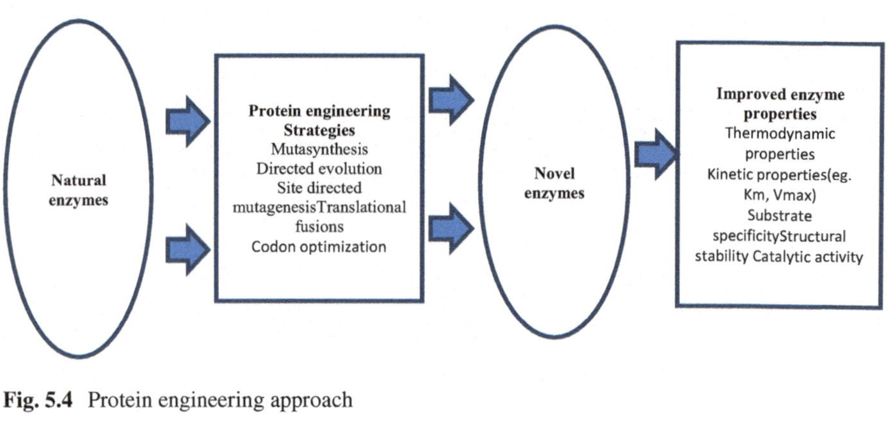

Fig. 5.4 Protein engineering approach

neering resulted in the production of 1.06 mg/L of resveratrol after 48 h of incubation without the addition of tyrosine whereas the titer increased to 1.90 mg/L after supplementing L-tyrosine in the medium (Wang et al. 2011b). Similarly, the metabolic pathway catalyzing the conversion of (+)-catechin to cyanidin was optimized by creating translational fusions between *A. thaliana* At3GT and *P. hybrida* PhANS. Two different types of translational fusions were created by varying order of the two genes placed next to each other and were connected via a mini-linker coding for Ser-Ser-Gly-Ser-Gly. The fusion protein, possessing At3GT at N-terminus of PhANS, produced 45.5 mg/L cyanidin 3-*O*-glucoside improving the production by approximately 17%. Furthermore, 70.7 mg/L of cyanidin 3-*O*-glucoside and 78.9 mg L^{-1} of pelargonidin 3-*O*-glucoside from (+)-catechin and (+)-afzelechin, respectively were obtained by co-expression of the fusion protein with galU and pgm (Yan et al. 2008). The production of kaempferol and astragalin from naringenin in *E. coli* was enhanced by designing a synthetic fusion enzyme and increasing the gene copy number (Pei et al. 2018). The catalytic efficiency of tyrosine ammonia-lyase (TAL), responsible for catalyzing the conversion of L-tyrosine to naringenin, was significantly increased by codon optimization involving resynthesis of the enzyme by changing the bacterial codons into yeast-preferred codons (Wang et al. 2011a, b). In a similar way, the production of resveratrol and pinocembrin in *E. coli* was enhanced upto 35–40 mg/L respectively by codon optimization of phenylalanine/tyrosine ammonia lyase (PAL/TAL), 4-coumarate:CoA ligase (4CL), chalcone synthase (CHS), and chalcone isomerase (CHI) (Wu et al. 2013). In addition to the above mentioned enzymes, the expression of stilbene synthase (STS) was also increased by mutating and optimizing its codons (Wu et al. 2013). Mutasynthesis is another semisynthetic tool which explores substrate promiscuity of the natural enzymes by feeding non-natural substrate analogs for the production of non-natural products possessing unique medicinal properties. Mutasynthesis or substrate feeding has been used for the production of various natural and unnatural flavonoids (Chemler et al. 2007, 2010; Katsuyama et al. 2007; Horinouchi 2008). The chemical diversity of polyketide (PK) compounds was increased by creating structure-guided mutants of *Vitis vinifera* STS. The substrate specificity of mutant STS was expanded by using different unnatural substrates which resulted in the synthesis of 15 novel polyketide molecules (Bhan et al. 2015). The cytochrome P450 enzymes, involved in flavonoid synthesis require NADPH dependent P450 oxidoreductase for electron transfer, a major bottleneck in prokaryotic host lacking endoplasmic reticulum. Such enzymes can be functionally expressed in the heterologous host by transmembrane engineering which involves replacement of their original ER N-terminal anchor with a modified version that targets the protein to the cell membrane. The efficiency of the heterologous pathway can be enhanced by translational fusion of the modified P450 enzymes with CPR (Leonard and Koffas 2007; Ajikumar et al. 2010; Zhu et al. 2014; Stahlhut et al. 2015). This type of translational fusions increase the proximity between the proteins resulting in enhanced electron transfer from one protein to another. Similarly, in the case of biosynthetic enzymes, translational fusions increase the flavonoid production by increasing the efficiency of transfer of reaction intermediates from one enzyme to the other by bringing them together, a phenomenon known as substrate channeling. It has been found that although translational fusions increase the efficiency of the heterologous pathways, it is something that must be carefully planned as the fused proteins may get misfolded or aggregated if the number of fused protein is high, leading to metabolic burden and hence to decreased production (Siddiqui et al. 2012). Another alternative is the formation of synthetic scaffolds, on which the enzymes can be anchored. The close proximity between the functionally related proteins has also been found in natural pathways where the proteins interact with each other by either noncovalent interactions or anchoring with the membrane. Scaffolds increase the efficiency of the metabolic pathway by minimizing the diffusion of reaction intermediates, efficient channeling of the substrate and reaction intermediates from one enzyme to the other thus preventing the accumulation of toxic intermediates and byproducts in the cell (Moon et al. 2010). Such synthetic scaffolds can be either made of DNA, RNA or proteins (Dueber et al. 2009; Delebecque et al. 2011; Conrado et al. 2012). The production of flavonoids can also be enhanced by the application of scaffolding strategy (Zhao et al. 2015). The production of resveratrol in *S. cerevisiae* has been enhanced by five-fold using scaffolding strategy in comparison to the non-scaffolded control (Wang and Yu 2011).

5.9 Producing Non-natural Derivatives of Flavonoids

The chemical synthesis of non-natural derivatives of flavonoids is complex and expensive task due to the involvement of a large number of steps and low final yields. Application of the recent molecular biology and system/synthetic biology tools in metabolic engineering has led to the synthesis of a diverse range of natural flavonoids and their unnatural derivatives in heterologous production hosts (such as *E. coli*, Streptomyces and *S. cerevisiae*). Combinatorial biosynthesis carried out the designing of new sets of gene clusters by combining genes from different organisms for the diversification of natural and non-natural product libraries, thus making efficient use of the potential of heterologous host. The strategy of combinatorial biosynthesis establishes novel enzyme-substrate combinations *in vivo* by combining enzymes that apparently do not function together in nature (Fukushima et al. 2013). Metabolic engineering in combination with protein engineering, mutasynthesis (precursor directed synthesis) and bio-transformation can result in the production of a multitude of both natural and non-natural flavonoid analogs without relying on expensive precursors and cofactors (Chemler et al. 2007, 2010; Horinouchi 2008, 2009; Fowler et al. 2011; Mora-Pale et al. 2013). The approaches for production on non-natural flavonoid derivatives are depicted in Fig. 5.5.

Mutasynthesis combined with metabolic engineering serves as an attractive alternative for creation of flavonoid analogues in contrast to the chemical synthesis. Mutasynthesis involves the chemical synthesis of substrate analogs having structural similarity to natural substrates and the conversion of these non-natural substrate analogs into novel non-natural compounds by reactions catalyzed by mutated

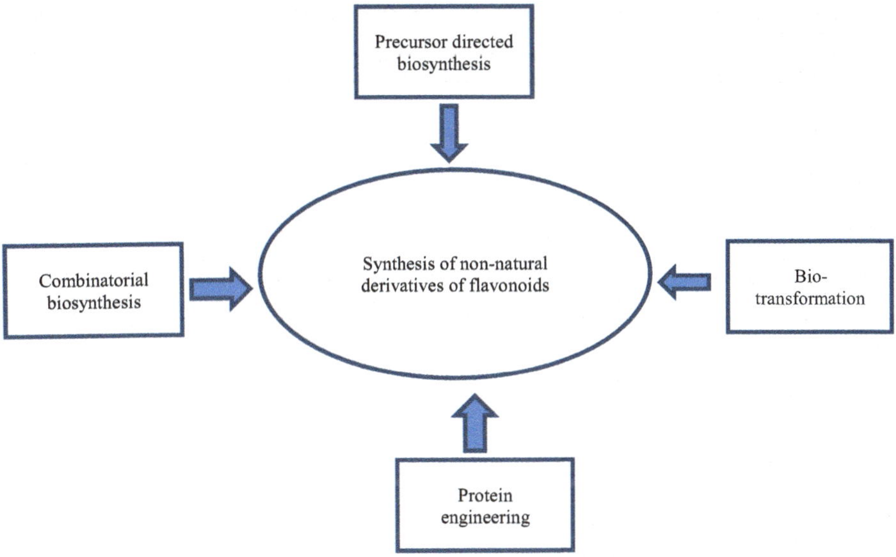

Fig. 5.5 Strategies for the synthesis of non-natural derivatives of flavonoids

or recombinant enzymes in the recombinant hosts. This combination of chemical synthesis and biosynthesis has been applied for the production of non-natural flavonoids and isoflavonoids (Cress et al. 2013; Bhan et al. 2014). Unnatural polypropanoids were synthesized using the chalcone synthase from *Scutellaria baicalensis*, possessing broad substrate specificity towards *p*-coumaroyl-CoA analogues (Abe et al. 2000; Morita et al. 2000).

Mutagenesis of type III polyketide synthases also resulted in the synthesis of a variety of natural and non-natural flavonoid molecules. The homology modeling, with previously resolved crystal structure of STS from *A. hypogaea*, resulted in the creation of wild type *V. vinifera* STS (WtVvSTS) variants which were able to accept various CATL and BYN-type pyrones (Abe et al. 2004; Morita et al. 2001). The substrate promiscuity of VvSTS was further diversified by using various substrate analogs such as propionyl-CoA, myristoyl-CoA, octanoyl-CoA and methylmalonyl-CoA instead of natural substrate (malonyl-CoA). WtVvSTS accepted all the non-natural substrates to produce various non-natural derivatives of aromatic polyketides, however the generation of polyketides by mutant VvSTSs involved coupling of non-natural CoAs with malonyl-CoA as an extender unit (Bhan et al. 2015).

For the precursor directed biosynthesis of unnatural flavonoids, the production medium is often exogenously supplemented with non-natural chemically synthesized phenylpropanoic acid precursors (structurally similar to natural substrates) in contrast to the natural substrates used by the plants and heterologous hosts carrying out the synthesis of natural flavonoids. Furthermore, production of non-natural flavonoid derivatives can also be achieved by using the substrate analogs that are entirely different from the natural substrates (Bhan et al. 2015). Non-natural or synthetic flavonoids were synthesized by Koffas and his colleagues by using combinatorial engineering approach assisted with precursor directed synthesis. Assembly of the flavonoid biosynthetic genes (4CL, CHS, and CHI) from different genetic sources resulted in the production of diverse kinds of synthetic flavonoids in *S. cerevisiae* by feeding the engineered strain with natural (*p*-coumaric acid, *m*-coumaric acid, *o*-coumaric acid) and chemically synthesized phenylpropanoic acids (*p*-fluorocinnamic acid, *o*-fluorocinnamic acid, *p*-aminocinnamic acid) (Chemler et al. 2007). The combination of both approaches resulted in the production of various flavonoid derivatives including 6.54 mg L^{-1} of (2S)-3′,5,7- trihydroxyflavanone, 2.81 mg/Lof (2*S*)-4′-fluoro-5,7-dihydroxyflavanone, 6.36 mg/L of (2*S*)-2′,5,7-trihydroxyflavanone and 15.82 mg/L of (2*S*)-4′-amino-5,7-dihydroxyflavanone. Chemler et al. further explored the substrate promiscuity of the flavonoid synthesizing enzymes of the engineered strain by supplementing with acrylic acid which accepted it as substrate and produced novel flavonoid derivative, (2E)-[(2*S*)-5, 7-dihydroxy-4-chromanone] propenoic acid (Chemler et al. 2007). Furthermore, Chemler and colleagues evaluated the substrate specificity and product diversity of IFS enzymes from five different plant species (*G. max*, *T. pratense*, *G. echinata*, *P. sativum*, and *M. truncatula*) on the basis of their potential of converting genistein from naringenin and applied precursor directed biosynthesis for the synthesis of natural and unnatural isoflavonoids in *S. cerevisiae* using three enzymes-IFS, CPR, and 2-hydroxyisoflavanone dehydratase (HID), screened from different plants (Chemler et al. 2010). In a similar manner, Horinuchi and his group used precursor-directed strategy combined with metabolic engineering for the synthesis of natural/non-natural flavonoids and stilbenoids by using three modules in *C. glutamicum* engineered to overexpress two subunits of ACC (*accBC*) for increasing the intracellular pool of malonyl-CoA. Firstly, Le4CL from *Lithospermum erythrohizon* carried out the catalysis of substrate synthesis step, secondly, the catalysis of polyketide synthesis step was carried out by either GeCHS from *G. echinata*, PlCHI from *P. lobata* or AhSTS from *A. hypogaea* leading to the synthesis of chalcone, flavonone or stilbene respectively. And finally in the third module, flavones and flavonols were derived from their precursor compounds by the post-polyketide modification reactions catalyzed by FSI from *P. crispum*, and F3H and FLS from Citrus respectively. The engineered strains were fed with 14 different chemically synthesized and naturally available carboxylic acid derivatives resulting in the production of 87 different aromatic polyketides of which 36 were novel compounds including triketide and tetraketide pyrones (Horinuchi 2008). Precursor directed synthesis of synthetic/non-natural flavanones, pyrones and stilbenes was carried out by various substrate analogs such as cinnamic acid, *p*-coumaric acid, fluorocinnamic acid, furyl and thienyl cinnamic acid resulting in the production of 70–90 mg/L of natural and 50–100 mg/L of synthetic flavanones, as well as 3.6 ± 1.1 mg/L and 2.7 ± 0.8 mg/L of natural and synthetic pyrones (Katsuyama et al. 2007).

Structure-guided enzyme engineering often manipulates at or around the active site or binding pocket region of enzyme and result in the alteration of substrate specificity, enzymatic activity, and product region-selectivity. Enzyme UGT71G1 from *M. truncatula* was mutated by replacing tyrosine to alanine at 202 position which lead to the conversion of genistein to both 7-*O*-glucoside and 5-*O*-glucoside, as compared to the native enzyme which only enables conversion of genistein to 7-*O*-glucoside (He et al. 2006). Structure-guided mutation of *Vitis vinifera* stilbene synthase (STS), involving replacement of threonine 197 with glycine (T197G), combined with precursor directed biosynthesis resulted in the synthesis of a novel C17 resorcylic acid (Bhan et al. 2014). Similarly, the precursor directed biosynthesis of 7-*O*-methyl aromadendrin was carried out by exogenous supplementation of *p*-coumaric acid precursor in recombinant *E. coli* harboring flavonoid pathway genes (Pc4Cl2, PhCHS, MsCHI, AtF3H) from different plant sources and a post modification gene 7-*O*-methyltransferase (SaOMT) from *Streptomyces avermitilis*. 2.7 mg/Lof 7-*O*-methyl aromadendrin was produced in the recombinant strain by the introduction of ACC subunits, biotin ligase, and acetyl-CoA synthase genes from *Nocardia farcinica*. The production was enhanced upto 30 mg L^{-1} on feeding naringenin (Malla et al. 2012). Non-natural derivatives of flavonoids can also be obtained by altering the regioselectivity of enzymes. *O*-methyltransferase isolated from poplar, POMT7, was altered by carrying out an error-prone polymerase chain reaction. The mutant POMT-7 having novel regioselectivity was screened from more than 100 mutants. The selected mutant (POMT-M1) Asp257Gly, possessed glycine at 257 position instead of an asparagine residue and methylated the 3-hydroxyl group of flavonols in addition to 7-hydrdoxyl group resulting in the production of 58 μM of 3, 7-*O*-dimethylquercetin and 70 μM of 3, 7-*O*-dimethylkaempferol in recombinant *E. coli* host harboring the novel enzyme (Joe et al. 2010).

Bio-transformation can also result in the synthesis of non-natural compounds by modifying the existing compounds by different types of biocatalysts (whole cells, cell extracts, or purified enzymes). This approach of biotransformation in combination with metabolic engineering and proteomic tools can be used for the production of flavonoid derivatives in model organisms (*E. coli*, *S. cerevisiae*) engineered by introducing partial or entire biosynthetic pathway gene clusters of intracellular cofactors/precursors and modification enzymes for production of derivatives of basic flavonoid compounds (Luo et al. 2015; Sun et al. 2015). Most of the modifications of flavonoids are carried out by enzymes such as BAHD acyltransferases, SCP acyltransferases, glycosyltransferases (GTs), *O*-methyltransferases (OMTs), cytochrome P450s (CYP) and prenyltransferases (PTs) (Shimoda et al. 2010; Wu et al. 2015a, b). The chemical diversity of flavonoids can be expanded by the addition of decorating or modification enzymes to the artificial pathway. *E. coli* strains produced 109.3 mg/L astragalin from naringenin (Malla et al. 2013), 13.56 mg/L from kaempferol (He et al. 2008) and 23.78 mg/L of 3-*O*-xylosyl quercetin from quercetin by attaching the sugars to the hydroxyl groups of the flavonoid backbone through the action of glycosyl transferases, a common biotransformation. Similarly, 3-*O*-, 7-*O*-, or 4-*O*-glucosides derivatives of apigenin, chrysin, luteolin, kaempferol, and quercetin were obtained by biotransformation using glycosyltransferases from various sources (He et al. 2008; Choi et al. 2012). Kaempferol and quercetin were converted into their corresponding 3-*O*-rhamnosides by the action of a rhamnose flavonol glycosyltransferase (Kim et al. 2012).

The development of methods for the in vitro biosynthesis of flavonoids holds great promise in further expanding the chemical space of these molecules by either precursor feeding experiments or by protein engineering (Zang et al. 2019).

5.10 Commentary on Future Trends

Flavonoids are valuable compounds having numerous health benefits and broad utility in pharmaceutical and food industries and so their increasing global demand cannot be met by natural production. Metabolic engineering is an alternative approach for the production of flavonoids at a large scale in heterologous hosts, which involves either the manipulation of natural pathways or the de novo construction of artificial pathways. The potential of metabolic engineering in the production of natural compounds in microbial hosts has increased due to recent developments in molecular, systems and synthetic biology which have effectively improved the product titers and significantly reduced the constraints or bottlenecks. The combination of metabolic engineering and synthetic/system biology has created novel approaches for design, construction, and optimization of artificial metabolic pathways as well as novel enzymes and microorganisms by providing a better understanding of the cellular metabolism and its regulation. Besides, the regulation of gene expression at the transcription, translation, and post translation levels by the application of synthetic regulatory tools like RNAi, has resulted in better fine tuning of the gene expression. This expression fine tuning has further been improved by using synthetic molecular sensors, such as for example a sensor for malonyl CoA which is the most important precursor metabolite in flavonoid biosynthesis (Xu et al. 2013, 2014). The regulatory tools also provide useful information for identifying the desired mutant from the library of mutants by rapid screening. Furthermore, novel computational tools like CiED and Optforce based on cellular modeling in combination with combinatorial engineering have been used to optimize strain and heterologous expression by redirecting carbon flux with minimum genetic interventions. Also, the innovative tools of protein engineering have not only improved the existing enzymes but also have created novel enzymes with entirely new properties and substrate specificities for the production of both natural and non-natural flavonoids. Non-natural derivatives of flavonoids can potentially have health benefits and improved properties (solubility and stability) allowing them to be used in food and pharmaceutical industries. Recently the synergistic effects of microbial consortia in co-cultures have been explored and exploited for the synthesis of valuable compounds (Jones et al. 2016, 2017; Fang et al. 2018). Also, various strategies like efflux-pump engineering and overexpression of membrane transporters have been developed to increase the strain robustness.

References

Abe I, Morita H, Nomura A, Noguchi H. Substrate specificity of chalcone synthase: enzymatic formation of unnatural polyketides from synthetic cinnamoyl-CoA analogues. J Am Chem Soc. 2000;122:11242–3.

Abe I, Watanabe T, Noguchi H. Enzymatic formation of long-chain polyketide pyrones by plant type III polyketide synthases. Phytochemistry. 2004;65:2447–53.

Ae Park S, Choi MS, Cho SY, Seo JS, Jung UJ, Kim MJ, Sung MK, Park YB, Lee MK. Genistein and daidzein modulate hepatic glucose and lipid regulating enzyme activities in C57BL/KsJ-db/db mice. Life Sci. 2006;79:1207–13.

Agar OT, Dikmen M, Ozturk N, Yilmaz MA, Temel H, Turkmenoglu FP. Comparative studies on phenolic composition, antioxidant, wound healing and cytotoxic activities of selected Achillea L. species growing in Turkey. Molecules. 2015;20:17976–8000.

Ajikumar PK, Xiao W-H, Tyo KEJ, Wang Y, Simeon F, Leonard E, Mucha O, Phon TH, Pfeifer B, Stephanopoulos G. Isoprenoid pathway optimization for Taxol precursor overproduction in *Escherichia coli*. Science. 2010;330:70–4.

Assini JM, Mulvihill EE, Huff MW. Citrus flavonoids and lipid metabolism. Curr Opin Lipidol. 2013;24:34–40.

Ayabe S, Akashi T. Cytochrome P450s in flavonoid metabolism. Phytochem Rev. 2006;5:271–82.

Bahaudin KNAK, Sabri S, Ramzi AB, Chor ALT, Tencomnao T, Baharum SN. Current progress in production of flavonoids using systems and synthetic biology platforms. Sains Malaysiana. 2018;47:3077–84.

Bailey JE. Towards a science of metabolic engineering. Science. 1991;252:1668–75.

Bakhtiari M, Panahi Y, Ameli J, Darvishi B. Protective effects of flavonoids against Alzheimer's disease related neural dysfunctions. Biomed Pharmother. 2017;93:218–29.

Baptista FI, Henriques AG, Silva AM, Wiltfang J, Da Cruz E Silva OA. Flavonoids as therapeutic compounds targeting key proteins involved in Alzheimer's disease. ACS Chem Neurosci. 2014;5:83–92.

Bhan N, Xu P, Khalidi O, Koffas MAG. Redirecting carbon flux into malonyl-CoA to improve resveratrol titers: proof of concept for genetic interventions predicted by Opt Force computational framework. Chem Eng Sci. 2012;103:109–14.

Bhan N, Xu P, Koffas MAG. Pathway and protein engineering approaches to produce novel and commodity small molecules. Curr Opin Microbiol. 2013;24(6):1137–43.

Bhan N, Li L, Cai C, Xu P, Linhardt RJ, Koffas MAG. Enzymatic formation of a resorcylic acid by creating a structure-guided single point mutation in stilbene synthase. Protein Sci. 2014;24:167–73.

Bhan N, Cress BF, Linhardt RJ, Koffas M. Expanding the chemical space of polyketides through structure guided mutagenesis of *Vitis vinifera* stilbene synthase. Biochimie. 2015;115:136–43.

Boock JT, Gupta A, Prather KLJ. Screening and modular design for metabolic pathway optimization. Curr Opin Biotechnol. 2015;36:189–98.

Bottcher D, Bornscheuer UT. Protein engineering of microbial enzymes. Curr Opin Microbiol. 2010;13:274–82.

Boyle PM, Silver PA. Parts plus pipes: synthetic biology approaches to metabolic engineering. Metab Eng. 2012;14:223–32.

Chemler JA, Yan Y, Leonard E, Koffas MAG. Combinatorial mutasynthesis of flavonoid analogues from acrylic acids in microorganisms. Org Lett. 2007;9:1855–8.

Chemler JA, Lim CG, Daiss JL, Koffas MAG. A versatile microbial system for biosynthesis of novel polyphenols with altered estrogen receptor binding activity. Chem Biol. 2010;17:392–401.

Chen AY, Chen YC. A review of the dietary flavonoid, kaempferol on human health and cancer chemoprevention. Food Chem. 2013;138:2099–107.

Chen X, Yang X, Shen Y, Hou J, Bao X. Increasing Malonyl-CoA derived product through controlling the transcription regulators of phospholipid synthesis in *Saccharomycescerevisiae*. ACS Synth Biol. 2017; https://doi.org/10.1021/acssynbio.6b00346.

Chen X, Yang X, Shen Y, Hou J, Bao X. Screening phosphorylation site mutations in yeast acetyl-CoA carboxylase using malonyl-CoA sensor to improve malonyl-CoA-derived product. Front Microbiol. (2018.); www.frontiersin.org. 47.

Choi S, Ryu M, Yoon Y, Kim D-M, Lee E. Glycosylation of various flavonoids by recombinant oleandomycin glycosyltransferase from *Streptomyces antibioticus* in batch and repeated batch modes. Biotechnol Lett. 2012;34:499–505. https://doi.org/10.1007/s10529-011-0789-z.

Chouhan S, Sharma K, Zha J, Guleria S, Koffas MAG. Recent advances in the recombinant biosynthesis of polyphenols. Front Microbiol. 2017;8:2259.

Conrado RJ, Wu GC, Boock JT, Xu H, Chen SY, Lebar T, Turnsˇek J, Tomšič N, Avbelj M, Gaber R, Koprivnjak T, Mori J, Glavnik V, Vovk I, Bencˇina M, Hodnik V, Anderluh G, Dueber JE, Jerala R, Delisa MP. DNA guided assembly of biosynthetic pathways promotes improved catalytic efficiency. Nucleic Acids Res. 2012;40:1879–89.

Costa SL, Silva VDA, Dos Santos Souza C, Santos CC, Paris I, Muñoz P, Segura-Aguilar J. Impact of plant-derived flavonoids on neurodegenerative diseases. Neurotox Res. 2016;30:41–52.

Cress BF, Linhardt RJ, Koffas MA. Isoflavonoid production by genetically engineered microorganisms. In: Ramawat KG, Merillon JM, editors. Natural products. Berlin/Heidelberg: Springer; 2013.

Cress BF, Trantas EA, Ververidis F, Linhardt RJ, Koffas MAG. Sensitive cells: enabling tools for static and dynamic control of microbial metabolic pathways. Curr Opin Biotechnol. 2015;36:205–14.

Davis MS, Solbiati J, Cronanjr JE. Overproduction of acetyl-CoA carboxylase activity increases the rate of fatty acid biosynthesis in *Escherichia coli*. J Biol Chem. 2000;275:28593–8.

De Luca V, Salim V, Atsumi SM, Yu F. Mining the biodiversity of plants: a revolution in the making. Science. 2012;336:1658–61.

Delebecque CJ, Lindner AB, Silver PA, Aldaye FA. Organization of intracellular reactions with rationally designed RNA assemblies. Science (New York, NY). 2011;333(6041):470–4.

Delmulle T, De Maeseneire SL, De Mey M. Challenges in the microbial production of flavonoids. Phytochem Rev. 2017;17:229–47.

Djouossi MG, Ngnokam D, Kuiate JR, Tapondjou LA, Harakat D, Voutquenne-Nazabadioko L. Antimicrobial and antioxidant flavonoids from the leaves of *Oncoba spinosa* Forssk. (Salicaceae). BMC Complement Altern Med. 2015;15:134.

Duan L, Ding W, Liu X, Cheng X, Cai J, Hua E, Jiang H. Biosynthesis and engineering of kaempferol in *Saccharomyces cerevisiae*. Microb Cell Factories. 2017;16:165.

Dueber JE, Wu GC, Malmirchegini GR, Moon TS, Petzold CJ, Ullal AV, Prather KLJ, Keasling JD. Synthetic protein scaffolds provide modular control over metabolic flux. Nat Biotechnol. 2009;27:753–9.

Etemadi A, Sinha R, Ward MH, Graubard BI, Inoue-Choi M, Dawsey SM, Abnet CC. Mortality from different causes associated with meat, heme iron, nitrates, and nitrites in the NIH-AARP diet and health study: population based cohort study. BMJ. 2017;357:j1957.

Fang Z, Jones JA, Zhu J, Koffas MAG. Engineering *Escherichia coli* co-cultures for production of curcuminoids from glucose. Biotechnol J. 2018;13(5) https://doi.org/10.1002/biot.201700576.

Fowler ZL, Gikandi WW, Koffas MAG. Increased malonyl coenzyme A biosynthesis by tuning the *Escherichia coli* metabolic network and its application to flavanone production. Appl Environ Microbiol. 2009;75:5831–9.

Fowler ZL, Shah K, Panepinto JC, Jacobs A, Koffas MAG. Development of nonnatural flavanones as antimicrobial agents. PLoS One. 2011;6:e25681.

Frabasile S, Koishi AC, Kuczera D, Silveira GF, Verri WA, Dos Santos CND, Bordignon J. The citrus flavanone naringenin impairs dengue virus replication in human cells. Sci Rep. 2017;7.

Fukushima EO, Seki H, Sawai S, Suzuki M, Ohyama K, Saito K, et al. Combinatorial biosynthesis of legume natural and rare triterpenoids in engineered yeast. Plant Cell Physiol. 2013;54:740–9. https://doi.org/10.1093/pcp/pct015.

Gassara F, Kouassi AP, Brar SK, Belkacemi K. Green alternatives to nitrates and nitrites in meat-based products – a review. Crit Rev Food Sci Nutr. 2016;56:2133–48.

He X-Z, Wang X, Dixon RA. Mutational analysis of the Medicago glycosyltransferase UGT71G1 reveals residues that control regioselectivity for (iso)flavonoid glycosylation. J Biol Chem. 2006;281:34441–7.

He XZ, Li WS, Blount JW, Dixon RA. Regioselective synthesis of plant (iso)flavone glycosides in *Escherichia coli*. Appl Microbiol Biotechnol. 2008;80:253–60. https://doi.org/10.1007/s00253-008-1554-7.

He L, Xiu Y, Jones JA, Baidoo EE, Keasling JD, Tang YJ, Koffas MAG. Deciphering flux adjustments of engineered *E. coli* cells during fermentation with changing growth conditions. Metab Eng. 2017;39:247–56.

Horinouchi S. Combinatorial biosynthesis of non-bacterial and unnatural flavonoids, Stilbenoids and Curcuminoids by microorganisms. J Antibiot. 2008;61:709–28.

Horinouchi H. Combinatorial biosynthesis of plant medicinal polyketides by microorganisms. Curr Opin Chem Biol. 2009;13:197–204.

Hwang EI, Kaneko M, Ohnishi Y, Horinouchi S. Production of plant specific flavanones by *Escherichia coli* containing an artificial gene cluster. Appl Environ Microbiol. 2003;69:2699–706.

Ibrahim A, Sobeh M, Ismail A, Alaa A, Sheashaa H, Sobh M, Badria F. Free-B-ring flavonoids as potential lead compounds for colon cancer therapy. Mol Clin Oncol. 2014;2:581–5.

Iwashina T. Contribution of flower colors of flavonoids including anthocyanins: a review. Nat Prod Commun. 2015;10:529–44.

Jiang H, Wood KV, Morgan JA. Metabolic engineering of the phenyl propanoid pathway in *Saccharomyces cerevisiae*. Appl Environ Microbiol. 2005;71:2962–9.

Joe EJ, Kim B-G, An B-C, Chong Y, Ahn JH. Engineering of flavonoid *O*-methyltransferase for a novel regioselectivity. Mol Cells. 2010;30:137–41.

Jones JA, Vernacchio VR, Sinkoe AL, Collins SM, Ibrahim MHA, Mlachance DM, et al. Experimental and computational optimization of an *Escherichia coli* co-culture for the efficient production of flavonoids. Metab Eng. 2016;35:55–63.

Jones JA, Vernacchio VR, Collins SM, Shirke AN, Xiu Y, Englaender JA, Cress BF, Mccutcheon CC, Linhardt RJ, Gross RA, Koffas MAG. Complete biosynthesis of anthocyanins using *E. coli* polycultures. MBio. 2017;8:e00617–21.

Juminaga D, Baidoo EE, Redding-Johanson AM, Bath TS, Burd H, Mukhopadhyay A, Petzold CJ, Keasling JD. Modular engineering of L-tyrosine production in *Escherichia coli*. Appl Environ Microbiol. 2012;78:89–98.

Kabera JN, Semana E, Mussa AR, He X. Plant secondary metabolites: biosynthesis, classification, function and pharmacological properties. J Pharm Pharmacol. 2014;2:377–92.

Karla RE, Heriberto VL, Diego H, Elisabeth M, Marta G, Rosa MC, Palazon J. Elicitation, an effective strategy for the biotechnological production of bioactive high-added value compounds in plant cell factories. Molecules. 2016;21:182.

Katsuyama Y, Funa N, Miyahisa I, Horinouchi S. Synthesis of unnatural flavonoids and stilbenes by exploiting the plant biosynthetic pathway in *Escherichia coli*. Chem Biol. 2007;14:613–21.

Keasling JD. Manufacturing molecules through metabolic engineering. Science. 2010;330:1355–8.

Kim BG, Ahn JH. Biosynthesis of pinocembrin from glucose using engineered *Escherichia coli*. J Microbiol Biotechnol. 2014;24:1536–41.

Kim BG, Kim HJ, Ahn JH. Production of bioactive flavonol rhamnosides by expression of plant genes in *Escherichia coli*. J Agric Food Chem. 2012;60:11143–8.

Ko KP, Kim CS, Ahn Y, Park SJ, Kim YJ, Park JK, Lim YK, Yoo KY, Yoo KY, Kim SS. Plasma isoflavone concentration is associated with decreased risk of type 2 diabetes in Korean women but not men: results from the Korean genome and epidemiology study. Diabetologia. 2015;58:726–35.

Kolewe ME, Gaurav V, Roberts SC. Pharmaceutically active natural product synthesis and supply via plant cell culture technology. Mol Pharm. 2008;5:243–56.

Koopman F, Beekwilder J, Crimi B, Van Houwelingen A, Hall RD, Bosch D, Van Maris AJ, Pronk JT, Daran JM. De novo production of the flavonoid naringenin in engineered *Saccharomyces cerevisiae*. Microb Cell Factories. 2012;11:155.

Kotopka BJ, Li Y, Smolke CD. Synthetic biology strategies toward heterologous phytochemical production. Nat Prod Rep. 2018;35:902.

Lani R, Hassandarvish P, Shu MH, Phoon WH, Chu JJH, Higgs S, Vandalingham D, Abu Bakar S, Zandi K. Antiviral activity of selected flavonoids against chikungunya virus. Antivir Res. 2016;133:50–61.

Lanza AM, Crook NC, Alper HS. Innovation at the intersection of synthetic and systems biology. Curr Opin Biotechnol. 2012;23:712–7.

Lee PC, Holtzapple E, Schmidt-Dannert C. Novel activity of *Rhodobacter sphaeroides* spheroidene monooxygenase CrtA expressed in *Escherichia coli*. Appl Environ Microbiol. 2010;76:7328–31.

Lee JW, Na D, Park JM, Lee J, Choi S, Lee SY. Systems metabolic engineering of microorganisms for natural and non-natural chemicals. Nat Chem Biol. 2012;8:536–46.

Lee H, Kim BG, Kim M, Ahn JH. Biosynthesis of two flavones, apigenin and genkwanin, in *Escherichia coli*. J Microbiol Biotechnol. 2015;25:1442–8.

Leiherer A, Mundlein A, Drexel H. Phytochemicals and their impact on adipose tissue inflammation and diabetes. Vasc Pharmacol. 2013;58:3–20.

Leonard E, Koffas MAG. Engineering of artificial plant cytochrome P450 enzymes for synthesis of isoflavones by *Escherichia coli*. Appl Environ Microbiol. 2007;73:7246–51.

Leonard E, Yan Y, Lim KH, Koffas MAG. Investigation of two distinct flavone synthases for plant-specific flavone biosynthesis in *Saccharomyces cerevisiae*. Appl Environ Microbiol. 2005;71:8241–8.

Leonard E, Lim KH, Saw PN, Koffas MAG. Engineering central metabolic pathways for high-level flavonoid production in *Escherichia coli*. Appl Environ Microbiol. 2007;73:3877–86.

Leonard E, Yan Y, Fowler ZL, Li Z, Lim CG, Lim KH, Koffas MAG. Strain improvement of recombinant *Escherichia coli* for efficient production of plant flavonoids. Mol Pharm. 2008;5:257–65.

Levisson M, Patinios C, Hein S, De Groot PA, Daran JM, Hall RD, Martens S, Beekwilder J. Engineering de novo anthocyanin production in *Saccharomyces cerevisiae*. Microb Cell Factories. 2018;17:103.

Li S, Si T, Wang M, Zhao H. Development of a synthetic Malonyl-CoA sensor in Saccharomyces cerevisiae for intracellular metabolite monitoring and genetic screening. ACS Synth Biol. 2015;4:1308–15.

Lia J, Tiana C, Xiaa Y, Mutandaa I, Wanga K, Wang Y. Production of plant-specific flavones baicalein and scutellarein in an engineered *E. coli* from available phenylalanine and tyrosine. Metab Eng. 2019;52:124–33.

Lim CG, Fowler ZL, Hueller T, Schaffer S, Koffas MA. High-yield resveratrol production in engineered *Escherichia coli*. Appl Environ Microbiol. 2011;77:3451–60. https://doi.org/10.1128/AEM.02186-10.

Lim CG, Wong L, Bhan N, Dvora H, Xu P, Venkiteswaran S, Koffas MAG. Development of a recombinant *Escherichia coli* strain for overproduction of plant pigment anthocyanin. Appl Environ Microbiol. 2015;81:6276–84.

Lovegrove JA, Stainer A, Hobbs DA. Role of flavonoids and nitrates in cardiovascular health. Proc Nutr Soc. 2017;76:83–95.

Luo Y, Li BZ, Liu D, Zhang L, Chen Y, Jia B, Zeng BX, Zhao H, Yuan YJ. Engineered biosynthesis of natural products in heterologous hosts. Chem Soc Rev. 2015;44:5265–90.

Malla S, Koffas MAG, Kazlauskas RJ, Kim BG. Production of 7-*O*-methyl aromadendrin, a medicinally valuable flavonoid, in *Escherichia coli*. Appl Environ Microbiol. 2012;78:684–94.

Malla S, Pandey RP, Kim BG, Sohng JK. Regiospecific modifications of naringenin for astragalin production in *Escherichia coli*. Biotechnol Bioeng. 2013;110:2525–35.

Marienhagen J, Bott M. Metabolic engineering of microorganisms for the synthesis of plant natural products. J Biotechnol. 2013;163:166–78.

Mcnerney MP, Watstein DM, Styczynski MP. Precision metabolic engineering: the design of responsive, selective, and controllable metabolic systems. Metab Eng. 2015;31:123–31.

Medema MH, Van Raaphorst R, Takano E, Breitling R. Computational tools for the synthetic design of biochemical pathways. Nat Rev Microbiol. 2012;10:191–202.

Miyahisa I, Kaneko M, Funa N, Kawasaki H, Kojima H, Ohnishi Y, Horinouchi S. Efficient production of (2S)-flavanones by *Escherichia coli* contain in g an artificial biosynthetic gene cluster. Appl Microbiol Biotechnol. 2005;68:498–504.

Miyahisa I, Funa N, Ohnishi Y, Martens S, Moriguchi T, Horinouchi S. Combinatorial biosynthesis of flavones and flavonols in *Escherichia coli*. Appl Microbiol Biotechnol. 2006;71:53–8.

Moon TS, Dueber JE, Shiue E, Prather KLJ. Use of modular, synthetic scaffolds for improved production of glucaric acid in engineered *E. coli*. Metab Eng. 2010;12:298–305.

Mora-Pale M, Sanchez-Rodriguez SP, Linhardt RJ, Dordick JS, Koffas MAG. Metabolic engineering and in vitro biosynthesis of phytochemicals and non-natural analogues. Plant Sci. 2013;210:10–24.

Morita H, Takahashi Y, Noguchi H, Abe I. Enzymatic formation of unnatural aromatic polyketides by chalcone synthase. Biochem Biophys Res Commun. 2000;279:190–5.

Morita H, Noguchi H, Schröder J, Abe I. Novel polyketides synthesized with a higher plant stilbene synthase. Eur J Biochem. 2001;268:3759–66.

Ng KR, Lyu X, Mark R, Chen WN. Antimicrobial and antioxidant activities of phenolic metabolites from flavonoid-producing yeast: potential as natural food preservatives. Food Chem. 2019;270:123–9.

Ong KL, Kaur G, Pensupa N, Uisan K, Lin CSK. Trends in food waste valorization for the production of chemicals, materials and fuels: case study south and Southeast Asia. Bioresour Technol. 2017:100–12.

Ortsater H, Grankvist N, Wolfram S, Kuehn N, Sjoholm A. Diet supplementation with green tea extract epigallocatechin gallate prevents progression to glucose intolerance in db/db mice. Nutr Metab. 2012;9:11.

Panche AN, Diwan AD, Chandra SR. Flavonoids: an overview. J Nutr Sci. 2016;5:47.

Pandey RP, Parajuli P, Koffas MAG, Sohng JK. Microbial production of natural and non-natural flavonoids: pathway engineering, directed evolution and systems/synthetic biology. Biotechnol Adv. 2016;34:634–62.

Pandey RP, Jung HY, Parajuli P, Nguyen THT, Bashyal P, Sohng JAEK. A synthetic approach for biosynthesis of Miquelianin and Scutellarin A in *Escherichia coli*. Appl Sci. 2019;9:215.

Park HJ, Choi YJ, Lee JH, Nam MJ. Naringenin causes ASK1-induced apoptosis via reactive oxygen species in human pancreatic cancer cells. Food Chem Toxicol. 2017;99:1–8.

Paterson I, Anderson EA. Chemistry. The renaissance of natural products as drug candidates. Science. 2005;310:451–3.

Pei J, Chen A, Dong P, Shi X, Zhao L, Cao F, Tang F. Modulating heterologous pathways and optimizing fermentation conditions for biosynthesis of kaempferol and astragalin from naringenin in *Escherichia coli*. J Ind Microbiol Biotechnol. 2018; https://doi.org/10.1007/s10295-018-02134-6.

Petrovska BB. Historical review of medicinal plants' usage. Pharmacogn Rev. 2012;6:1–5.

Prior RL, Lazarus SA, Cao G, Muccitelli H, Hammerstone JF. Identification of procyanidins and anthocyanins in blueberries and cranberries (Vaccinium spp.) using high-performance liquid chromatography/mass spectrometry. J Agric Food Chem. 2001;49:1270–6.

Ranganathan S, Suthers PF, Maranas CD. OptForce: an optimization procedure for identifying all genetic manipulations leading to targeted overproductions. PLoS Comput Biol. 2010;6:e1000744.

Rasines-Perea Z, Teissedre PL. Grape polyphenols' effects in human cardiovascular diseases and diabetes. Molecules. 2017;22:68.

Saewan N, Jimtaisong A. Natural products as photoprotection. J Cosmet Dermatol. 2015;14:47–63.

Santos CNS, Koffas M, Stephanopoulos G. Optimization of a heterologous pathway for the production of flavonoids from glucose. Metab Eng. 2011;13:392–400.

Sawayama AM, Chen MMY, Kulanthaivel P, Kuo M-S, Hemmerle H, Arnold FH. A panel of cytochrome P450 BM3 variants to produce drug metabolites and diversify lead compounds. Chem Eur J. 2009;15:11723–9.

Schujman GE, Guerin M, Buschiazzo A, Schaeffer F, Llarrull LI, Reh G, Vila AJ, Alzari PM, De Mendoza D. Structural basis of lipid biosynthesis regulation in gram-positive bacteria. EMBO J. 2006;25:4074–83.

Schujman GE, Altabe S, De Mendoza D. A malonyl-CoA-dependent switch in the bacterial response to a dysfunction of lipid metabolism. Mol Microbiol. 2008;68:987–96.

Shimoda K, Kubota N, Taniuchi K, Sato D, Nakajima N, Hamada H, Hamada H. Biotransformation of naringin and naringenin by cultured *Eucalyptus perriniana* cells. Photochemistry. 2010;71:201–5.

Si W, Gong J, Tsao R, Kalab M, Yang R, Yin Y. Bioassay-guided purification and identification of microbial components in Chinese green tea extracts. J Chromatogr A. 2006;1125:204–10.

Siddiqui MS, Thodey K, Trenchard I, Smolke CD. Advancing secondary metabolite biosynthesis in yeast with synthetic biology tools. FEMS Yeast Res. 2012;12:144–70.

Silva MM, Lidon FC. Food preservatives-An overview on applications and side effects. Emirates J Food Agric. 2016;28:366.

Stahlhut SG, Siedler S, Malla S, Harrison SJ, Maury J, Neves AR, Forster J. Assembly of a novel biosynthetic pathway for production of the plant flavonoid fisetin in *Escherichia coli*. Metab Eng. 2015;31:84–93.

Stefan L. Beyond directed evolution semi-rational protein engineering and design. Curr Opin Biotechnol. 2010;21:734–43.

Stephanopoulos G, Alper H, Moxley J. Exploiting biological complexity for strain improvement through systems biology. Nat Biotechnol. 2004;22:1261–7.

Sun H, Liu Z, Zhao H, Ang EL. Recent advances in combinatorial biosynthesis for drug discovery. Drug Des Devel Ther. 2015;9:823–33.

Tee KL, Wong TS. In directed enzyme evolution: advances and applications. Alcalde M, editors. Cham: Springer International Publishing; 2017. p. 201–27, https://doi.org/10.1007/978-3-319-50413-1_8.

Tokuyama K, Toya Y, Matsuda F, Cress B, Koffas MAG, Shimizu H. Magnesium starvation improves production of malonyl-CoA-derived metabolites in *Escherichia coli*. Metab Eng. 2018; https://doi.org/10.1016/j.ymben.2018.12.002.

Trantas E, Panopoulos N, Ververidis F. Metabolic engineering of the complete pathway leading to heterologous biosynthesis of various flavonoids and stilbenoids in *Saccharomyces cerevisiae*. Metab Eng. 2009;11:355–66.

Trantas E, Koffas M, Xu P, Ververidis F. When plants produce not enough or at all: metabolic engineering of flavonoids in microbial hosts. Front Plant Sci. 2015;6:7.

Treutter D. Significance of flavonoids in plant resistance: a review. Environ Chem Lett. 2006;4:147–57.

Tsao R. Chemistry and biochemistry of dietary polyphenols. Nutrients. 2010;2:1231–46.

Tsuda T. Dietary anthocyanin-rich plants: biochemical basis and recent progress in health benefits studies. Mol Nutr Food Res. 2012;56:159–70.

Veitch NC, Grayer RJ. Flavonoids and their glycosides, including anthocyanins. Nat Prod Rep. 2011;28:1626–95.

Vemuri GN, Aristidou AA. Metabolic engineering in the –omics era: elucidating and modulating regulatory networks. Microbiol Mol Biol Rev. 2005;69:197–216.

Vila-Real H, Alfaia AJ, Bronze MR, Calado AR, Ribeiro MH. Enzymatic synthesis of the flavone glucosides, Prunin and Isoquercetin, and the Aglycones, Naringenin and quercetin, with selective alpha-L-Rhamnosidase and beta-D-glucosidase activities of Naringinase. Enzym Res. 2011;373:692618.

Wang YC, Yu O. Synthetic scaffolds increased resveratrol biosynthesis in engineered yeast cells. J Biotechnol. 2011;157:258–60.

Wang Y, Halls C, Zhang J, Matsuno M, Zhang Y, Yu O. Stepwise increase of resveratrol biosynthesis in yeast *Saccharomyces cerevisiae* by metabolic engineering. Metab Eng. 2011a;13:455–63.

Wang YC, Chen S, Yu O. Metabolic engineering of flavonoids in plants and microorganisms. Appl Microbiol Biotechnol. 2011b;91:949–56.

Williams RJ, Spencer JP, Rice-Evans C. Flavonoids: antioxidants or signaling molecules? Free Radic Biol Med. 2004;36:838–49.

Wilson SA, Roberts SC. Recent advances towards development and commercialization of plant cell culture processes for the synthesis of biomolecules. Plant Biotechnol J. 2012;10:249–68.

Winkel-Shirley B. Flavonoid biosynthesis. A colorful model for genetics, biochemistry, cell biology, and biotechnology. Plant Physiol. 2001;126:485–93.

Wu J, Du G, Zhou J, Chen J. Metabolic engineering of *Escherichia coli* for (2S)-pinocembrin production from glucose by a modular metabolic strategy. Metab Eng. 2013;16:48–55.

Wu J, Zhou T, Du G, Zhou J, Chen J. Modular optimization of heterologous pathways for de novo synthesis of (2S)-naringenin in *Escherichia coli*. PLoS One. 2014;9(7):e101492.

Wu C, Zacchetti B, Ram A, Wezel GPV. Expanding the chemical space for natural products by Aspergillus-Streptomyces co-cultivation and biotransformation. Sci Rep. 2015a;5:10868.

Wu J, Du G, Chen J, Zhou J. Enhancing flavonoid production by systematically tuning the central metabolic pathways based on a CRISPR interference system in *Escherichia coli*. Sci Rep. 2015b;5:13477.

Wu J, Zhang X, Zhou J, Dong M. Efficient biosynthesis of (2S)-pinocembrin from D-glucose by integrating engineering central metabolic pathways with a pH-shift control strategy. Bioresour Technol. 2016;218:999–1007.

Xiu Y, Jang S, Jones JA, Zill NA, Linhardt RJ, Yuan Q, Jung GY, Koffas MAG. Naringenin-responsive riboswitch-based fluorescent biosensor module for *Escherichia coli* co-cultures. Biotechnol Bioeng. 2017;114:2235–44.

Xu P, Ranganathan S, Fowler ZL, Maranas CD, Koffas MA. Genome-scale metabolic network modeling results in minimal interventions that cooperatively force carbon flux towards malonyl- CoA. Metab Eng. 2011a;13:578–87.

Xu P, Ranganathan S, Maranas CD, Koffas MAG. An integrated computational and experimental study to increase the intra-cellular malonyl-CoA: application to flavanone synthesis. 2011b; 978-1-61284-8928-0/11.

Xu P, Gu Q, Wang WY, Wong L, Bower AGW, Collins CH, Koffas MAG. Modular optimization of multi-gene pathways for fatty acids production in *Escherichia coli*. Nat Commun. 2013;4:1–8.

Xu P, Wang W, Li L, Bhan N, Zhang F, Koffas MAG. Design and kinetic analysis of a hybrid promoter-regulator system for malonyl-CoA sensing in *Escherichia coli*. ACS Chem Biol. 2014;9:451–8.

Yan Y, Chemler J, Huang L, Martens S, Koffas MAG. Metabolic engineering of anthocyanin biosynthesis in *Escherichia coli*. Appl Environ Microbiol. 2005a;71:3617–23.

Yan Y, Kohli A, Koffas MAG. Biosynthesis of natural flavanones in *Saccharomyces cerevisiae*. Appl Environ Microbiol. 2005b;71:5610–3.

Yan Y, Li Z, Koffas MA. High-yield anthocyanin biosynthesis in engineered *Escherichia coli*. Biotechnol Bioeng. 2008;100:126–40.

Yang Y, Lin Y, Li L, Linhardt RJ, Yan Y. Regulating malonyl-CoA metabolism via synthetic antisense RNAs for enhanced biosynthesis of natural products. Metab Eng. 2015;29:217–26.

Yang D, Kim WJ, Yoo SM, Choi JH, Ha SH, Lee MH, Lee SY. Repurposing type III polyketide synthase as a malonyl- CoA biosensor for metabolic engineering in bacteria. PNAS. 2018.; www.pnas.org/cgi/doi/10.1073/pnas.1808567115.

Youns M, Hegazy WAH. The natural flavonoid fisetin inhibits cellular proliferation of hepatic, colorectal, and pancreatic cancer cells through modulation of multiple signalling pathway. PLoS One. 2017;12:e0169335.

Zamora-Ros R, Knaze V, Rothwell JA, Hémon B, Moskal A, Overvad K, Boutron-Ruault MC. Dietary polyphenol intake in Europe: the European prospective investigation into Cancer and nutrition (EPIC) study. Eur J Nutr. 2016;55:1359–75.

Zang Y, Zha J, Wu X, Zheng Z, Ouyang J, Koffas MAG. In vitro naringenin biosynthesis from *p*-coumaric acid using recombinant enzymes. J Agric Food Chem. 2019; https://doi.org/10.1021/acs.jafc.9b00413.

Zha J, Koffas MAG. Anthocyanin production in engineered microorganisms. In: Schwab W, Lange BM, Wüst M, editors. Biotechnology of natural products. Cham: Springer International Publishing; 2018. p. 81–97.

Zha W, Rubin-Pitel SB, Shao Z, Zhao H. Improving cellular malonyl-CoA level in *Escherichia coli* via metabolic engineering. Metab Eng. 2009;11:192–8.

Zhang Z, He Y, Huang Y, Ding L, Chen L, Liu Y, Nie Y, Zhang X. Development and optimization of an *in vitro* multienzyme synthetic system for production of kaempferol from naringenin. J Agric Food Chem. 2018; https://doi.org/10.1021/acs.jafc.8b01299.

Zhao S, Jones JA, Lachance DM, Bhan N, Khalidi O, Venkataraman S, Wang Z, Koffas MAG. Improvement of catechin production in *Escherichia coli* through combinatorial metabolic engineering. Metab Eng. 2015;28:43–53.

Zhu SJ, Wu J, Du GC, Zhou JW, Chen J. Efficient synthesis of eriodictyol from L-tyrosine in *Escherichia coli*. Appl Environ Microbiol. 2014;80:3072–80.

Chapter 6
Microbial Production of Natural Food Colorants

Lei Chen and Bobo Zhang

Colorants are mainly employed in food industry to improve the sensory attribute of food. Owing to the bright color, addition of certain colorants brings pleasing preference for food. Additionally, natural colorants generally have nutritional value, directly improving the market value of the colored food product (Narsing et al. 2017; Sen et al. 2019). Most of these compounds have significant health benefits, such as antioxidative, antibacterial, and anticancer activity, that they have drawn extensive attention by the nutraceutical and pharmaceutical industries (Celli et al. 2019; Lila et al. 2016).

The initial food colorants were mainly extracted from natural plant with special colors, such as paprika, turmeric, indigo, saffron, and various flowers (Aberoumand 2011). However, these natural pigments have some defects in chemical stability, high cost, and low yield. With the advance of science and technology, researchers attempted to synthesize colors to overcome the disadvantages of natural sources and larger ranges of hue and shade. Although most of the existed problems has been solved, new issues appeared, including allergenicity, toxicity, and carcinogenicity, astricting the application of synthetic food colorants (Oplatowska-Stachowiak and Elliott 2017; Potera 2010; Sen et al. 2019). Therefore, people's eyes turned back to natural food colorants from synthetic ones.

Under this background, microbial fermentation becomes an alternative way for colorants production for its intrinsic characters, such as fast growth and easy cultivation. In addition, sophisticated microbial manipulation techniques promote microbial production of food colorants facile, controllable, and cost-effective (Pandey et al. 2016; Staniek et al. 2014). Microbial production of natural food colorants is higher yields, lower cost, easier extraction, without seasonal limitation, and

L. Chen · B. Zhang (✉)
Key Laboratory of Carbohydrate Chemistry and Biotechnology, Ministry of Education, Jiangnan University, Wuxi, China

Key Laboratory of Industrial Biotechnology, Ministry of Education, Jiangnan University, Wuxi, China

L. Liu, J. Chen (eds.), *Systems and Synthetic Biotechnology for Production of Nutraceuticals*, https://doi.org/10.1007/978-981-15-0446-4_6

environmentally friendly (Kaur 2015). Some microbes could produce a variety of plant-derived food colorants, such as carotenoids, flavonoids, terpenoids, and alkaloids in prokaryotic and eukaryotic microorganisms (Chemler and Koffas 2008; Li and Smolke 2016).Among the biosynthetic pigments, carotenoids, lycopene, anthocyanins, and monascus pigments are the commonly used food colorants. The present review focuses on the strategies for the microbial production of these four types of natural food colorants. In general, the producing microorganisms are metabolically engineered for overproduction of the target colorants, with the benefits of genome sequencing, enzymology, metabolic engineering and fermentation engineering (Woolston et al. 2013; Yadav et al. 2012).

6.1 Carotenoids

Carotenoids can be applied as natural coloring agents and they are distributed in a wide range of organisms (plants, microalgae, fungi and bacteria) (Delgado-Vargas et al. 2000; Walter and Strack 2011). Carotenoids pigments occur universally in photosynthetic systems. On the other hand, in non-photosynthetic organisms, carotenoids are important in protecting against photooxidative damage (Mata-Gomez et al. 2014).

In general, carotenoids are derivatives of tetraterpenes, which share a common C40 backbone structure of eight isoprenoid units (Delgado-Vargas et al. 2000). Besides, some bacteria can produce C30 and C50 carotenoids *via* different intermediates (Walter and Strack 2011).

According to the chemical structure, carotenoids can be classified into two categories: the first class are carotenes, which contain only polyunsaturated hydrocarbons, and the second class are oxycarotenoids or xanthophylls, which contain polyunsaturated hydrocarbons with some oxygen functional groups (Sigurdson et al. 2017). According to the number of rings, carotenoids can be classified as acyclic, monocyclic or dicyclic. The representative compounds are lycopene, γ-carotene, and lutein, respectively. The chemical structure of some typical carotenoids is shown in Fig. 6.1.

Carotenoids can be classified as primary or secondary according to their functions. The primary carotenoids refer to the compounds such as lutein which involved in photosynthesis. They act as the structural and functional constituents in photosynthesis, which are essential for transferring and absorbing light energy (Ye et al. 2008). In comparison, secondary carotenoids are usually synthesized when responded to stress conditions, acting as protective compounds. Some typical secondary carotenoids such as astaxanthin and canthaxanthin can be largely accumulated by microalgae in stress conditions (Begum et al. 2016; Wang et al. 2015).

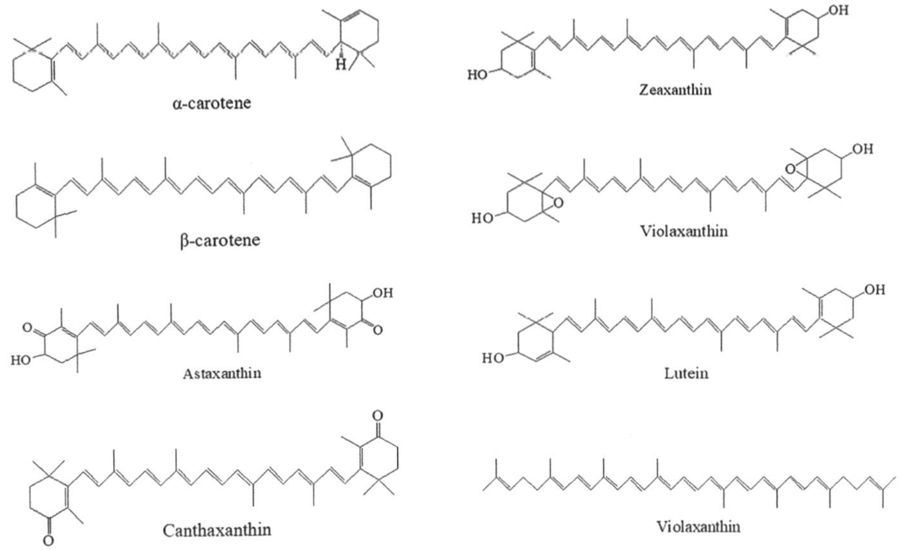

Fig. 6.1 The chemical structures of some typical carotenoids in microorganisms

6.1.1 Biosynthesis Pathway

Although the detailed regulatory mechanism is slightly different in various species, the metabolic pathway of carotenoid in most of the photosynthetic plant and algae species is highly conserved and ubiquitous. The carotenoid pathways begin from the plastidic methylerythritol 4-phosphate (MEP) pathway or from the cytosolic mevalonic acid pathway (MVA) pathway. After synthesizing the same C5 building block dimethylallyl diphosphate (DMAPP) or isopentenyl pyrophosphate (IPP), a C10-geranyl pyrophosphate (GPP) is further synthesized by condensation of IPP and DMAPP. Afterwards, by successive chain elongation, GPP is used for the synthesis of C15-farnesyl pyrophosphate (FPP) and C20-geranylgeranyl diphosphate (GGDP) (Liang et al. 2018). Subsequently, the C40 carotenoid phytoene is synthesized by the head to head condensation style. Then lycopene, the first important colored carotenoid is formed. Afterwards, *α*-carotene and *β*-carotene are synthesized through different cyclization styles. Finally, the commonly recognized carotenoids including lutein, zeaxanthin, astaxanthin and canthaxanthin are synthesized through a series of biochemical reactions (Gong and Bassi 2016) (Fig. 6.2).

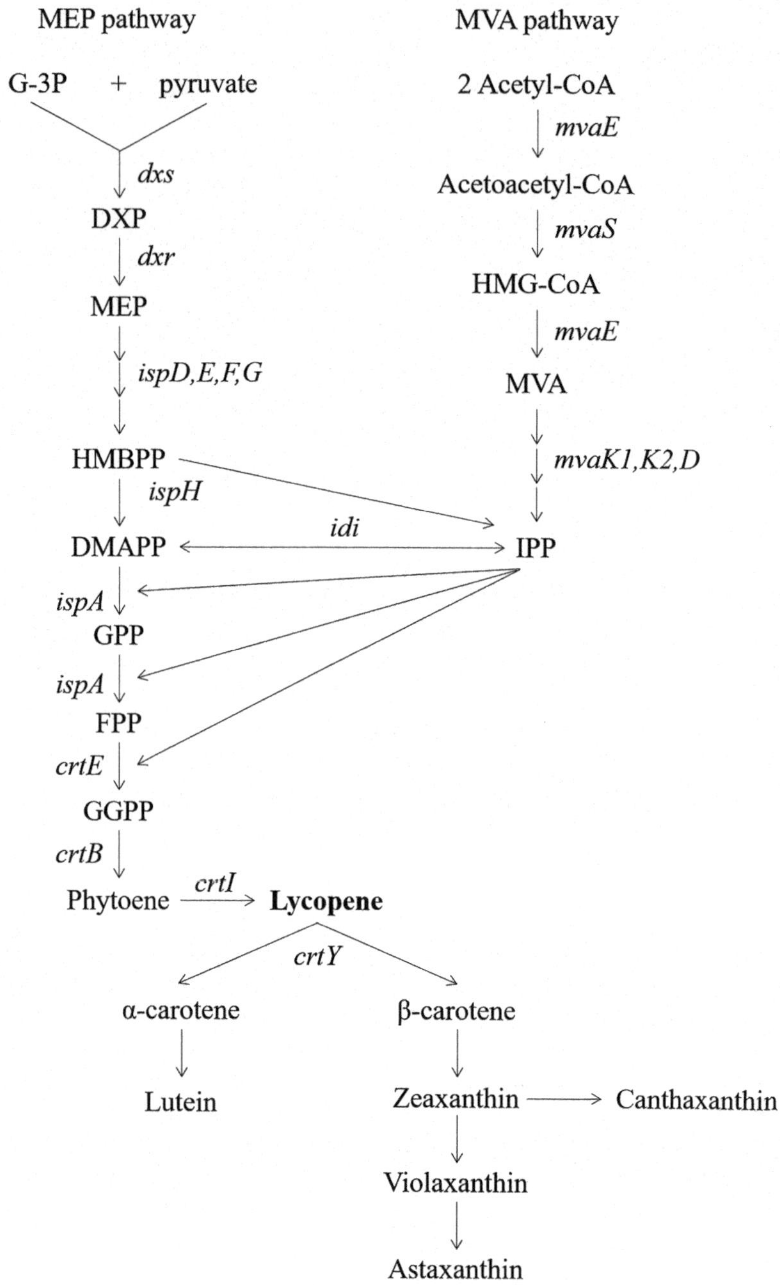

Fig. 6.2 Metabolic pathway for carotenoid biosynthesis in most microbial species. Metabolite abbreviations: *G-3P* D-glyceraldehyde-3-phosphate; *HMG-CoA* (S)-3-hydroxy-3-methylglutaryl-CoA; *DXP* 1-deoxy-D-xylulose-5-phosphate; *MVA* mevalonate; *MEP* 2-C-methyl-D-erythritol-4-phosphate; *MVP* mevalonate-5-phosphate; *HMBPP* 1-hydroxy-2-methyl-2-(E)-butenyl-4-diphosphate; *DMAPP* dimethylallyl diphosphate; *IPP* isopentenyl diphosphate; *GPP* geranyl diphosphate; *FPP* farnesyl diphosphate; *GGPP* geranylgeranyl diphosphate

6.1.2 Fermentation & Production

The commercial carotenoids are usually produced by extraction from plants (Dufossé et al. 2005) or chemical synthesis (Coulson 1980). However, the main challenges of producing carotenoids from plants are the vagaries of nature. The chemical synthesis generates hazardous wastes that can affect the environment. Compared to these conventional methods, the production of carotenoids originated from microorganisms obtains increasing attentions and safety to use. Therefore, microbial fermentation provides a promising approach for producing carotenoids in large scale (Mata-Gomez et al. 2014).

Carotenoids can be biosynthesized by different microbial sources. The biomass, the specific type and yield of carotenoids are highly dependent by many factors, including the nutritional composition and the culture conditions such as illumination intensity, pH, temperature and aeration rate (Mussagy et al. 2019; Mezzomo and Ferreira 2016).

6.1.2.1 Microalgae

Some important commercial carotenoids such as β-carotene, astaxanthin, canthaxanthin and lutein can be massively produced by microalgae, especially the strains in Chlorophyceae (Del Campo et al. 2007; Prommuak et al. 2013; Zhang et al. 2014b). Due to their versatility in adapting to a wide range of growth conditions and climates, (e.g., glacial to tropic, fresh water to hyper-saline), and varied pH, microalgae display a clear advantage over higher plants.

Compared with plants, the advantages of using microalgae as producers of carotenoids are their versatility in adapting different cultural conditions, even in hash and stress environments. Besides, the specific carotenoid content (mg/g) in microalgae is also higher than that of plants. For instance, the content of lutein in microalgae can reach as high as 4 mg/g, which is much higher than that of marigold flowers (0.3 mg/g) (Del Campo et al. 2007; Prommuak et al. 2013; Zhang et al. 2014b). Moreover, some pigments like astaxanthin are rarely found in higher plants, making microalgae can produce some unique carotenoids such as astaxanthin which are rarely synthesized by plants (Gong and Bassi 2016). However, the main issue of producing carotenoids by microalgae is the relatively higher cost, which make it prohibitive in commercial application (Gong and Bassi 2016).

Open pond and closed photo-bioreactors (PBRs) are the two common cultivation method for producing carotenoids by microalgae (Gong and Bassi 2016). Although the cost of open ponds system is much lower than that of PBRs, some critical disadvantages including uneven illumination, easy contamination, and poor mass transfer are the major concerns for using open ponds in large scale application (Singh and Sharma 2012; Rogers et al. 2014). Hence, the closed PBRs are preferred in recent studies due to high efficiency, higher biomass productivity, modernized operation mode and accurate control.

Generally, different kinds of PBRs including flat, tubular and stirred tank bioreactors can be applied for culturing microalgae. However, further reduction on the equipment and operation cost is required for PBRs. (Gong and Bassi 2016; Gupta et al. 2015; Olivieri et al. 2014).

Many variables affect the microalgae biomass and the productivity of carotenoid, such as the type of algal species, the nutritional composition, temperature, pH, light intensity, photoperiod, salinity, and aeration rate (Mezzomo and Ferreira 2016).

Temperature exert significant influence on the cell growth and carotenoids accumulation. It was found that 28 °C is the suitable temperature for both the cell growth and the accumulation of lutein is 28 °C, while the concentration of lutein dramatically decrease when the temperature increase to 32 °C (Fernandez-Sevilla et al. 2010). Light is also an important factor which greatly affects the production of carotenoids. For example, when increase the light intensity from186 to 460 μmol photon $m^{-2} \cdot s^{-1}$, the cell growth and lutein productivity could be simultaneously enhanced (Xie et al. 2013; Huang et al. 2008). It was found that the production of lutein was benefit from the full white light, when compared to the light source of monochromatic LED (Ho et al. 2014). The pH can affect the CO_2 availability, which is critical for the microalgae growth and the productivity of carotenoids.

Besides the microalgae strains and culture conditions, various technologies and strategies can be applied for the efficient production of carotenoids (Mussagy et al. 2019). For the cultivation strategies, two-stage cultivation and stress conditions are usually employed for the accumulation of high carotenoid content in microalgae. However, the biosynthesis of different carotenoids was affected by stress conditions to different degrees (Minhas et al. 2016).

For instance, oxidative stress can significantly stimulate the content of lutein by the addition of a suitable degree of oxidizing compounds (Guedes et al. 2011). The yield of astaxanthin increase with the combine use of nutrient reduction, salt introduction, and light stress (Harker et al. 1996; Orosa et al. 2000).

Ferrous salts can be applied for stimulating the production of canthaxanthin in *C. zofingiensis*, and thus it is regarded as an alternative oxidative stress method instead of strong light (Dan Pelah 2004). By combined use of salt stress and light limiting, the yield of canthaxanthin can reach 8.5 mg/g in *C. zofingiensis* (Gong and Bassi 2016). A high level of astaxanthin and β-carotene (over 50 mg/g) could be achieved by applying proper stress conditions (Kyriakopoulou et al. 2015; Suh et al. 2006).

The strategy of two-stage cultivation is proved to be an effective method for simultaneous production of microalgae biomass and carotenoids (Wan et al. 2014). For example, two cultivation phases were developed for biomass accumulation and astaxanthin biosynthesis, respectively, resulting in a high astaxanthin yield (57.9 mg/g) in a double layer bioreactor (Suh et al. 2006).

In order to enhance the accumulation of carotenoids in microalgae, investigation has been carried out not only in optimizing the fermentation conditions but also applying more advanced and efficient biology approaches, e.g. metabolic engineering and synthetic biology. The major method to enhance the yield of carotenoids by metabolic engineering is to provide sufficient isoprenoid precursor. It can be achieved through overexpressing the critical enzymes or suppressing the branch pathways (Varela et al. 2015).

The critical enzymes PSY, PDS, BKT are generally used for improving the yield of carotenoids. For example, the yield of carotenoids could be enhanced by twofold through overexpressing the related genes in the microalgae *Chlamydomonas* (Huang et al. 2008). PDS gene plays an important role in the biosynthesis of astaxanthin in *C. zofingiensis* (Liu et al. 2014). Both DXS and PSY genes were reported to associate with the production of fucoxanthin in *P. tricornutum*. Their transformants can produce 2.8-fold and 1.8-fold of fucoxanthin when compared with the wild type (Eilers et al. 2016).

6.1.2.2 Fungi (Yeast)

Fungi have been demonstrated to play important roles in producing high levels of carotenoids. Several studies have reported the production of carotenoids by fungus. *Blakeslea trispora* and *Phycomyces blakesleeanus* have been suggested as excellent producers for β-carotene at large scale (Almeida and Cerda-Olmedo 2008; Papaioannou and Liakopoulou-Kyriakides 2010). The biomass and the yield of astaxanthin could be largely boosted in *P. rhodozyma* through adding some chemical elicitors (ethanol and acetic acid). The yeast *Xanthophyllomyces dendrorhous* has been reported to be a suitable producer for producing astaxanthin (Buzzini et al. 2007).

Many studies have shown that the genus *Rhodotorula* and *Sporobolomyces* are excellent producers for specific carotenoids (Maldonade et al. 2008). By regulating the medium and culture conditions, *Rhodotorula* spp. can produce various carotenoids including β-carotene, torulene and torularhodin (Aksu and Eren 2007; Davoli et al. 2004; Sarada et al. 2002; Tinoi et al. 2005).

For instance, the most suitable temperature for the biomass and the yield of *β*-carotene by *Rhodotorula glutinis* was reported to be 30 °C (Malisorn and Suntornsuk 2009). Shaking rate is another critical factor which plays important role in both the cell growth and carotenoids production. In the case of *R. glutinis*, the shaking rate was critical. When the shaking rate was too low, the nutrients was not feasible for the cell growth. However, too high shaking rate was detrimental for the cell viability (Tinoi et al. 2005). Metal ions and salts are also important factors for the biosynthesis of carotenoids in yeast (Mata-Gomez et al. 2014). Some kinds of specific chemical elicitors have demonstrated their important roles in the carotenoids production by *Rhodotorula* spp. One powerful evidence was mevalonic acid, which could significantly stimulate the yield of carotenoids when added at proper concentration (Valduga et al. 2008).

6.1.2.3 Bacteria

Recently, more and more bacteria strains have been regarded as potential producers for carotenoids (Nasri Nasrabadi and Razavi 2010a, b). It was reported that yellowish-orange carotenoids was produced by *Chryseobacterium artocarpi*. The production of carotenoids was significantly enhanced about 7.2-fold in a 50-L bioreactor using Box Behnken design and RSM analysis (Venil et al. 2015). Immobilization approach was successfully applied for the production of zeaxanthin using the fluidized bed bioreactor. The maximal zeaxanthin concentration of 3.16 g/L was achieved with *Flavobacterium* sp. immobilized cells using orthogonal experimental design, which was tenfolds higher than previously reported yield (Chávez-Parga et al. 2012).

6.1.3 Function and Application

Carotenoids show two major functions in photosynthesis: (1) as accessory components in light harvesting, and (2) as photoprotectors against oxidative stress (Delgado-Vargas et al. 2000). For instance, xanthophylls are produced as an adaptative function, protecting the photosynthetic apparatus from stress conditions (Phillips et al. 1995). Carotenoid functions are tightly connected with their associated proteins, which are mainly membranal and hydrophobic. These proteins are usually bound carotenoids through noncovalent bonds (Gottfried et al. 1991).

The photoprotective function is suggested to link with the antioxidant activity of carotenoids by in vivo and in vitro studies. Moreover, it was found that there is a close relationship between the antioxidant activity and the chemical structure of different carotenoids. For instance, canthaxanthin and astaxanthin display better antioxidant activity than β-carotene or zeaxanthin. It is believed that the number of double bonds, keto groups, and cyclopentane rings are important parameters for the antioxidant activity of carotenoid. This finding is useful for the selection of carotenoids as food antioxidant (Nielsen et al. 1996). Carotenoids are also important for the integrity and fluidity of cell membranes. It is essential for the cell viability when exposed to stress conditions (Camejo et al. 2006). Carotenoids have been regarded as beneficiary substances in age-related diseases, against certain kinds of cancer (in especial lung cancer), strokes, macular degeneration, and cataracts (Delgado-Vargas et al. 2000).

Carotenoids can be widely applied in the area of colorants, nutraceuticals, cosmetics, and feed (Ye et al. 2008). At present, β-carotene and astaxanthin are two of the most popular carotenoids in the global market, which occupied nearly half of the carotenoid market. Other attentions are mainly paid on lutein (with zeaxanthin), lycopene, and canthaxanthin (Gong and Bassi 2016). Astaxanthin not only shows distinguish antioxidant activity, but also exhibits significant anti-inflammatory activity when supplied with aspirin (Li et al. 2011). β-carotene has been widely used in preventing night blindness and cataract. Moreover, it has been demonstrated to improve the immune system (Dufossé et al. 2005).

Lutein has gained increasing attentions as it is beneficial for preventing some kinds of eye diseases such as cataract and macular degeneration (Gong and Bassi 2016). Another important carotenoid is fucoxanthin, which can be supplied as an anti-obesity functional food. It has also exhibited anti-cancer and anti-inflammatory activities (Heo et al. 2010). Although not a major market occupier, canthaxanthin has been reported to prevent some blood disorder diseases (Clinton 1998).

6.2 Lycopene

Lycopene is a tetraterpenoid with C_{40} backbone and 13 double bonds, including 11 conjugated and 2 unconjugated double bonds, as shown in Fig. 6.3, which make contribution for its sensitivity to temperature or light (Shi et al. 2008). As a carotenoid pigment, lycopene is a potent antioxidant and food colorant naturally produced by tomatoes with content of 3–14 mg/100 g (Story et al. 2010). Lycopene is one of the most powerful singlet oxygen quenchers and plays a vital role in antioxidant, anticancer, prevention of diabetes, cardiovascular protection (Rissanen et al. 2002), and alleviation of osteoporosis (Rao et al. 2007). Therefore, lycopene has been broadly applied in food, medicine, and cosmetic industries, with increasing global market requirement (Su et al. 2018). Lycopene was originally derived from tomatoes and other red fruits, followed by chemical synthesis. However, the source availability and intensive labor limited the large-scale extraction from plant materials, while chemical synthesis confronted with safe, economic and environmental issues. Under this background, microbial fermentation has gained a growing number of interests as an economic, environmentally friendly, and sustainable technique (Hernández-Almanza et al. 2016).

6.2.1 *Metabolic Pathways for Producing Lycopene*

As a terpene compounds, lycopene is derived from two common precursors in living organisms: isopentenyl pyrophosphate (IPP) and its isomer dimethylallyl diphosphate (DMAPP), which could be synthesized through either mevalonate pathway (MVA) or 2-methyl-D-erythritol-4-phosphate pathway (MEP) (Ma et al.

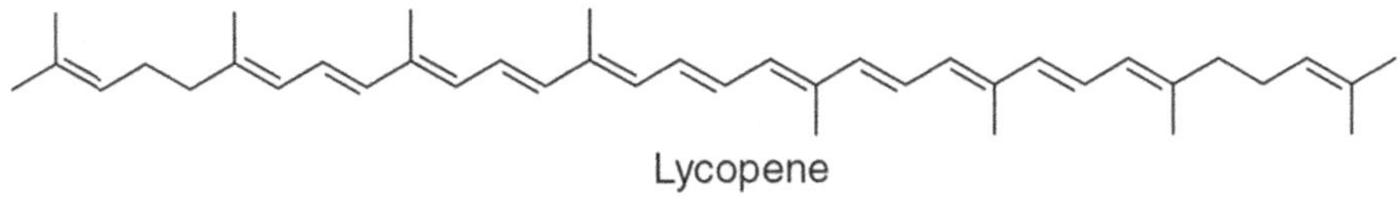

Fig. 6.3 Chemical structure of lycopene

2016). The MVA pathway is generally found in eukaryotes, which firstly convert acetyl-CoA to 3-hydroxy-3-methyl glutaryl-CoA (HMG-CoA) and subsequently synthesize mevalonic acid (MVA) (Fig. 6.2). After two steps of kinase reaction, MVA is converted into MVA-5-PP and then decarboxylation produces the precursor substance IPP, subsequently isomerized to generate DMAPP.

Meanwhile, the MEP pathway is commonly found in most bacteria, some eukaryotic parasites, and plastids of plant cells (Boucher and Doolittle 2000) as a source of IPP and DMAPP (Fig. 6.2). In general, with the help of enzyme 1-deoxy-D-xylulose-5-phosphate synthase (DXS), pyruvate and D-glyceraldehyde 3-phosphate (GAP) are firstly condensed to 1-deoxy-D-xylulose 5-phosphate (DXP), which is then converted to MEP by the enzyme 1-deoxy-D-xylulose-5-phosphate reductoisomerase (DXR). Through three successive enzymatic reactions, MEP is converted to methylerythritol 2,4-cyclodiphosphate (ME-cPP), a cyclic intermediate to form 1-hydroxy-2-methyl-2-(E)-butenyl-4-diphosphate (HMBPP) in the next step. HMBPP is then reduced to the precursors, IPP and DMAPP (Banerjee and Sharkey 2014).

The following steps for lycopene comprise three successive condensations from DMAPP to from geranyl pyrophosphate (GPP), farnesyl pyrophosphate (FPP) and finally geranylgeranyl pyrophosphate (GGPP) with a molecule of IPP in each step. Two molecules of GGPP condense to form phytoene, which is then desaturated to produce lycopene (Hernández-Almanza et al. 2016). Lycopene could be further converted to cyclic carotenoids by cyclization (Srivastava and Srivastava 2015). Therefore, the commonly used method for accumulation of lycopene was the addition of different inhibitors to prevent the biosynthesis of β-carotene. For example, bacterium *Dietzia natronolimnaea* HS-1 has been fermented in beet molasses to produce up to 8.26 mg/L of lycopene by adding enzyme inhibitors, including imidazole, nicotinic acid, piperidine, pyridine and triethylamine (within the range 0–50 ppm) (Nasri Nasrabadi and Razavi 2010a, b).

Inhibitors of lycopene cyclase have also been employed in culture of *Rhodotorula* strains to increase lycopene production. Yields of lycopene could be increased to 77% of total carotenoids by adding 20 mM of nicotine to the culture medium of *R. glutinis* and *R. rubra* (Squina and Mercadante 2005). Imidazole has been added to the fermentation medium of *R. glutinis* YB-252 in a concentration of 250 mg/L for improvement of the lycopene production up to 6.82 mg/L (Hernández-Almanza et al. 2014).

Blakeslea trispora is the widely utilized microorganism to achieve industrial production of lycopene by adding different inhibitors, as well as optimizing culture medium compositions and growth parameters. The increment of lycopene could be up to 113% by adding imidazole or pyridine in culture medium of *B. trispora* F-816 (+) and F-744 (−) at a concentration of 0.8 g/L (López-Nieto et al. 2004). The other study has increased the yields of lycopene to 779 ± 3 mg/L in fermentation of *B. trispora* NRRL 2895 and 2896 after 96 h of fermentation, with the addition of both vitamin A acetate at the beginning of the fermentation and piperidine at 500 mg/L (after 48 h) (Choudhari et al. 2008). Furthermore, with addition of 2-isopropylimidazole (300 mg/L) as lycopene cyclase inhibitor and ketoconazole

(20 mg/L) as ergosterol synthesis inhibitor, the content of lycopene from *B. trispora* could be up to 18.9 mg/g and 288.7 mg/L, respectively (Wang et al. 2014). Although the present production of lycopene is mainly achieved by adding cyclase inhibitors to fermentation of *B. trispora*, these inhibitor substances are expensive and may cause food safety issues.

6.2.2 Metabolic Regulation for Lycopene Production

With advances in metabolic engineering, metabolic regulation has been widely applied in the biotechnological production of lycopene within various microorganisms as shown in Table 6.1. The improvement of lycopene yields through metabolic engineering was mainly achieved by genetically engineering microorganisms. A plasmid carrying the lycopene synthesis gene should be firstly constructed and transformed into the host microorganisms to affect the biosynthesis of lycopene (Wang et al. 2015).

E. coli has been commonly used as microbial cell factory for the synthesis of lycopene by overexpressing the *idi*, *dxs*, *ispD*, and *ispF* genes in MEP pathway to elevate isoprenoid accumulation (Lv et al. 2013), or expressing heterologous *crtE*,

Table 6.1 Production of lycopene by engineered microorganisms

Microorganism	Metabolic engineering approaches	Carbon source	Lycopene production	References
M. circinelloides	Disruption of negative regulator gen crgA.	Saccharose	54 mg/L	Nicolás-Molina et al. (2008)
P. pastoris	Plasmid pGAPZB-EbI∗ and pGAPZB-EpBpi∗p	Glucose	73.9 mg/L	Bhataya et al. (2009)
E. coli	Genes of *B. licheniformis* as the host (iBl and iEc)	Citrate	198 mg/g	Rad et al. (2012)
E. coli	Triclosan-induced chromosomal evolution	Glucose	33.43 mg/g	Chen et al. (2013)
E. coli	Genes crtE, crtB and CrtI of *P. ananatis*, plasmid pMH1 and T7 promoter.	Glycerol	1234 mg/L	Zhu et al. (2015)
E. coli	Engineering of global regulator cAMP receptro protein	Glucose	18.48 mg/L	Huang et al. (2015)
E. coli	*E. coli* DH5α containing the plasmid pTrc99A-EBI	Glycerol	925 mg/L	Xu et al. (2018)
Y. lipolytica	Adjusting the copy numbers of crtE, crtB and crtI and overexpressing AMPD	Glucose	60 mg/g	Zhang et al. (2019)
S. cerevisiae	Overexpressed key genes associated with fatty acid synthesis, TAG production, and *ole1* and deletion of *fld1*	Glucose	2370 mg/L	Ma et al. (2019)

crtB and *crtI* to convert FPP into lycopene (Xu et al. 2018). The recombinant *E. coli* strain 99DH produced 925 mg/L of lycopene in 2 × YT medium with glycerol as carbon source under bright conditions (Xu et al. 2018). *Yarrowia lipolytica* was considered to be a convincing producer of lycopene, or other carotenoids. Zhang et al. has enhanced lycopene content to highest level (60 mg/g cdw) in *Y. lipolytica* by altering the copy numbers of three heterologous lycopene biosynthesis genes (crtE, crtB and crtI) and overexpressing AMP deaminase-encoding gene (AMPD) in a 5-L fermenter cultivation (Zhang et al. 2019).

A nature-inspired strategy has recently been established by improving lipid oil–triacylglycerol (TAG) metabolism to increase lycopene accumulation of *Saccharomyces cerevisiae* (Ma et al. 2019). The production of lycopene was increased to the highest level in *S. cerevisiae*, 2.37 g/L and 73.3 mg/g cell dry weight (cdw), by overexpressing key genes of fatty acid synthesis, TAG production, and fatty acid desaturase (OLE1), followed by deletion of Seipin (FLD1).

With benefit of metabolic engineering, microbial production of lycopene is an environmentally friendly, sustainable, and economical production approach. These strategies mainly involve the introduction of heterologous pathways, reengineering of regulatory networks through rational or random approaches, cofactor tuning, and transport of the accumulation of toxic intermediates. Moreover, the optimization of fermentation conditions is an efficient method to improve the yields in large-scale fermentation by using recombinant microbes.

6.3 Anthocyanins

Anthocyanins are an important sub-class of flavonoids in plants with the role of pigments, antioxidants, and antimicrobials (Chouhan et al. 2017; Zha and Koffas 2017). Owing to their diverse colors and nutritional characters, anthocyanins are extensively applied as food colorants (Aberoumand 2011). Most anthocyanins have been reported for their bioactivities and pharmaceutical functions. The polyphenol structure gives some anthocyanins strong ability for absorbing visible and ultraviolet (UV) spectra (Giusti and Wrolstad 2001), which could be used to protect human skin from aging and UV-induced damage (Gonsalves et al. 2016). With powerful antioxidant potency, anthocyanins are compelling in the prevention and treatment of some diseases. The bioactivities were mainly proved through *in vitro* and *in vivo* trials on diseases of antitumor, anti-cardiovascular, anti-neurodegenerative, anti-obesity and anti-diabetes (Kolakul and Sripanidkulchai 2017; Lila et al. 2016).

Anthocyanins widely exist in various tissues of plants, which are primarily used to obtain anthocyanins by extraction and purification (Mora-Pale et al. 2014). These traditional methods have long been utilized mainly due to diverse sources of availably cheap feedstocks. Advances on genetic engineering stimulated the production of anthocyanins through metabolic regulation of anthocyanin biosynthesis in plants (Shi and Xie 2014; Zhang et al. 2014a). However, anthocyanins sourced from plants still have several limitations, such as long growth cycle, depending on season and environment, sophisticated extraction and purification techniques (Santos-Buelga and González-Paramás 2019; Zhu et al. 2017).

With elucidation of biosynthetic pathways on anthocyanins (Fig. 6.4), microbial production provides a dramatic alternative to plant extraction. At present, various flavonoid compounds, including anthocyanins, have been produced by engineered microbes (Pandey et al. 2016). Meanwhile, some non-natural anthocyanins have been generated by feeding of specific substrates in microbial production (Pontrelli et al. 2018). *E. coli* is the most commonly studied microbial platform for production of natural flavonoids, including naringenin, kaempferol, and catechin (Jones et al. 2016). Afterwards, the range of heterologous host has been expanded to *Lactococcus lactis* (Solopova et al. 2019), *Saccharomyces cerevisiae* (Levisson et al. 2018), and *Corynebacterium glutamicum* (Zha et al. 2018). The metabolic engineering strategy on improving microbial production of anthocyanins mainly classifies in four aspects: heterologous expression of related enzymes, optimization of co-factor/co-substrate supply, construction of specific transporters and optimization of the production condition (Lim et al. 2015; Zha and Koffas 2017). The production of several anthocyanin compounds by microorganism is listed in Table 6.2.

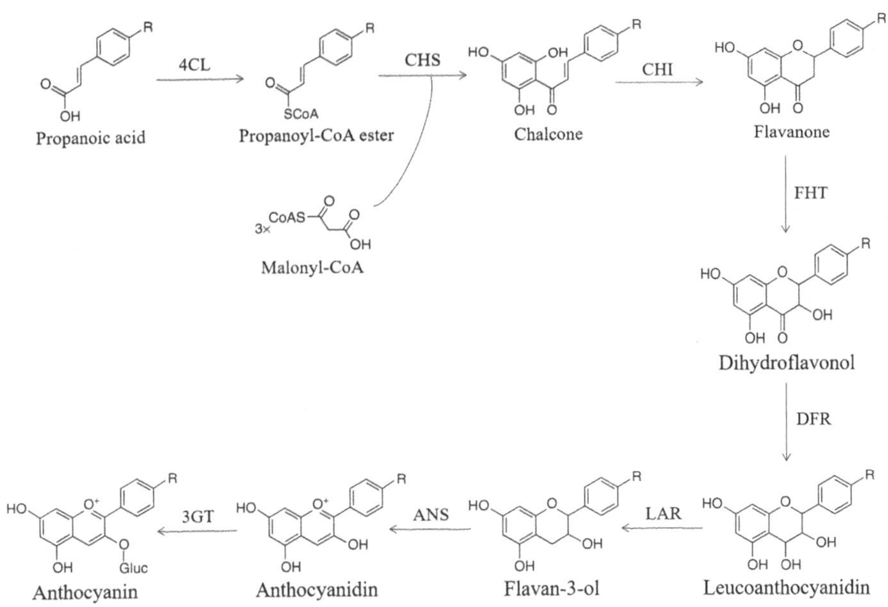

Fig. 6.4 Biosynthetic pathway of flavonoid compounds. *4CL* 4-coumarate: coenzyme A ligase, *CHS* chalcone synthase, *CHI* chalcone isomerase, *FHT* flavanone 3β-hydroxylase, *DFR* dihydroflavonol 4-reductase, *LAR* leucoanthocyanidin reductase, *ANS* anthocyanidin synthase, *3GT* UDP-glucose: flavonoid 3-O-glucosyltransferase

Table 6.2 Production of anthocyanin in engineered microorganisms

Product	Microorganism	Genetic modifications	Substrate	Yield/ μM	References
Cyanidin 3-*O*-glucoside	*E. coli* JM109	MdF3H/AaDFR/MdANS/PhF3GT	0.1 mM Eriodictyol	0.012	Yan et al. (2005)
	E. coli BL21* (DE3)	MdF3H/AaDFR/At3GT/PhANS	0.2 mM Eriodictyol	3.88	Yan et al. (2008)
		MdF3H/AaDFR/At3GT/PhANS/DuLAR	0.2 mM Eriodictyol	4.27	
		At3GT/PhANS/galU/pgm	0.75 mM Catechin	127	
		Fusion of At3GT and PhANS/galU/pgm	0.75 mM Catechin	146	
		Fusion of At3GT and PhANS/galU/pgm/ndk/ Dudg/galE/ T(inactive)	Catechin	215	Leonard et al. (2008)
		At3GT/PhANS/galU/pgm/cmk/ndk/YadH/ΔtolC	2.5 mM Catechin	260	Lim et al. (2015)
		At3GT/PhANS/cmk/ndk/ycjU	2.5 mM Catechin	421	
		At3GT/PhANS/galU/pgm/cmk/ndk/ycjU	2.5 mM Catechin	722	
	C. glutamicum GPAG	Fusion of At3GT and PhANS	1.72 mM Catechin	~89	Zha et al. (2018)
Pelargonidin 3-*O*-glucoside	*E. coli* JM109	MdF3H/AaDFR/MdANS/PhF3GT	0.25 mM Naringenin	0.012	Yan et al. (2005)
	E. coli BL21* (DE3)	MdF3H/AaDFR/At3GT/PhANS	0.2 mM Naringenin	1.34	Yan et al. (2008)
		MdF3H/AaDFR/At3GT/PhANS/DuLAR	0.2 mM Naringenin	2.09	
		Fusion of At3GT and PhANS/galU/pgm	0.75 mM afzelechin	168	

		Fusion of At3GT and PhANS/galU/pgm/ndk/ Dudg/galE/ T(inactive)	Afzelechin	241	Leonard et al. (2008)
Pelargonidin 3-*O*-6″-*O*-malonylglucoside	*E. coli* BL21* (DE3)	MdF3H/AaDFR/At3GT/PhANS/DuLAR/ Dv3MaT	0.2 mM Naringenin	0.18	Yan et al. (2008)
Cyanidin 3-*O*-6″-*O*-malonylglulcoside	*E. coli* BL21* (DE3)	MdF3H/AaDFR/At3GT/PhANS/DuLAR/ Dv3MaT	0.2 mM Eriodictyol	0.21	
Peonidin 3-*O*-glucoside	*E. coli* BL21* (DE3)	MBP-At3GT/MBP-PhANS/VvAOMT/ MetJ↓(CRISPRi)	3.44 mM Catechin	112	Cress et al. (2017)

6.3.1 Heterologous Expression of Plant-Derived Enzymes

Selection of plant enzymes from diverse species and heterologous expression in microorganisms is a primary method to increase the yield of anthocyanins (Zhao et al. 2015), because gene orthologs encode enzymes can improve yields of the target components. Several enzymes have been screened to produce anthocyanins, such as ANS from *P. hybrida* to produce cyanidin (Yan et al. 2008), 4-coumaroyl-CoA ligase (4CL), CHS and CHI to produce naringenin (Jones et al. 2016). To heterologously express these enzymes in microorganisms, engineering modification are generally required for relevant encoding genes. In microbial production of quercetin, a new hydroxylase was constructed to catalyze the formation of quercetin in *E. coli* by deletion and replacement of several groups on P450 F3′5′H and fusion with a shortened P450 reductase from *Catharanthus roseus* (Leonard et al. 2006).

In some cases, the contributing enzymes from different plant sources should be combined to perform their roles in successive steps (Jones et al. 2016). A chimerical enzyme has been constructed by fusing F3GT from *Arabidopsis thaliana* to the *N*-terminus of ANS from *P. hybrida* with a pentapeptide linker (Yan et al. 2008). The chimerical enzyme exhibited higher yield of cyanidin 3-*O*-glucoside than individual ANS & F3GT.

6.3.2 Regulation of Co-factor/Co-substrate Supply

To perform the enzymatic reaction for anthocyanin production, co-factors and/or co-substrates are normally required for electron transfer and enzyme activation/stabilization. For instance, malonyl-CoA is a crucial co-factor for flavonoid biosynthesis. A study has overexpressed the enzyme acetyl-CoA carboxylase (ACC) to increase the conversion rate from acetyl-CoA to malonyl-CoA in the fatty acid biosynthesis pathway (Miyahisa et al. 2005). The productions of naringenin and pinocembrin were up to a threefold and fourfold increments when using tyrosine and phenylalanine as substrates, respectively.

UDP-glucose is an indispensable co-factor for biosynthesis of glycosylated anthocyanins. The frequently-used strategy is to overexpress the biosynthetic genes. In general, relative genes on UDP-glucose biosynthesis from orotic acid (*pyrE*, *pyrR*, *cmk*, *ndk*, *pgm*, *galU*) could be selected to overexpress in microbes. By using *E. coli* strain without genes *galE* and *galT* (converting UDP-glucose to UDP-galactose) and gene *udg* (converting UDP-glucose to UDP-glucuronate), anthocyanin production was observed additional improvement with overexpression of *ndk* and supplementation of orotic acid (Leonard et al. 2008). In addition, S-Adenosyl-L-methionine and sodium ascorbate are necessary co-substrates for overproduction of certain anthocyanins (Cress et al. 2017).

6.3.3 Construction of Specific Transporters

Some biosynthetic products accumulate in the microorganisms and often cause toxic and side effect to host strains, which limiting their high-yield production. Therefore, transport of theses metabolites to extracellular matrix is considered as an effective method for continuous biosynthesis. This strategy has been attempted by several studies to improve the production of anthocyanin compounds. For example, YadH, a efflux pump on cyanidin 3-*O*-glucoside, has been characterized and over-expressed to increase over 15% yield of anthocyanins (Lim et al. 2015). The other efflux pump TolC has been deleted to promote the production of cyanidin 3-*O*-glucoside, probably due to its regulation of catechin secretion. Furthermore, researchers have proposed to combine transporters from plant with that in microbe hosts (Zha and Koffas 2017).

6.3.4 Optimization of Culture Conditions

Similar with other metabolites, production of anthocyanins is greatly affected by the cultural conditions, including induction time-point, temperature, and pH. For example, the stationary phase was found as the optimal for cyanidin 3-*O*-glucoside production in engineered *E. coli*, which promoting the production of anthocyanin (Lim et al. 2015). In a study on afzelechin fermentation, the yield was up to 22.9 mg/L at induction temperature of 20 °C, while the yield was only 6.1 mg/L at 10 °C (Jones et al. 2016). Additionally, pH is considered as one of the most important factors affecting anthocyanins production, because the synthesized anthocyanins is extremely unstable under the intracellular pH (~7) for microorganisms. Based on this problem, the host cells are firstly cultured to certain growth stage at pH 7, and then inoculated to fresh medium at pH~5 to reduce anthocyanin decomposition. This method has elevated the yield of cyanidin 3-*O*-glucoside in *E. coli* by ~15-fold to 38.9 mg/L against traditional production (2.5 mg/L) (Yan et al. 2008).

Microbial production of anthocyanin is an intricate and time-consuming process. Although the above factors have been identified and studied, there is still a variety of unsolved problems. Therefore, future work should be done by evaluating multiple aspects simultaneously.

6.4 *Monascus* Pigments

Monascus pigments have been traditionally used in Asian countries, which are produced by the fermentation of edible and medicinal fungi *Monascus* spp. *Monascus* pigments have been produced in large scale and successfully applied as colorants and additives in food industry (Srianta et al. 2014).

The genus *Monascus* belonging to the family *Monascaceae* and to the class *Ascomyceta*, whose most important characteristic is the ability to produce natural pigments of polyketidic structure (Jůzlová et al. 1996). Among *Monascus* species, *M. purpureus, M. ruber* and *M. pilosus* are the most common species used in industrial applications (Vendruscolo et al. 2016).

Monascus spp. can produce three kinds of pigments (yellow, orange, and red), which are determined by the used species and the employed cultivation conditions. Chemical structures of six well-known *Monascus* pigments were shown in Fig. 6.5, including two red ones (rubropunctamine and monascorubramine), two orange ones (rubropunctatin and monascorubrin) and two yellow ones (monascin and ankaflavin). The yellow, orange, and red pigments are traditionally determined by the maximum absorbance wavelength at 330–450, 460–480, and 490–530 nm, respectively. To the best of our knowledge, more than 94 *Monascus* pigments have been reported up to now, which include 42 red, 8 orange, and 44 yellow pigments (Chen et al. 2017; Feng et al. 2012).

The discovery of citrinin (a type of mycotoxin) production initiated a controversy over the safety of *Monascus* pigments (Blanc et al. 1995). Hence, it is critical to eliminate or reduce the concentration of citrinin in *Monascus* products. The conventional methods for controlling citrinin concentration are to optimize the fermentation conditions and screen citrinin-free strains. Advanced approaches have been developed like blocking or suppressing the citrinin biosynthetic pathway by genetic modification (Chen et al. 2015; Kang et al. 2014).

6.4.1 Biosynthetic Pathway

Although *Monascus* pigments have been produced in commercial scale and widely applied in food industry, the biosynthetic pathway of *Monascus* pigments are still remained incomplete. As shown in Fig. 6.6, it has been generally recognized that the biosynthetic pathway of *Monascus* pigments initiates from the fatty acid synthase pathway and polyketide synthase pathway, which generate β-ketoacid and the chromophore, respectively. Then, the orange pigments rubropunctatin and monascorubrin

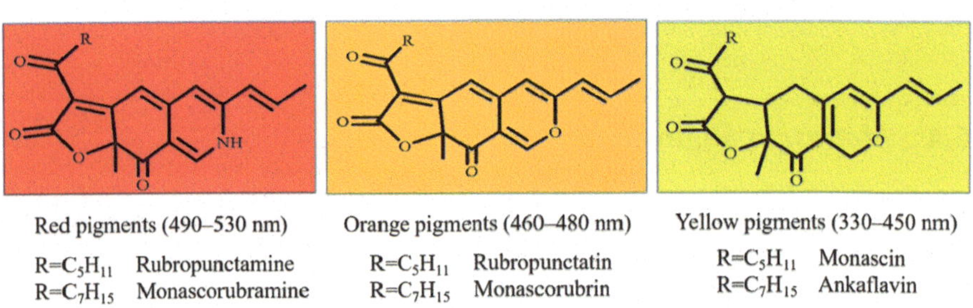

Fig. 6.5 Chemical structures of six well-known *Monascus* pigments

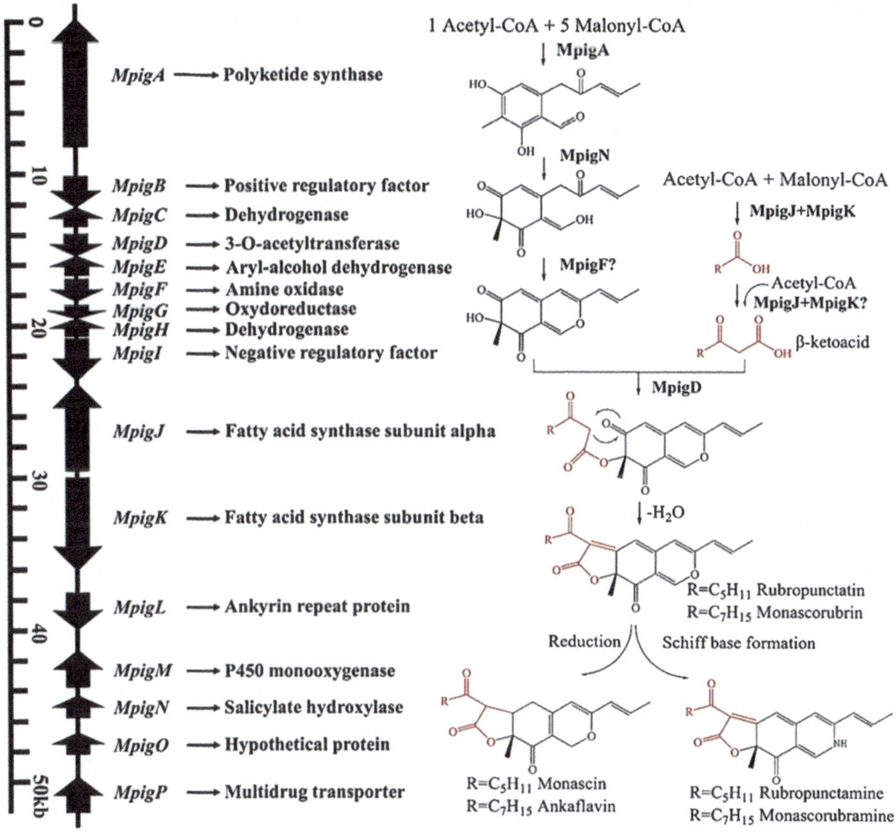

Fig. 6.6 The proposed biosynthetic pathways of *Monascus* pigments and the related gene cluster. (Chen et al. 2015)

are biosynthesized through the esterification reaction ofβ-ketoacid and the chromophore. Afterwards, the *Monascus* yellow pigments monascin and ankaflavin are formed by the reduction of the orange pigments. Comparatively, amination of the orange pigments with NH_3 leads to the red pigments rubropunctamine and monascorubramine. The gene cluster and their related functions for the biosynthesis of Monascus pigments were also indicted in Fig. 6.6 (Chen et al. 2015).

Nevertheless, the generally recognized biosynthetic pathway still remained assumptions based on chemical principles, which may not accurately describe the actual situation. Besides the generally recognized pathway, some studies suggested that the yellow *Monascus* pigments were the primary products of the shunt pathway, and then the orange pigments were formed by enzymatic transformation. Subsequently, red pigments were biosynthesized from the orange pigments through non-enzymatic conversion with amines (Chen et al. 2017). Further investigations are still required to illuminate the biosynthetic pathway of *Monascus* pigments.

6.4.2 Fermentation & Production

The color value and color tune of *Monascus* pigments are depended on the fermentation mode, the nutrients and the operational conditions.

6.4.2.1 Fermentation Mode

The two main approaches for producing *Monascus* pigments are solid-state fermentation (SSF) and submerged fermentation (SMF). SSF are traditional method that the *Monascus* seed culture are inoculated on the solid substrates (e.g. rice). After that, the fermented mixture are applied as food colorant or further used for pigment extraction. In contrast, SMF are relatively modern process that the fermentation process can be conducted in large scale bioreactors, in which the final product of *Monascus* pigments must be extracted (Feng et al. 2012; Liu et al. 2010).

SSF is the process conducted on solid substrate with little or no free water. The solid substrates can be grain substrate (rice, millet, barley, wheat, etc.) or the low-cost agricultural by-product (Babitha et al. 2007; Kantifedaki et al. 2018; Kaur 2015; Zhang et al. 2018). In general, the advantages of SSF are low water demand and sterility requirement, low cost, low product repression and high concentration of the end-product. However, the detailed parameters such as temperature, pH and aeration are difficult to control. Compared to SSF, use of SMF for the production of *Monascus* pigments has been widely applied in the food industry, due to the advantages such as easy for operation at a large scale and avoiding contamination (Vendruscolo et al. 2016).

6.4.2.2 Key Factors for the Production of *Monascus* Pigments

Due to the lack of knowledge and of scalable bioreactor technologies, SSF are less considered as the main approach for the production of *Monascus* pigments. Therefore, the factors for efficient production of *Monascus* pigments by SMF have been extensively investigated.

The nutrients (carbon source, nitrogen source and minerals, etc.), and the operational conditions (pH, temperature, dissolved oxygen, light, etc.) exert critical effects on the quantity and quality of *Monascus* pigments in SMF (Buhler et al. 2015; Lv et al. 2017, 2018; Shi et al. 2015; Yang et al. 2015).

Many kinds of carbon sources can be used for *Monascus* pigments in SMF, including starch, oligo- and polysaccharides, various monosaccharides, and even glycerol and ethanol, etc. Different carbon sources have different and complex effects on both *Monascus* growth and its pigments production. The nutrients not only affect the yield of Monascus pigments, but also the intracellular or extracellular distribution of the pigments. For instance, the production of intracellular/extracellular *Monascus* yellow pigment were affected by the nitrogen sources. In addition, some kinds of food industry wastes have been applied as nutrients for the efficient biosynthesis of *Monascus* pigments. For instance, a brewery waste hydrolysate, brewer's spent grain was used for the production of red pigments in SMF of *Monascus purpureus* fermentation system (Silbir and Goksungur 2019). Another low-cost substrate sweet potato could be used in culture broth for producing water-soluble *Monascus* red pigments (Srivastav et al. 2015).

Culture conditions can significantly influence the production of *Monascus* pigments. For example, it was found that low pH or amino ion content were beneficial for the production of yellow and orange pigments. Comparatively, red pigments were easily accumulated at neutral pH or higher nitrogen concentration. Light intensity has been recognized as a critical parameter for the biosynthesis of various *Monascus* pigments. When exposed to direct illumination, the cell growth and pigment yield were both inhibited during the cultivation of *M. ruber*. In contrast, the production of *Monascus* pigments can be stimulated by red light or in dark (Buhler et al. 2015). Another study indicated that the colony morphology, the composition and permeability of the *Monascus* mecelial cell wall were obviously influenced by blue light in static liquid culture, suggesting that blue light may be beneficial for the secretion of pigments from aerial mycelia to culture broth (Zhang et al. 2017). The differential gene expression was related to the light-cause *Monascus* pigment production. It was found that although the cell growth was inhibited by low intensity of blue light (500 lx), increased the production of *Monascus* pigment was stimulated through upregulation of *MpigA*, *MpigB*, and *MpigJ* genes expression (Wang et al. 2016a).

Other studies also suggested the relationship between fermentation conditions, mycelial morphology and the biosynthesis of *Monascus* pigments. Microscopic images revealed that a high *Monascus* yellow pigment yield was associated with the formation of freely dispersed small mycelial pellets with shorter, thicker and multi-branched hyphae. Furthermore, the hyphal diameter was suggested to be highly correlated with the prodcution of the *Monascus* yellow pigments (Lv et al. 2017).

Some newly developed technology were applied for producing highly colored *Monascus* pigments. For example, in order to alleviate the phenomenon of product inhibition, a novel integrated fermentation system consisting of surfactant and *in situ* extractant was established, in which the production of *Monascus* yellow pigments was greatly improved. Critical factors such as alleviating the product inhibition, increasing the membrane permeability, changing the hyphal morphology, and influencing the cell activity have been suggested as the underlying mechanisms (Lv et al. 2018). Other new method such as Micelle aqueous solution was applied for facilitating the intracellular conversion of *Monascus* orange pigments to yellow pigments through adding nonionic surfactant in the culture medium (Xiong et al. 2015). Moreover, *Monascus* orange pigments was successfully produced by using the resting cells in submerged fermentation. The method exhibited several advantages over the normal fermentation process using growing cells, such as non-sterilization operation, high cell density, high productivity, and high product yield (Wang et al. 2016b).

6.4.3 Function & Application

Monascus pigments have been widely applied in Asian countries as food colorants in different kinds of traditional foods (yoghurt, sausages, tofu, hams, meats, etc.). Nowadays, the application has been expanded to other areas including textile, cosmetic, and pharmaceutical industries (Vendruscolo et al. 2016).

Besides the application as coloring agents, *Monascus* pigments possess notable functions such as antimicrobial activity, antioxidant activity, and anticancer activity to a certain extent. The biological activities of *Monascus* pigments vary according to their diverse component structures. It was found that the *Monascus* orange pigments exhibited antimicrobial activity against *Staphylococcus aureus* while the red pigments against *S. aureus* and *E. coli* (Vendruscolo et al. 2014).

It was found that the *Monascus*-fermented soybean can be supplied as functional food additives due to its antioxidant activity (Pyo and Lee 2007). Researches proved that the crude extracts of *Monascus* rice from solid-state fermentation possessed potential anti-mutagenic activities (Hsu and Pan 2012). Two *Monascus* yellow pigments monascin and ankaflavin have been proved to improve memory and learning abilities of Ab40-icv-infused rats through suppressing Alzheimer's disease risk factors (Lee et al. 2015).

Based on their beneficial effects, *Monascus* pigments are promising for the development as functional additives in dietary food. Moreover, the *Monascus* pigments may be selected as an inhibitor to preserve food products where a natural preservative is required.

6.5 Conclusion and Perspectives

Nowadays, there is a significant increasing demand on safe and natural food pigments. Toxicological concerns on certain synthetic pigments and the increased consumer awareness on health have promoted the food colorant market for use of natural pigments.

Natural pigments produced by microbial fermentation possesses high economic value and has attracted more and more attention due to the advantages of productivity and versatility. There are many advantages using microbial fermentation for natural pigment production, including short harvest period, efficient process control, easily production on low-cost substrates, numbers of bioactive components, and completely safe under specific conditions.

However, there are several limitations in the natural pigments production by microbial fermentation. (1) It is lack of microorganism strains with high pigment concentration; (2) Current fermentation process is not efficient in terms of final yield; (3) The cost and price is still too high for common application in food industry; (4) Some natural pigments produced oriented from microorganisms are unstable at high temperatures, strong illumination, presence of oxygen, metal ions, and pH changes, etc.; (5) The application area of natural pigment is limited in food product.

Correspondingly, to overcome the limitation listed above and effectively promote the production of natural pigments by microbial fermentation, some useful suggestions are proposed as follows: (1) The screening of microorganism strains can be initiated by traditional or novel mutation methods. However, in order to overcome the time-, cost- and labor-intensive processes of strain development, high throughput screening approaches are useful for obtaining excellent microbial strain for industrial application; (2) Because the fundamentals of the biosynthesis and metabolism pathways of microbial pigments remain unclear, methods for efficiently producing high yield of microbial pigments still require intensive investigation. The employment of more complex but state-of-the-art approaches of metabolic engineering and synthetic biology will contribute to the advance of this promising research area; (3) The cost reduction during industrial production and downstream processes can be overcome by optimizing the fermentation conditions with intelligent control technology, as well as the use of low-cost substrate as nutrients for microbial growth and pigment production; (4) Structure modification is useful for improving the stability of natural pigments. Besides, combination or encapsulation with other food-grade materials can effectively enhance the stability of microbial pigments; (5) It is possible to verify the potentiality of various natural pigments that beyond the application in food industry, but also use these microbial biomolecules in the functional supplementary, pharmaceutical, textile, and cosmetic industries.

It is promising that microbial pigments have quite good future prospects for robust industrial production of various colors. Microorganisms can serve as sustainable cell factories for producing natural pigments that are economical and human-friendly.

References

Aberoumand A. A review article on edible pigments properties and sources as natural biocolorants in foodstuff and food industry. World J Dairy Food Sci. 2011;6:71–8.

Aksu Z, Eren AT. Production of carotenoids by the isolated yeast of Rhodotorula glutinis. Biochem Eng J. 2007;35(2):107–13.

Almeida ER, Cerda-Olmedo E. Gene expression in the regulation of carotene biosynthesis in Phycomyces. Curr Genet. 2008;53(3):129–37.

Babitha S, Soccol CR, Pandey A. Solid-state fermentation for the production of *Monascus* pigments from jackfruit seed. Bioresour Technol. 2007;98(8):1554–60.

Banerjee A, Sharkey TD. Methylerythritol 4-phosphate (MEP) pathway metabolic regulation. Nat Prod Rep. 2014;31(8):1043–55.

Begum H, Yusoff FM, Banerjee S, Khatoon H, Shariff M. Availability and utilization of pigments from microalgae. Crit Rev Food Sci Nutr. 2016;56(13):2209–22.

Bhataya A, Schmidt-Dannert C, Lee PC. Metabolic engineering of *Pichia pastoris* X-33 for lycopene production. Process Biochem. 2009;44(10):1095–102.

Blanc PJ, Laussac JP, Le Bars J, Le Bars P, Loret MO, Pareilleux A, Prome D, Prome JC, Santerre AL, Goma G. Characterization of monascidin A from *Monascus* as citrinin. Int J Food Microbiol. 1995;27(2–3):201–13.

Boucher Y, Doolittle WF. The role of lateral gene transfer in the evolution of isoprenoid biosynthesis pathways. Mol Microbiol. 2000;37(4):703–16.

Buhler RM, Muller BL, Moritz DE, Vendruscolo F, de Oliveira D, Ninow JL. Influence of light intensity on growth and pigment production by *Monascus ruber* in submerged fermentation. Appl Biochem Biotechnol. 2015;176(5):1277–89.

Buzzini P, Innocenti M, Turchetti B, Libkind D, van Broock M, Mulinacci N. Carotenoid profiles of yeasts belonging to the genera *Rhodotorula*, *Rhodosporidium*, *Sporobolomyces*, and *Sporidiobolus*. Can J Microbiol. 2007;53(8):1024–31.

Camejo D, Jimenez A, Alarcon JJ, Torres W, Gomez JM, Sevilla F. Changes in photosynthetic parameters and antioxidant activities following heat-shock treatment in tomato plants. Funct Plant Biol. 2006;33(2):177–87.

Celli GB, Tan C, Selig MJ. Anthocyanidins and anthocyanins. In: Melton L, Shahidi F, Varelis P, editors. Encyclopedia of food chemistry. Oxford: Academic; 2019. p. 218–23.

Chávez-Parga MDC, Munguía-Franco A, Aguilar-Torres M, Escamilla-Silva EM. Optimization of zeaxanthin production by immobilized *Flavobacterium sp.* cells in fluidized bed bioreactor. Adv Microbiol. 2012;2(4):598.

Chemler JA, Koffas MA. Metabolic engineering for plant natural product biosynthesis in microbes. Curr Opin Biotechnol. 2008;19(6):597–605.

Chen YY, Shen HJ, Cui YY, Chen SG, Weng ZM, Zhao M, Liu JZ. Chromosomal evolution of *Escherichia coli* for the efficient production of lycopene. BMC Biotechnol. 2013;13(1):6.

Chen W, He Y, Zhou Y, Shao Y, Feng Y, Li M, Chen F. Edible filamentous fungi from the species *Monascus*: early traditional fermentations, modern molecular biology, and future genomics. Compr Rev Food Sci Food Saf. 2015;14(5):555–67.

Chen W, Chen R, Liu Q, He Y, He K, Ding X, Kang L, Guo X, Xie N, Zhou Y, Lu Y, Cox RJ, Molnar I, Li M, Shao Y, Chen F. Orange, red, yellow: biosynthesis of azaphilone pigments in *Monascus* fungi. Chem Sci. 2017;8(7):4917–25.

Choudhari SM, Ananthanarayan L, Singhal RS. Use of metabolic stimulators and inhibitors for enhanced production of beta-carotene and lycopene by *Blakeslea trispora* NRRL 2895 and 2896. Bioresour Technol. 2008;99(8):3166–73.

Chouhan S, Sharma K, Zha J, Guleria S, Koffas M. Recent advances in the recombinant biosynthesis of polyphenols. Front Microbiol. 2017;8:2259.

Clinton SK. Lycopene: chemistry, biology, and implications for human health and disease. Nutr Rev. 1998;56(2):35–51.

Coulson J. Miscellaneous naturally occurring colouring materials for foodstuff. Dev Food Colour. 1980;189:218.

Cress BF, Leitz QD, Kim DC, Amore TD, Suzuki JY, Linhardt RJ, Koffas MA. CRISPRi-mediated metabolic engineering of *E. coli* for *O*-methylated anthocyanin production. Microb Cell Factories. 2017;16(1):10.

Dan Pelah ASAE. The effect of salt stress on the production of canthaxanthin and astaxanthin by *Chlorella zofingiensis* grown under limited light intensity. World J Microbiol Biotechnol. 2004;20(5):483–6.

Davoli P, Mierau V, Weber RW. Carotenoids and fatty acids in red yeasts Sporobolomyces roseus and Rhodotorula glutinis. Prikl Biokhim Mikrobiol. 2004;40(4):460–5.

Del Campo JA, Garcia-Gonzalez M, Guerrero MG. Outdoor cultivation of microalgae for carotenoid production: current state and perspectives. Appl Microbiol Biotechnol. 2007;74(6):1163–74.

Delgado-Vargas F, Jimenez AR, Paredes-Lopez O. Natural pigments: carotenoids, anthocyanins, and betalains-characteristics, biosynthesis, processing, and stability. Crit Rev Food Sci Nutr. 2000;40(3):173–289.

Dufossé L, Galaup P, Yaron A, Arad SM, Blanc P, Chidambara Murthy KN, Ravishankar GA. Microorganisms and microalgae as sources of pigments for food use: a scientific oddity or an industrial reality? Trends Food Sci Technol. 2005;16(9):389–406.

Eilers U, Bikoulis A, Breitenbach J, Büchel C, Sandmann G. Limitations in the biosynthesis of fucoxanthin as targets for genetic engineering in *Phaeodactylum tricornutum*. J Appl Phycol. 2016;28(1):123–9.

Feng Y, Shao Y, Chen F. *Monascus* pigments. Appl Microbiol Biotechnol. 2012;96(6):1421–40.

Fernandez-Sevilla JM, Acien FF, Molina GE. Biotechnological production of lutein and its applications. Appl Microbiol Biotechnol. 2010;86(1):27–40.

Giusti MM, Wrolstad RE. Characterization and measurement of anthocyanins by UV-visible spectroscopy. Curr Protocol Food Anal Chem. 2001;00(1):F1.2.1–F1.2.13.

Gong M, Bassi A. Carotenoids from microalgae: a review of recent developments. Biotechnol Adv. 2016;34(8):1396–412.

Gonsalves J, Divya AJ, Lekha G. Study of anthocyanin content, antioxidant property, UV absorbance & SPF analysis of a few petals. J Adv Appl Sci Res. 2016;1(3):1–6.

Gottfried D, Steffen M, Boxer S. Large protein-induced dipoles for a symmetric carotenoid in a photosynthetic antenna complex. Science. 1991;251(4994):662–5.

Guedes AC, Amaro HM, Malcata FX. Microalgae as sources of carotenoids. Mar Drugs. 2011;9(4):625–44.

Gupta PL, Lee SM, Choi HJ. A mini review: photobioreactors for large scale algal cultivation. World J Microbiol Biotechnol. 2015;31(9):1409–17.

Harker M, Tsavalos AJ, Young AJ. Factors responsible for astaxanthin formation in the Chlorophyte *Haematococcus pluvialis*. Bioresour Technol. 1996;55(3):207–14.

Heo SJ, Yoon WJ, Kim KN, Ahn GN, Kang SM, Kang DH, Affan A, Oh C, Jung WK, Jeon YJ. Evaluation of anti-inflammatory effect of fucoxanthin isolated from brown algae in lipopolysaccharide-stimulated RAW 264.7 macrophages. Food Chem Toxicol. 2010;48(8–9):2045–51.

Hernández-Almanza A, Montañez-Sáenz J, Martínez-Ávila C, Rodríguez-Herrera R, Aguilar CN. Carotenoid production by *Rhodotorula glutinis* YB-252 in solid-state fermentation. Food Biosci. 2014;7:31–6.

Hernández-Almanza A, Montañez J, Martínez G, Aguilar-Jiménez A, Contreras-Esquivel JC, Aguilar CN. Lycopene: progress in microbial production. Trends Food Sci Technol. 2016;56:142–8.

Ho SH, Chan MC, Liu CC, Chen CY, Lee WL, Lee DJ, Chang JS. Enhancing lutein productivity of an indigenous microalga *Scenedesmus obliquus* FSP-3 using light-related strategies. Bioresour Technol. 2014;152:275–82.

Hsu WH, Pan TM. Monascus purpureus-fermented products and oral cancer: a review. Appl Microbiol Biotechnol. 2012;93(5):1831–42.

Huang J, Liu J, Li Y, Chen F. Isolation and characterization of the phytoene desaturase gene as a protentíal selective marker for genetic engineering of the astaxanthin-producing green alga *Chlorella zofingiensis* (Chlorophyta). J Phycol. 2008;44(3):684–90.

Huang L, Pu Y, Yang X, Zhu X, Cai J, Xu Z. Engineering of global regulator cAMP receptor protein (CRP) in *Escherichia coli* for improved lycopene production. J Biotechnol. 2015;199:55–61.

Jones JA, Vernacchio VR, Sinkoe AL, Collins SM, Ibrahim M, Lachance DM, Hahn J, Koffas M. Experimental and computational optimization of an *Escherichia coli* co-culture for the efficient production of flavonoids. Metab Eng. 2016;35:55–63.

Jûzlová P, Martinkova L, Křen V. Secondary metabolites of the fungus *Monascus*: a review. J Ind Microbiol. 1996;16(3):163–70.

Kang B, Zhang X, Wu Z, Wang Z, Park S. Production of citrinin-free *Monascus* pigments by submerged culture at low pH. Enzym Microb Technol. 2014;55:50–7.

Kantifedaki A, Kachrimanidou V, Mallouchos A, Papanikolaou S, Koutinas AA. Orange processing waste valorisation for the production of bio-based pigments using the fungal strains *Monascus purpureus* and *Penicillium purpurogenum*. J Clean Prod. 2018;185:882–90.

Kaur S. Production of microbial pigments utilizing agro-industrial waste: a review. Curr Opin Food Sci. 2015;1(1):70–6.

Kolakul P, Sripanidkulchai B. Phytochemicals and anti-aging potentials of the extracts from *Lagerstroemia speciosa* and *Lagerstroemia floribunda*. Ind Crop Prod. 2017;109:707–16.

Kyriakopoulou K, Papadaki S, Krokida M. Life cycle analysis of β-carotene extraction techniques. J Food Eng. 2015;167:51–8.

Lee C, Lin P, Hsu Y, Pan T. Monascus-fermented monascin and ankaflavin improve the memory and learning ability in amyloid β-protein intracerebroventricular-infused rat via the suppression of Alzheimer's disease risk factors. J Funct Foods. 2015;18:387–99.

Leonard E, Yan Y, Koffas MA. Functional expression of a P450 flavonoid hydroxylase for the biosynthesis of plant-specific hydroxylated flavonols in *Escherichia coli*. Metab Eng. 2006;8(2):172–81.

Leonard E, Yan Y, Fowler ZL, Li Z, Lim CG, Lim KH, Koffas MA. Strain improvement of recombinant *Escherichia coli* for efficient production of plant flavonoids. Mol Pharm. 2008;5(2):257–65.

Levisson M, Patinios C, Hein S, de Groot PA, Daran JM, Hall RD, Martens S, Beekwilder J. Engineering de novo anthocyanin production in *Saccharomyces cerevisiae*. Microb Cell Factories. 2018;17(1):103.

Li Y, Smolke CD. Engineering biosynthesis of the anticancer alkaloid noscapine in yeast. Nat Commun. 2016;7:12137.

Li J, Zhu D, Niu J, Shen S, Wang G. An economic assessment of astaxanthin production by large scale cultivation of *Haematococcus pluvialis*. Biotechnol Adv. 2011;29(6):568–74.

Liang MH, Zhu J, Jiang JG. Carotenoids biosynthesis and cleavage related genes from bacteria to plants. Crit Rev Food Sci Nutr. 2018;58(14):2314–33.

Lila MA, Burton-Freeman B, Grace M, Kalt W. Unraveling anthocyanin bioavailability for human health. Annu Rev Food Sci Technol. 2016;7(1):375–93.

Lim CG, Wong L, Bhan N, Dvora H, Xu P, Venkiteswaran S, Koffas MA. Development of a recombinant *Escherichia coli* strain for overproduction of the plant pigment anthocyanin. Appl Environ Microbiol. 2015;81(18):6276–84.

Liu D, Wu S, Tan F. Effects of addition of anka rice on the qualities of low-nitrite Chinese sausages. Food Chem. 2010;118(2):245–50.

Liu J, Sun Z, Gerken H, Huang J, Jiang Y, Chen F. Genetic engineering of the green alga *Chlorella zofingiensis*: a modified norflurazon-resistant phytoene desaturase gene as a dominant selectable marker. Appl Microbiol Biotechnol. 2014;98(11):5069–79.

López-Nieto MJ, Costa J, Peiro E, Méndez E, Rodríguez-Sáiz M, de la Fuente JL, Cabri W, Barredo JL. Biotechnological lycopene production by mated fermentation of *Blakeslea trispora*. Appl Microbiol Biotechnol. 2004;66(2):153–9.

Lv X, Xu H, Yu H. Significantly enhanced production of isoprene by ordered coexpression of genes dxs, dxr, and idi in *Escherichia coli*. Appl Microbiol Biotechnol. 2013;97(6):2357–65.

Lv J, Zhang B, Liu X, Zhang C, Chen L, Xu G, Cheung PCK. Enhanced production of natural yellow pigments from *Monascus purpureus* by liquid culture: the relationship between fermentation conditions and mycelial morphology. J Biosci Bioeng. 2017;124(4):452–8.

Lv J, Qian G, Chen L, Liu H, Xu H, Xu G, Zhang B, Zhang C. Efficient biosynthesis of natural yellow pigments by *Monascus purpureus* in a novel integrated fermentation system. J Agric Food Chem. 2018;66(4):918–25.

Ma T, Deng Z, Liu T. Microbial production strategies and applications of lycopene and other terpenoids. World J Microbiol Biotechnol. 2016;32(1):15.

Ma T, Shi B, Ye Z, Li X, Liu M, Chen Y, Xia J, Nielsen J, Deng Z, Liu T. Lipid engineering combined with systematic metabolic engineering of *Saccharomyces cerevisiae* for high-yield production of lycopene. Metab Eng. 2019;52:134–42.

Maldonade IR, Rodriguez-Amaya DB, Scamparini ARP. Carotenoids of yeasts isolated from the Brazilian ecosystem. Food Chem. 2008;107(1):145–50.

Malisorn C, Suntornsuk W. Improved β-carotene production of *Rhodotorula glutinis* in fermented radish brine by continuous cultivation. Biochem Eng J. 2009;43(1):27–32.

Mata-Gomez LC, Montanez JC, Mendez-Zavala A, Aguilar CN. Biotechnological production of carotenoids by yeasts: an overview. Microb Cell Factories. 2014;13:12.

Mezzomo N, Ferreira SRS. Carotenoids functionality, sources, and processing by supercritical technology: a review. J Chem. 2016;2016:1–16.

Minhas AK, Hodgson P, Barrow CJ, Adholeya A. A review on the assessment of stress conditions for simultaneous production of microalgal lipids and carotenoids. Front Microbiol. 2016;7:546.

Miyahisa I, Kaneko M, Funa N, Kawasaki H, Kojima H, Ohnishi Y, Horinouchi S. Efficient production of (2S)-flavanones by *Escherichia coli* containing an artificial biosynthetic gene cluster. Appl Microbiol Biotechnol. 2005;68(4):498–504.

Mora-Pale M, Sanchez-Rodriguez SP, Linhardt RJ, Dordick JS, Koffas MA. Biochemical strategies for enhancing the in vivo production of natural products with pharmaceutical potential. Curr Opin Biotechnol. 2014;25:86–94.

Mussagy CU, Winterburn J, Santos-Ebinuma VC, Pereira J. Production and extraction of carotenoids produced by microorganisms. Appl Microbiol Biotechnol. 2019;103(3):1095–114.

Narsing RM, Xiao M, Li WJ. Fungal and bacterial pigments: secondary metabolites with wide applications. Front Microbiol. 2017;8:1113.

Nasri Nasrabadi MR, Razavi SH. High levels lycopene accumulation by *Dietzia natronolimnaea* HS-1 using lycopene cyclase inhibitors in a fed-batch process. Food Sci Biotechnol. 2010a;19(4):899–906.

Nasri Nasrabadi MR, Razavi SH. Use of response surface methodology in a fed-batch process for optimization of tricarboxylic acid cycle intermediates to achieve high levels of canthaxanthin from *Dietzia natronolimnaea* HS-1. J Biosci Bioeng. 2010b;109(4):361–8.

Nicolás-Molina FE, Navarro E, Ruiz-Vázquez RM. Lycopene over-accumulation by disruption of the negative regulator gene crgA in *Mucor circinelloides*. Appl Microbiol Biotechnol. 2008;78(1):131–7.

Nielsen BR, Mortensen A, Jørgensen K, Skibsted LH. Singlet versus triplet reactivity in photodegradation of C_{40} carotenoids. J Agric Food Chem. 1996;44(8):2106–13.

Olivieri G, Salatino P, Marzocchella A. Advances in photobioreactors for intensive microalgal production: configurations, operating strategies and applications. J Chem Technol Biotechnol. 2014;89(2):178–95.

Oplatowska-Stachowiak M, Elliott CT. Food colors: existing and emerging food safety concerns. Crit Rev Food Sci Nutr. 2017;57(3):524–48.

Orosa M, Torres E, Fidalgo P, Abalde J. Production and analysis of secondary carotenoids in green algae. J Appl Phycol. 2000;12(3–5):553–6.

Pandey RP, Parajuli P, Koffas M, Sohng JK. Microbial production of natural and non-natural flavonoids: pathway engineering, directed evolution and systems/synthetic biology. Biotechnol Adv. 2016;34(5):634–62.

Papaioannou EH, Liakopoulou-Kyriakides M. Substrate contribution on carotenoids production in *Blakeslea trispora* cultivations. Food Bioprod Process. 2010;88(2–3):305–11.

Phillips LG, Cowan AK, Rose PD, Logie MRR. Operation of the xanthophyll cycle in non-stressed and stressed cells of *Dunaliella salina Teod.* In response to diurnal changes in incident irradiation: a correlation with intracellular β-carotene content. J Plant Physiol. 1995;146(4):547–53.

Pontrelli S, Chiu T, Lan EI, Chen FYH, Chang P, Liao JC. *Escherichia coli* as a host for metabolic engineering. Metab Eng. 2018;50:16–46.

Potera C. The artificial food dye blues. Environ Health Perspect. 2010;118(10):A428.

Prommuak C, Pavasant P, Quitain AT, Goto M, Shotipruk A. Simultaneous production of biodiesel and free lutein from *Chlorella vulgaris*. Chem Eng Technol. 2013;36(5):733–9.

Pyo YH, Lee TC. The potential antioxidant capacity and angiotensin I-converting enzyme inhibitory activity of *Monascus*-fermented soybean extracts: evaluation of *Monascus*-fermented soybean extracts as multifunctional food additives. J Food Sci. 2007;72(3):S218–23.

Rad SA, Zahiri HS, Noghabi KA, Rajaei S, Heidari R, Mojallali L. Type 2 IDI performs better than type 1 for improving lycopene production in metabolically engineered *E. coli* strains. World J Microbiol Biotechnol. 2012;28(1):313–21.

Rao LG, Mackinnon ES, Josse RG, Murray TM, Strauss A, Rao AV. Lycopene consumption decreases oxidative stress and bone resorption markers in postmenopausal women. Osteoporos Int. 2007;18(1):109–15.

Rissanen T, Voutilainen S, Nyyssonen K, Salonen JT. Lycopene, atherosclerosis, and coronary heart disease. Exp Biol Med (Maywood). 2002;227(10):900–7.

Rogers JN, Rosenberg JN, Guzman BJ, Oh VH, Mimbela LE, Ghassemi A, Betenbaugh MJ, Oyler GA, Donohue MD. A critical analysis of paddlewheel-driven raceway ponds for algal biofuel production at commercial scales. Algal Res. 2014;4(1):76–88.

Santos-Buelga C, González-Paramás AM. Anthocyanins. In: Melton L, Shahidi F, Varelis P, editors. Encyclopedia of food chemistry. Oxford: Academic; 2019. p. 10–21.

Sarada R, Tripathi U, Ravishankar GA. Influence of stress on astaxanthin production in Haematococcus pluvialis grown under different culture conditions. Process Biochem. 2002;37(6):623–7.

Sen T, Barrow CJ, Deshmukh SK. Microbial pigments in the food industry-challenges and the way forward. Front Nutr. 2019;6:7.

Shi MZ, Xie DY. Biosynthesis and metabolic engineering of anthocyanins in *Arabidopsis thaliana*. Recent Pat Biotechnol. 2014;8(1):47–60.

Shi J, Dai Y, Kakuda Y, Mittal G, Xue SJ. Effect of heating and exposure to light on the stability of lycopene in tomato purée. Food Control. 2008;19(5):514–20.

Shi K, Song D, Chen G, Pistolozzi M, Wu Z, Quan L. Controlling composition and color characteristics of *Monascus* pigments by pH and nitrogen sources in submerged fermentation. J Biosci Bioeng. 2015;120(2):145–54.

Sigurdson GT, Tang P, Giusti MM. Natural colorants: food colorants from natural sources. Annu Rev Food Sci Technol. 2017;8:261–80.

Silbir S, Goksungur Y. Natural red pigment production by *Monascus Purpureus* in submerged fermentation systems using a food industry waste: Brewer's spent grain. Foods. 2019;8(5):161.

Singh RN, Sharma S. Development of suitable photobioreactor for algae production-A review. Renew Sust Energ Rev. 2012;16(4):2347–53.

Solopova A, van Tilburg AY, Foito A, Allwood JW, Stewart D, Kulakauskas S, Kuipers OP. Engineering *Lactococcus lactis* for the production of unusual anthocyanins using tea as substrate. Metab Eng. 2019;54:160–9.

Squina FM, Mercadante AZ. Influence of nicotine and diphenylamine on the carotenoid composition of *rhodotorula* strains. J Food Biochem. 2005;29(6):638–52.

Srianta I, Ristiarini S, Nugerahani I, Sen SK, Zhang BB, Xu GR, Blanc PJ. Recent research and development of *Monascus* fermentation products. Int Food Res J. 2014;21(1):1–12.

Srivastav P, Yadav VK, Govindasamy S, Chandrasekaran M. Red pigment production by *Monascus purpureus* using sweet potato-based medium in submerged fermentation. Forum Nutr. 2015;14(3):159–67.

Srivastava S, Srivastava AK. Lycopene; chemistry, biosynthesis, metabolism and degradation under various abiotic parameters. J Food Sci Technol. 2015;52(1):41–53.

Staniek A, Bouwmeester H, Fraser PD, Kayser O, Martens S, Tissier A, van der Krol S, Wessjohann L, Warzecha H. Natural products – learning chemistry from plants. Biotechnol J. 2014;9(3):326–36.

Story EN, Kopec RE, Schwartz SJ, Harris GK. An update on the health effects of tomato lycopene. Annu Rev Food Sci Technol. 2010;1(1):189–210.

Su A, Chi S, Li Y, Tan S, Qiang S, Chen Z, Meng Y. Metabolic redesign of *Rhodobacter sphaeroides* for lycopene production. J Agric Food Chem. 2018;66(23):5879–85.

Suh IS, Joo HN, Lee CG. A novel double-layered photobioreactor for simultaneous *Haematococcus pluvialis* cell growth and astaxanthin accumulation. J Biotechnol. 2006;125(4):540–6.

Tinoi J, Rakariyatham N, Deming RL. Simplex optimization of carotenoid production by *Rhodotorula glutinis* using hydrolyzed mung bean waste flour as substrate. Process Biochem. 2005;40(7):2551–7.

Valduga E, Valério A, Tatsch PO, Treichel H, Furigo A, Luccio MD. Assessment of cell disruption and carotenoids extraction from *Sporidiobolus salmonicolor* (CBS 2636). Food Bioprocess Technol. 2008;2(2):234–8.

Varela JC, Pereira H, Vila M, Leon R. Production of carotenoids by microalgae: achievements and challenges. Photosynth Res. 2015;125(3):423–36.

Vendruscolo F, Tosin I, Giachini AJ, Schmidell W, Ninow JL. Antimicrobial activity of Monascus pigments produced in submerged fermentation. J Food Process Preserv. 2014;38(4):1860–5.

Vendruscolo F, Meinicke BR, Cesar DCJ, de Oliveira D, Moritz DE, Schmidell W, Ninow JL. *Monascus*: a reality on the production and application of microbial pigments. Appl Biochem Biotechnol. 2016;178(2):211–23.

Venil CK, Zakaria ZA, Ahmad WA. Optimization of culture conditions for flexirubin production by *Chryseobacterium artocarp*i CECT 8497 using response surface methodology. Acta Biochim Pol. 2015;62(2):185–90.

Walter MH, Strack D. Carotenoids and their cleavage products: biosynthesis and functions. Nat Prod Rep. 2011;28(4):663–92.

Wan M, Zhang J, Hou D, Fan J, Li Y, Huang J, Wang J. The effect of temperature on cell growth and astaxanthin accumulation of *Haematococcus pluvialis* during a light-dark cyclic cultivation. Bioresour Technol. 2014;167:276–83.

Wang H, He F, Lu M, Zhao C, Xiong L, Yu L. High-quality lycopene overaccumulation via inhibition of γ-carotene and ergosterol biosyntheses in *Blakeslea trispora*. J Funct Foods. 2014;7:435–42.

Wang Q, Feng LR, Luo W, Li HG, Zhou Y, Yu XB. Effect of inoculation process on lycopene production by *Blakeslea trispora* in a stirred-tank reactor. Appl Biochem Biotechnol. 2015;175(2):770–9.

Wang L, Dai Y, Chen W, Shao Y, Chen F. Effects of light intensity and color on the biomass, extracellular red pigment, and citrinin production of *Monascus ruber*. J Agric Food Chem. 2016a;64(50):9506–14.

Wang B, Zhang X, Wu Z, Wang Z. Biosynthesis of *Monascus* pigments by resting cell submerged culture in nonionic surfactant micelle aqueous solution. Appl Microbiol Biotechnol. 2016b;100(16):7083–9.

Woolston BM, Edgar S, Stephanopoulos G. Metabolic engineering: past and future. Annu Rev Chem Biomol Eng. 2013;4(1):259–88.

Xie Y, Ho SH, Chen CN, Chen CY, Ng IS, Jing KJ, Chang JS, Lu Y. Phototrophic cultivation of a thermo-tolerant *Desmodesmus sp.* for lutein production: effects of nitrate concentration, light intensity and fed-batch operation. Bioresour Technol. 2013;144:435–44.

Xiong X, Zhang X, Wu Z, Wang Z. Accumulation of yellow *Monascus* pigments by extractive fermentation in nonionic surfactant micelle aqueous solution. Appl Microbiol Biotechnol. 2015;99(3):1173–80.

Xu J, Xu X, Xu Q, Zhang Z, Jiang L, Huang H. Efficient production of lycopene by engineered *E. coli* strains harboring different types of plasmids. Bioprocess Biosyst Eng. 2018;41(4):489–99.

Yadav VG, De Mey M, Lim CG, Ajikumar PK, Stephanopoulos G. The future of metabolic engineering and synthetic biology: towards a systematic practice. Metab Eng. 2012;14(3):233–41.

Yan Y, Chemler J, Huang L, Martens S, Koffas MA. Metabolic engineering of anthocyanin biosynthesis in *Escherichia coli*. Appl Environ Microbiol. 2005;71(7):3617–23.

Yan Y, Li Z, Koffas MA. High-yield anthocyanin biosynthesis in engineered *Escherichia coli*. Biotechnol Bioeng. 2008;100(1):126–40.

Yang J, Chen Q, Wang W, Hu J, Hu C. Effect of oxygen supply on *Monascus* pigments and citrinin production in submerged fermentation. J Biosci Bioeng. 2015;119(5):564–9.

Ye ZW, Jiang JG, Wu GH. Biosynthesis and regulation of carotenoids in *Dunaliella*: progresses and prospects. Biotechnol Adv. 2008;26(4):352–60.

Zha J, Koffas MAG. Production of anthocyanins in metabolically engineered microorganisms: current status and perspectives. Synth Syst Biotechnol. 2017;2(4):259–66.

Zha J, Zang Y, Mattozzi M, Plassmeier J, Gupta M, Wu X, Clarkson S, Koffas M. Metabolic engineering of *Corynebacterium glutamicum* for anthocyanin production. Microb Cell Factories. 2018;17(1):143.

Zhang Y, Butelli E, Martin C. Engineering anthocyanin biosynthesis in plants. Curr Opin Plant Biol. 2014a;19:81–90.

Zhang J, Sun Z, Sun P, Chen T, Chen F. Microalgal carotenoids: beneficial effects and potential in human health. Food Funct. 2014b;5(3):413–25.

Zhang X, Liu W, Chen X, Cai J, Wang C, He W. Effects and mechanism of blue light on *Monascus* in liquid fermentation. Molecules. 2017;22(3):385.

Zhang B, Xing H, Jiang B, Chen L, Xu G, Jiang Y, Zhang D. Using millet as substrate for efficient production of monacolin K by solid-state fermentation of *Monascus ruber*. J Biosci Bioeng. 2018;125(3):333–8.

Zhang XK, Nie MY, Chen J, Wei LJ, Hua Q. Multicopy integrants of crt genes and co-expression of AMP deaminase improve lycopene production in *Yarrowia lipolytica*. J Biotechnol. 2019;289:46–54.

Zhao S, Jones JA, Lachance DM, Bhan N, Khalidi O, Venkataraman S, Wang Z, Koffas M. Improvement of catechin production in *Escherichia coli* through combinatorial metabolic engineering. Metab Eng. 2015;28:43–53.

Zhu F, Lu L, Fu S, Zhong X, Hu M, Deng Z, Liu T. Targeted engineering and scale up of lycopene overproduction in *Escherichia coli*. Process Biochem. 2015;50(3):341–6.

Zhu L, Huang Y, Zhang Y, Xu C, Lu J, Wang Y. The growing season impacts the accumulation and composition of flavonoids in grape skins in two-crop-a-year viticulture. J Food Sci Technol. 2017;54(9):2861–70.

Chapter 7
Microbial Production of Vitamins

Panhong Yuan, Shixiu Cui, Jianghua Li, Guocheng Du, Jian Chen, and Long Liu

7.1 Water-Soluble Vitamin

7.1.1 Vitamin B12

7.1.1.1 Structure and Functions of Vitamin B12

Vitamin B12 (cobalamin, Cbl) is an essential nutrient for human health, which is a cofactor required for methionine synthase (MS, EC 2.1.1.13) and mitochondrial methylmalonyl-CoA mutase (MUT, EC 5.4.99.2) (Fowler et al. 2008). MUT utilizes an adenylated form of Cbl to catalyze the conversion of L-methylmalonyl-CoA to succinyl-CoA, this is an important step in the catabolism of odd-chain fatty acids and the side chain of cholesterol (Froese et al. 2010). MS requires a methylated form of Cbl and uses 5-methyltetrahydrofolate as a methyl donor to catalyze the remethylation of homocysteine to methionine (Froese et al. 2018). The importance of this reaction is beyond the production of methionine, an essential amino acid capable of producing S-adenosylmethionine (Adomet, commonly known as sam). The methyl group of methionine is formed by donation to form various extremely

Panhong Yuan and Shixiu Cui have been equally contributed to this chapter.

P. Yuan · S. Cui · J. Li · G. Du · L. Liu (✉)
Key Laboratory of Carbohydrate Chemistry and Biotechnology, Ministry of Education, Jiangnan University, Wuxi, China

Key Laboratory of Industrial Biotechnology, Ministry of Education, Jiangnan University, Wuxi, China
e-mail: longliu@jiangnan.edu.cn

J. Chen
Key Laboratory of Industrial Biotechnology, Ministry of Education, Jiangnan University, Wuxi, China

L. Liu, J. Chen (eds.), *Systems and Synthetic Biotechnology for Production of Nutraceuticals*, https://doi.org/10.1007/978-981-15-0446-4_7

important methylated compounds such as adrenaline, sarcosine and creatine, as well as methylated DNA, RNA and protein (Froese et al. 2018).

Vitamin B12 is at least the most complex coenzyme (Roth et al. 1993) (Fig. 7.1). Vitamin B12 is used to describe compounds of the cobalt corrinoid family, in particular, those of the cobalamin group, which are composed of upper and lower ligands are respectively composed of a corrinoid ring, and the upper ligand may be a methyl group, an adenosine, a hydroxyl group or a cyano group (Roth et al. 1993). The focus of Cbl is a central cobalt atom, and may exist in a reduced state of cob(III), cob(II) or cob(I), and may form 4–6 bonds (Roth et al. 1996). There is a correlation between the oxidation state of cobalt and its preferred number of coordination elements, wherein cob(III) typically forms six bonds, cob(II) forms five bonds, and cob(I) forms four bonds. Cobalt may additionally be bonded to a lower axial ligand, such as a dimethylbenzimidazole (DMB) moiety attached to the corrin ring. When combined, Cbl is considered "base open" when it is not "base closed". Finally, the cobalt atom may also or alternatively be combined with an axial ligand (R-group) which may be composed of any number of compounds (Froese et al. 2018).

The structural complexity of the molecule is also reflected in the complex chemical synthesis, which requires more than 60 steps to catalyze the synthesis. In nature, cobalamin is synthesized by branching of a modified tetrapyrrole pathway involving about 30 enzyme-mediated steps (Raux et al. 2000). Cbl is close to 1300–1500 Da, the uniqueness of vitamin B12 in all vitamins seems to limit its de novo synthesis to only a few bacteria and archaea (Martens et al. 2002a). Therefore, the commercial production of vitamin B12 is achieved by a bacterial fermentation process, mainly

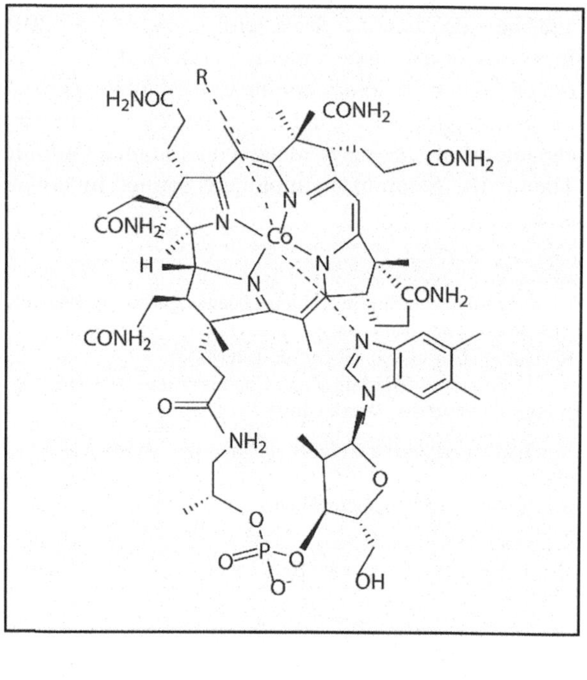

Fig. 7.1 The chemical structure of vitamin B12 (modified from Froese et al. 2018). The corrin ring and the dimethyl benzimidazole (DMB) moiety in the base configuration are indicated. Gray dashed lines indicate non-essential keys. R represents a variety of upper axial ligands, including: methyl, glutathione, adenosine, cyano and hydroxyl

using genetically modified strains of *Propionibacterium shermanii* and *Pseudomonas denitrificans*, while other organisms that utilize Cbl must modify Cbl obtained from other sources (Martens et al. 2002a). For humans, the source of B12 is limited to animal products, so there is a certain percentage of people, low-photographic animal products, moderate vitamin B12 deficiency, even if the daily diet requires only a few micrograms, even processing pre-synthesized Cbl is complicated and high metabolic cost (Banerjee and Matthews 1990). About 20 human genes are known to be involved in Cbl obtained from the diet. The functions of these genes include absorption, selection, transport, modification and utilization.

7.1.1.2 Metabolic Pathways of Vitamin B12

There are two major pathways for the biosynthesis of cobalamin coenzyme forms, called oxygen-dependent and oxygen-independent pathways, in bacteria and archaea, respectively (Fang et al. 2017). The main difference between these two pathways is the synthesis of the corrin ring components of cobalamin, in which they diverge on the demethylated derivative of uroporphyrinogen III and combine upon the formation of adenosine (Martens et al. 2002b). Some strains can also absorb porphyrins through salvage pathways to synthesize cobalamin (Fig. 7.2). Biosynthetic pathway of tetrapyrrole compounds: δ-aminolevulinic acid (ALA) is synthesized by the C4 or C5 pathway, and adenosine cobalamin is synthesized by de novo or salvage pathways (Avissar et al. 1989). The enzyme shown in the adenosine cobalamin biosynthesis pathway is derived from *P. denitrificans* or *Salmonella typhimurium* using an aerobic pathway or an anaerobic pathway, respectively (Blanche et al. 1989; Capozzi et al. 2012). ALA is the first committed tetrapyrrole synthesis precursor pathway that can be synthesized by the C4 pathway or the C5 pathway (Kang et al. 2011). In the C4 pathway, ALA synthase catalyzes the production of ALA by glycine and succinyl-CoA. In the C5 pathway, ALA is catalyzed by three enzymatic reactions by glutamate (Blanche et al. 1989). Two ALA molecules are condensed by porphobilinogen synthase to form monopyrrole porphobilinogen, and then Uroporphyrinogen III is polymerized by four porphobilinogen molecules and cyclized to form (Zappa et al. 2010). Catalyzed by enzyme porphobilinogen deaminase and uroporphyrinogen III synthase.

The aerobic and anaerobic pathways diverge at precorrin-2 and converge at coby (II) rinic acid a, c-diamide (Roth et al. 1993). Cob(I) yrinic acid a, c-diamide is adenylated to form adenosine cobyrinic acid a, c-diamide. Cob(I) yrinic acid a, c-diamide adenosyl transferase may also adenylate other porphyrins in which at least the a and c positions of the carboxyl group are amidated. A four-step stepwise amidation reaction of adenine cobyrinic acid a, c-diamide on the carboxyl groups at the b, d, e and g positions an adenosine-assisted acid is produced. Two separate methods have been developed, namely aerobic and anaerobic pathways, to be able to attach (f)-1-amino-2-propanol or (R)-1-amino-2 to the carboxyl group of adenos-

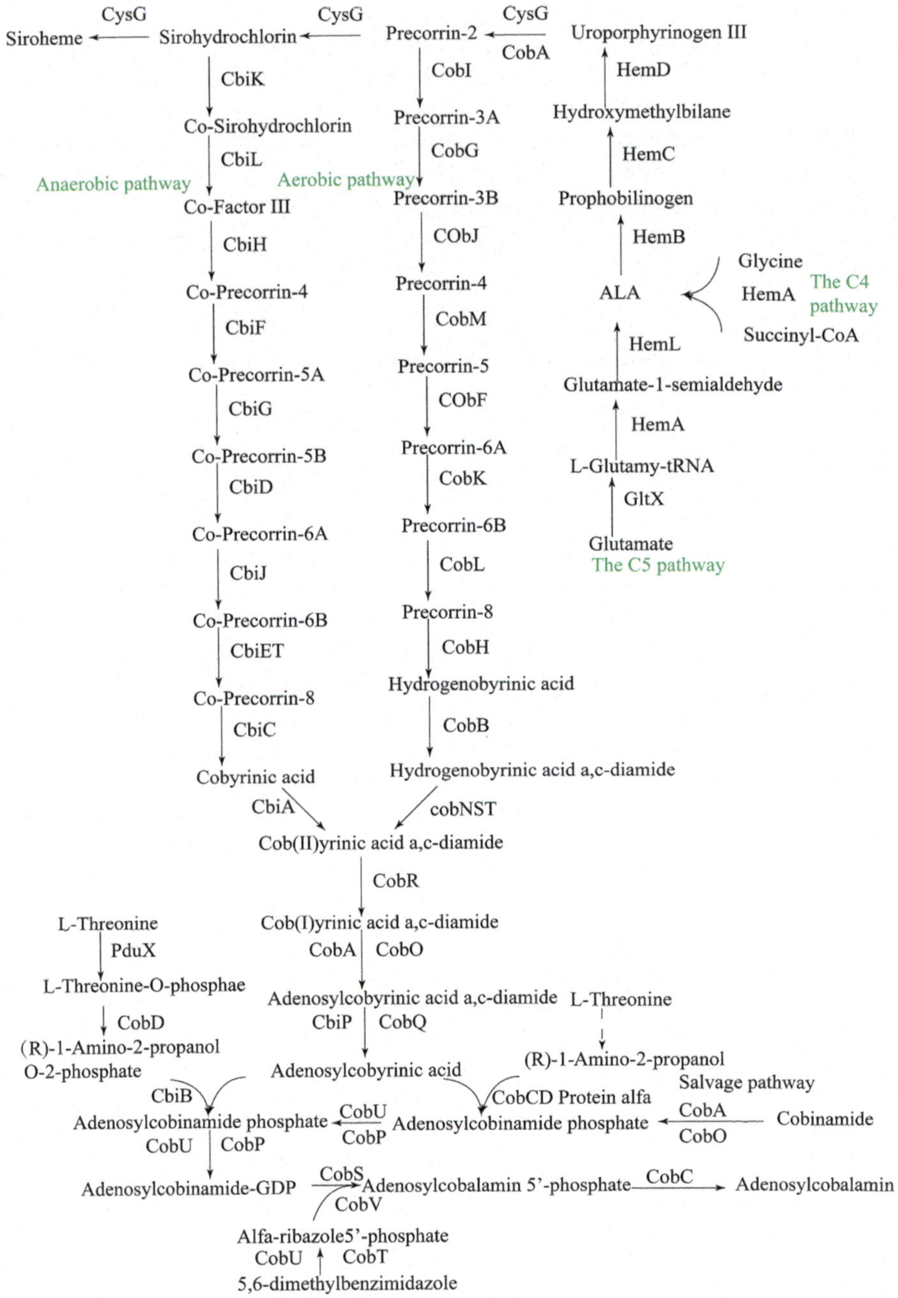

Fig. 7.2 Biosynthetic pathways of tetrapyrrole compounds. ALA is synthesized by either the C4 or the C5 pathway. (Modified from Kang et al. 2011)

ine glycine. In the anaerobic pathway, the linker between the porphyrin-like loop and the lower-axis ligand is phosphorylated prior to attachment to the porphyrin-like loop (Fan and Bobik 2008). The final step in vitamin B12 has two different biosynthetic perspectives. One view is that the last reaction of AdoCb1 biosynthesis involves the cocatalytic addition of α-oxazole to cobalamin synthase (Raux et al. 1996).

The synthesis of vitamin B12 is inhibited by post-transcriptional regulation mechanisms (Pfleger et al. 2006). Such as expression of the *Salmonella typhimurium* cob operon encoding the CBL biosynthetic pathway and the *btuB* gene encoding the vitamin B12 transporter *E. coli* and *S. typhimurium* (Fan and Bobik 2008; Brushaber et al. 1998; Raux et al. 1996). Studies have shown that this regulation requires an abnormally long 5′-untranslated leader sequence of the corresponding mRNA, which contains several conserved elements. The leader mRNA of the cob and btuB genes contains an evolutionarily conserved sequence called the B12-box (Vitreschak et al. 2003). Adenosine cobalamin (Ado-CBL) is an effector molecule involved in the regulation of CBL gene (Vitreschak 2003). Structurally dependent spontaneous cleavage of RNA technology was applied to the *E. coli btuB* leader sequence in the presence and absence of Ado-CBL, and it was shown that this sequence fragment can directly bind to Ado-CBL, thus conformation changes in secondary and tertiary structure of RNA (Nahvi et al. 2002).

Cobalamin riboswitches are the main metabolic regulatory form that controls the concentration of vitamin B12 in microorganisms (Vitreschak 2003). The riboswitch is an evolutionarily conserved non-coding region in which the 5′ untranslated region of the mRNA of the regulatory gene is expressed in response to the direct binding of the RNA itself to intracellular metabolites (Winkler et al. 2002). A riboswitch consists of two domains, one of which is used as an evolutionarily conserved natural aptamer, which is capable of binding to a target metabolite due to its high selectivity and affinity, while the other domain utilizes the formation of an aptamer-ligand complex. Allosteric changes in RNA structure to control the expression of adjacent genes or operons (Rodionov et al. 2002). In the case of high cobalamin concentration and transcriptional repression, alternative Rho-independent termination of the Rho binding site or hairpin results in premature transcription termination. The high cobalamin concentration also promotes the isolation of the ribosome binding site (RBS) and the blockade of translation initiation. While the concentration of cobalamin is low, an anti-terminating hairpin is formed, which enabling RNA polymerase to complete transcription of the downstream gene. Low-cobalamin concentration promotes the formation of anti-chelating hairpins and releases RBS for translation initiation (Vitreschak 2003).

7.1.2 Folates Acid

7.1.2.1 Structure and Functions of Folates Acid

The vitamin folates are a group of water-soluble compounds that are parts of B vitamin family (B9) (Crider et al. 2012). They act as coenzymes in the C1 transfer reaction involving purine, pyrimidine and methionine synthesis and amino acid metabolism (Tibbetts and Appling 2010). Since animals cannot produce folic acid (FA), these essential vitamins must be obtained from exogenous foods to prevent defects. The main dietary sources of folic acid are dairy products, cereal products and green vegetables, especially fortified bread and breakfast cereals (USDA National Nutrient Database for Standard Reference Legacy Release, April 2018). In the body, folic acid is converted to dihydrofolate, tetrahydrofolate, L-methyl folate, and other derivatives to participate in specific biological activities (Greenberg et al. 2011). The important folate derivative 5-methyltetrahydrofolate (5CH3-THF, 5CH3-H4 folic acid) plays a key role in the methyl donor (Botto and Yang 2000). Homocysteine (Hcy) produces methionine, a methionine cycle that produces carbon metabolism of proteins. Folic acid and vitamin B12 deficiency cause severe abnormalities in one carbon metabolism, which is considered to be risk factors for chronic diseases and developmental disorders, including autism Alzheimer's disease (Hinterberger and Fischer 2013) and neural tube defects (NTD) (Copp et al. 2013). NTD is a group of abnormalities in the brain, spine and spinal cord that are usually manifested during the first month of pregnancy, leading to the closure of neural tube failure during embryogenesis (Williams et al. 2015). Therefore, it is strongly recommended to use a folate-rich diet and folic acid supplements during pregnancy to prevent NTD and other chronic dysfunctions, such as congenital heart defects (CHD) (Czeizel et al. 2013).

They have a common chemical structure formed by a pteridine ring, a p-aminobenzoic acid (pABA), and one or more gamma-linked glutamates (Poe et al. 1979) (Fig. 7.3a). Food folates in different forms contain additional glutamic acid residues to form glutamic acid. Various forms of folic acid differ in a carbon unit attached to the N5- and/or N10-position of the pteridine ring, such as methyl (5-CH3), methylene (5,10-CH2), formyl Amino (5-CHNH), formyl (5- or 10-HCO) and methyl (5,10-CH) (Fig. 7.3b) (Hanson and Gregory 2011).

7.1.2.2 Metabolic Pathways of Folates Acid

FA is usually chemically synthesized in industry, because there are no biotechnological methods for mass production (Zhang et al. 2008). Although the synthetic form of vitamins does not exist in nature, it can be metabolized into a biologically active form by the action of dihydrofolate reductase (DHFR). However, human DHFR shows that the conversion of synthetic FA to bioactive glaze is extremely low, so the application of a high concentration of FA can lead to its accumulation in

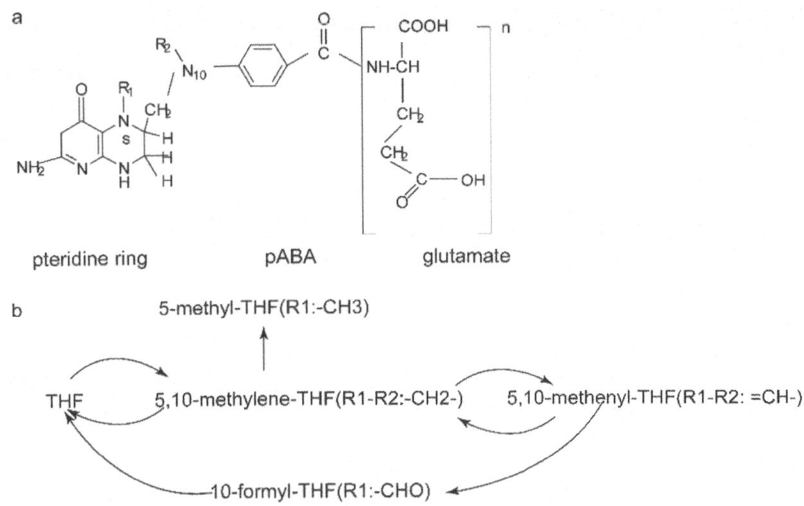

Fig. 7.3 Folic acid structure (modified from Poe et al. 1979). (**a**) Folates acid contains a pteridine ring, a pABA molecule and a gammalinked L-glutamic acid tail. (**b**) The different substituents on R1 and R2 characterize different vitamins that can be converted into each other

the blood (Bailey and Ayling 2009). Only plants and a few microorganisms have a complete synthesis pathway for folates biosynthesis, which is very conservative throughout the evolutionary process (Basset et al. 2005). It includes the synthesis of pteridine rings which is a common precursor of the riboflavin synthesis from GTP, condensation with pABA, synthesis from chorismate and glutamic acid moieties (Fig. 7.4) (Saini et al. 2016). The pterin branch begins with the action of GTP via GTP cyclohydrolase I (GTPCHI), converts GTP to 7,8-dihydropterin triphosphate, and then performs two dephosphorylation steps. The dihydropterin aldolase converted 7,8-dihydropterin to 2-amino-4-hydroxy-6-hydroxymethyl-7,8-dihydropterin (DHNA), followed by 2-amino 4-hydroxy-6-hydroxymethyldihydropterin pyrophosphate. The pABA was synthesized from chorismate, which is the product of the shikimate pathway (Lawrence et al. 2005). In the first step, it is a combination of two enzymes, namely transglutaminase and aminodeoxycholate synthase (ADCS), which produce ammonia from glutamine, and transfers the amino group to the chorismate acid ester to form 4-amino-4-deoxycholate (ADC). In the second step, converts the ADC to pABA was catalyzed by 4-amino-4-deoxycholate lyase. Dihydropropionate synthase (DHPS) was used to synthesize 7,8-dihydropropionic acid from 2-amino-4-hydroxy-6-hydroxymethyl dihydropterin diphosphate and PABA. Then, 7,8-dihydrofolate (DHF) was obtained by adding glutamic acid through dihydrofolate synthase (DHFS). The DHF is finally reduced by dihydrofolate reductase (DHFR) present in the animal to produce the first biological form of folic acid, tetrahydrofolate (THF) (Moretti et al. 2014). In bacteria, a gene encoding

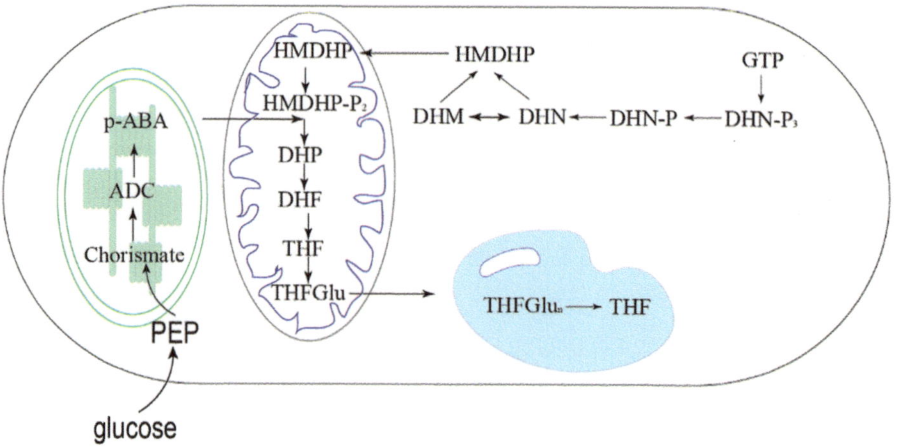

Fig. 7.4 Folate biosynthesis pathway (modified from Saini et al. 2016). Abbreviations of compounds: *ADC* Aminodeoxychorismate, *DHF* Dihydrofolate, *DHM* dihydromonapterin, *DHN* Dihydroneopterin, *DHP* Dihydropteroate, *Glu* Glucose ester, *Glu$_n$* Polyglutamate, *HMDHP* Hydroxymethyldihydropterin, *P* Phosphate, *THF* Tetrahydrofolate

an enzyme, and in fungi and plants, a fusion gene that results in a multidomain enzyme is usually found (Lawrence et al. 2005).

Microbial Metabolism Engineering - is a powerful tool for manipulating metabolic pathways through genetic engineering to enable microbial production beyond the natural synthesis capacity of high-value molecules. Taking folic acid production as an example, metabolic engineering can enhance the synthesis of folic acid from three aspects: (Fowler et al. 2008) enhancing the metabolic flux of folate production, increasing titer and yield, and (Froese et al. 2010) controlling folate distribution to maximize optimal (activity/stability) Form, and (Froese et al. 2018) maximizing folic acid stability, known to be an important issue in folate storage (Revuelta et al. 2018). In order to increase the synthesis of folic acid, metabolic engineering has performed a great deal of work in *Lactococcus lactis*, including the identification of gene clusters involved in folate synthesis followed by overexpression of some components. Increasing the activity of HPPK and GTPCHI can increase the extracellular folate production by nearly 10 times and the total amount of folic acid by 3 times (Sybesma et al. 2003a). Furthermore, overexpression of endogenous folKE with folC encoding FPGS increases the retention of folate in the cells. Overexpression of folC alone increases the polyglutamyl tail, resulting in the retention of all folate in the cell. In contrast, overexpression of folA encoding DHFR reduced folate production, indicating feedback inhibition mechanism (Sybesma et al. 2003b). In another work, the author expressed mammalian γ-glutamyl hydrolase in *Lactococcus lactis*, converted polyglutamyl folate to monoglutamyl FA, and improved bioavailable monoglutamyl folate Excreted into the fermentation broth (Sybesma et al. 2004). The foods were fermented by *Lactococcus lactis* indicates that these engineered strains can be used to enhance the synthesis of folic acid, although regulatory restrictions do not include the use of GMOs in food. However, the level achieved so far is still very low (200 μg/L), so more engineering methods are still needed (Sybesma et al. 2003a).

By combining theoretical metabolic flux analysis and metabolic engineering, the model organisms Bacillus subtilis and E. coli were designed to increase folic acid production (Zhu et al. 2003, 2005). The best strain produced in *B. subtilis* showed that pyruvate kinase was induced, In addition overexpressing *E. coli* aroH (2-dehydro-3-deoxyphosphonate heptanoic acid aldehydase, involved in pABA synthesis), and increasing the gene in the folate operon transcription and translation. This strain reached a yield of 163 μg/L folic acid (Zhu et al. 2005). By deleting the pyruvate kinase (PYK) gene and redirecting the metabolic flux to the synthesis of the basal metabolic precursor phosphoenolpyruvate (PEP) and erythrose-4-phosphate (E4P), the model organism *E. coli* was also designed to overproduce folic acid. This achieves a yield of 275 μg/L (Zhu et al. 2003).

Tetrahydrofolate (THF) polyglutamic acid acts as a coenzyme family in cells that activate at the N-5 and/or N-10 positions and carry a single carbon. THF can be activated for carbon transfer reaction in three different oxidation states (Revuelta et al. 2018). THF often cooperates with activated formate, which is 10-formyl-THF, 5-formyl-THF or 5,10-methyl-THF. In addition, THF transfers activated formaldehyde as 5,10-methylene-THF and activated methanol as 5-methyl-THF (Hoffbrand and Weir 2001). When transported into cells, the folate monoglutamate molecule is modified by a covalently bound glutamate polypeptide consisting of three to eight glutamic acid residues, which are polymerized by unusual gamma-linked peptide bonds (Wright et al. 2003). Addition of glutamate polypeptides is essential for the production of functional coenzymes; polyglutamate peptides are essential for high affinity binding of many folate-dependent and binding proteins as well as folate coenzymes in cell and intracellular compartments. The THF polyglutamate accepts and transfers a carbon in a network of biosynthesis and catabolic reactions called folate-mediated one-carbon metabolism (Shane 1989). The folate coenzyme and single carbon pathway play a role in three cellular compartments: cytoplasm, mitochondria and nucleus. Folic acid coenzyme performs specialized metabolic functions in each compartment and is difficult to exchange between cell compartments. Mitochondria - Carbon metabolism produces formic acid by catabolism of serine, glycine and choline (Christensen and MacKenzie 2006). Cytoplasmic – Carbon Metabolism Mitochondria-derived formic acid was used for nucleotide biosynthesis and remethylation of homocysteine to methionine. Nuclear folate metabolism produces thymidylate during DNA replication and repair.

7.1.3 Vitamin B1

Vitamin B1, also known as thiamin or thiamine, is a vitamin found in food as a dietary supplement and medication. Food sources of vitamin B1 include legumes, grains, dairy products, vegetables poultry and fish (Lonsdale 2006). When pH is more than 5, it is resistant to high temperatures (100 °C), while pH is less than 3, it is unstable. Therefore, vitamin B1 and phosphorylation thiamin all exist in the form of thiamine salt (thiamine hydrochloride) (Voelker et al. 2018).

7.1.3.1 Structure and Functions of Vitamin B1

Vitamin B1 is used to treat or prevent vitamin B1 deficiency. This medicine injection is used to treat beriberi, serious condition caused by prolonged lack of vitamin B1 (Attaluri et al. 2018). Its phosphate derivatives are involved in many cellular processes. The best-characterized form is thiamine pyrophosphate (TPP), a coenzyme in the catabolism of sugars and amino acids (Iosue et al. 2016).

Vitamin B1 is a colorless organic sulfur compounds, molecular formula $C_{12}H_{17}N_4OS$. Its structure consists of amino pyrimidine and methylene bridged up of thiazole ring (Fig. 7.5a). The thiazole is substituted with methyl and hydroxyethyl side chains. Thiamine and its phosphorylation can be oxidized to thiochrome (Fig. 7.5b), a fluorescent compound by potassium ferricyanide in alkaline solution, and this can be used to test vitamin B1 and other thiamine phosphates (Schyns et al. 2005).

A report said about vitamin B1 market forecast 2019–2024 by Brother Enterprises, Huazhong Pharma, Zhejiang Tianxin and DSM. They reported that thiamine prices at around \$20–30/kg in 2012–2014. Since 2015, manufacturers are under pressure from environmental policy. The price of vitamin B1 rose 32% in 2015 and 42% in 2016. Due to insufficient supply of vitamin B1, which prices will remain higher for years. Vitamin B1 has a capacity of 7791 metric ton (MT) in the world market in 2016 and it will be 9981 MT in 2023, with a growth rate of 3.6% (Rahul 2019).

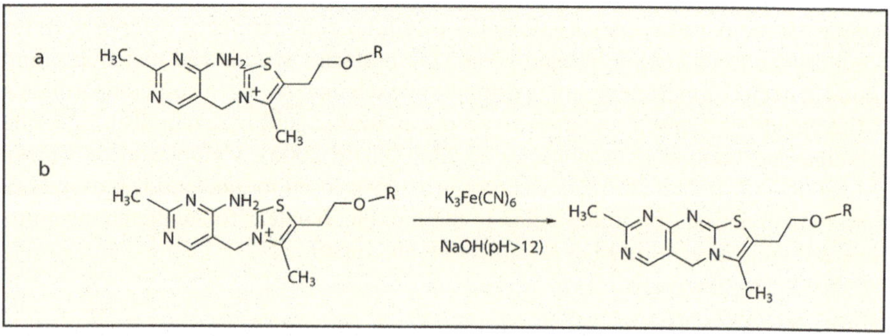

Fig. 7.5 Structure and derivative of vitamin B1 (modified from Jenkins et al. 2007). (**a**) The structure of vitamin B1; (**b**) Derivative of vitamin B1 to thiochrome. "R" represent hydroxyl radical or phosphate group(s)

7.1.3.2 Production, Advantage and Disadvantage

Vitamin B1 can be synthesized by means of chemical and biological way. 4-amino-5-aminomethyl-2-methylpyrimidine is the key intermediate of vitamin B1 by chemical synthesis (Fig. 7.6). There are many routes to synthesis the intermediate: carbonitrile pyrimidine approach (Hoffmann-La Roche Co. process Fig. 7.7a), formyl pyrimidine approach (UBE Co. process, Fig. 7.7b) and formamide pyrimidine approach (Chinese producers process, Fig. 7.7c) (Zhao et al. 2012; Létinois et al. 2013).

Thiamine pyrophosphate (TPP) is synthesized via thiamine monophosphate (TMP) by separate construction of the thiazole (THZ-P) and pyrimidine (HMP-P) heterocycles, followed by their coupling distinct routes exist (Fig. 7.8) (Jenkins et al. 2007). The biosynthetic pathways are negative feedback regulated by TPP riboswitches. If TPP is deficient in cells, TPP synthase is activated and TPP synthesis is stimulated. However, when the amount of TPP becomes excessive in cells, it binds directly to the riboswitch as a signaling molecule. Because TPP binding can alter the structure of the mRNA, it inhibits the translation of TPP synthase, which results in the inhibition of TPP synthesis. Therefore, TPP-dependent riboswitches maintain a constant amount of TPP in cells by controlling their own TPP synthase. In bacteria, the TPP riboswitch has already evolved and generated different gene-regulation mechanisms using the same binding domain. In Gram-positive bacteria, the TPP bind to the nascent RNA molecule and triggers the formation of a transcription terminator signal (Nudler 2006). In Gram-negative bacteria, the binding of TPP to the RNA molecule results in a structural rearrangement that masks the Shine–Dalgarno sequence (which helps to recruit ribosomes to the mRNA in prokaryotes) and, as a result, ribosomes fail to initiate translation (Bocobza and Aharoni 2008).

7.1.3.3 Biosynthesis and Regulation

Unlike other vitamin biosynthetic pathways in bacteria (for example, for riboflavin and biotin), thiamine is part of the salvage pathway and is not synthesized de novo. A key enzyme in the biosynthesis of vitamin B1 has somehow evolved the ability to perform a complex series of some 15–20 steps and the negative feedback regulation mechanism of TPP riboswitch successfully limit the vitamin B1 microbial fermentation (Acevedo-Rocha et al. 2019).

Thiamine pyrophosphate (TPP) is synthesized in bacteria through multiple complex metabolic pathways. Here, *Bacillus subtilis* will be selected as the representative

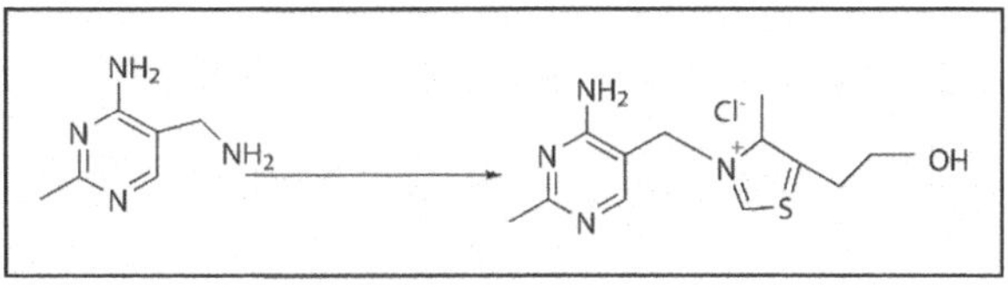

Fig. 7.6 Synthetic strategy of vitamin B1. (Modified from Rahul 2019)

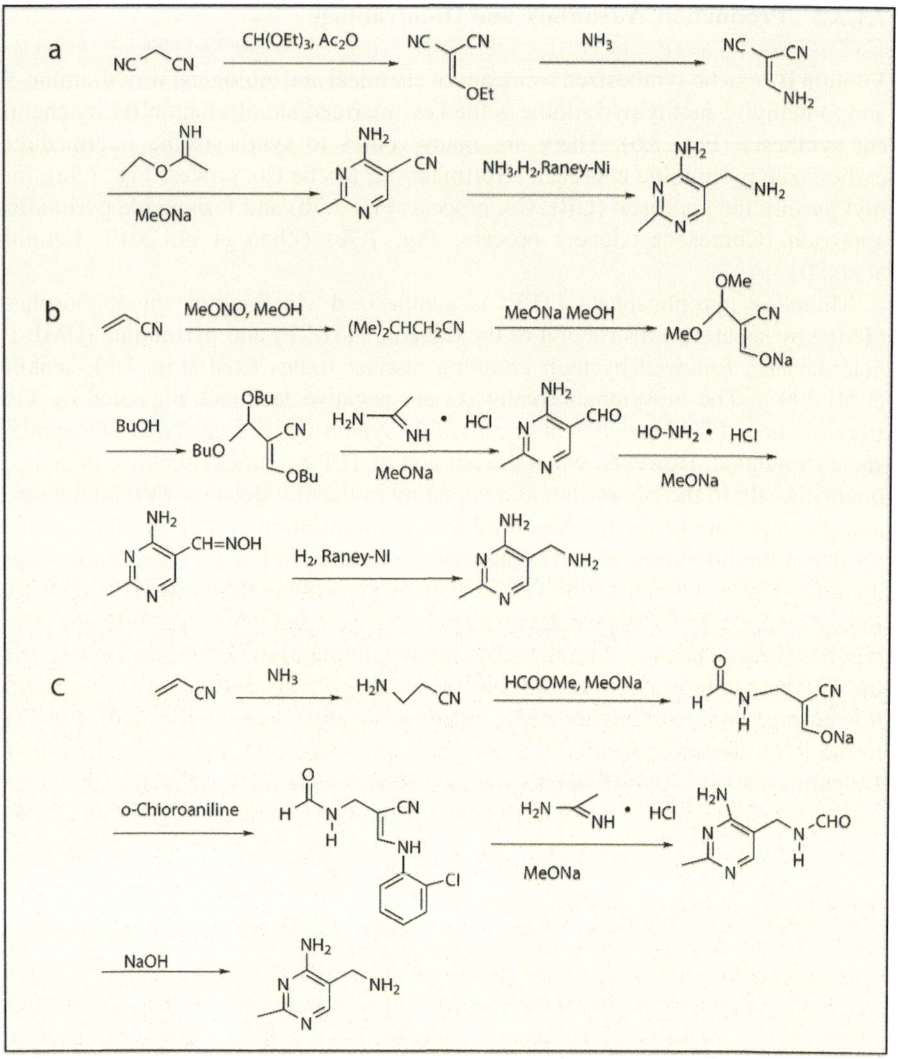

Fig. 7.7 Synthesis of 4-amino-5-aminomethyl-2-methylpyrimidine (modified from Zhao et al. 2012). (**a**) Carbonitrile pyrimidine approach; (**b**) Formyl pyrimidine approach; (**c**) Formamide pyrimidine approach

to explain the biosynthesis pathway of TPP (Fig. 7.8). The pyrimidine moiety, 4-amino-2-methyl-5-hydroxymethylpyrimidine pyrophosphate (HMP-PP), is derived from an intermediate in the indole biosynthetic pathway and is obtained by thiC enzyme catalysis 5-Aminoimidazole nucleotide (AIR) (Begley et al. 1999). HMP is phosphorylated to HMP-PP by catalysis of thiD and then coupled to a thiazole unit (Park et al. 2004). Derivatization of 1-deoxy-D-xylulose-5-phosphate (DXP) to form 5-(2-hydroxyethyl)-4-methylthiazole phosphate (HET-P), glycine

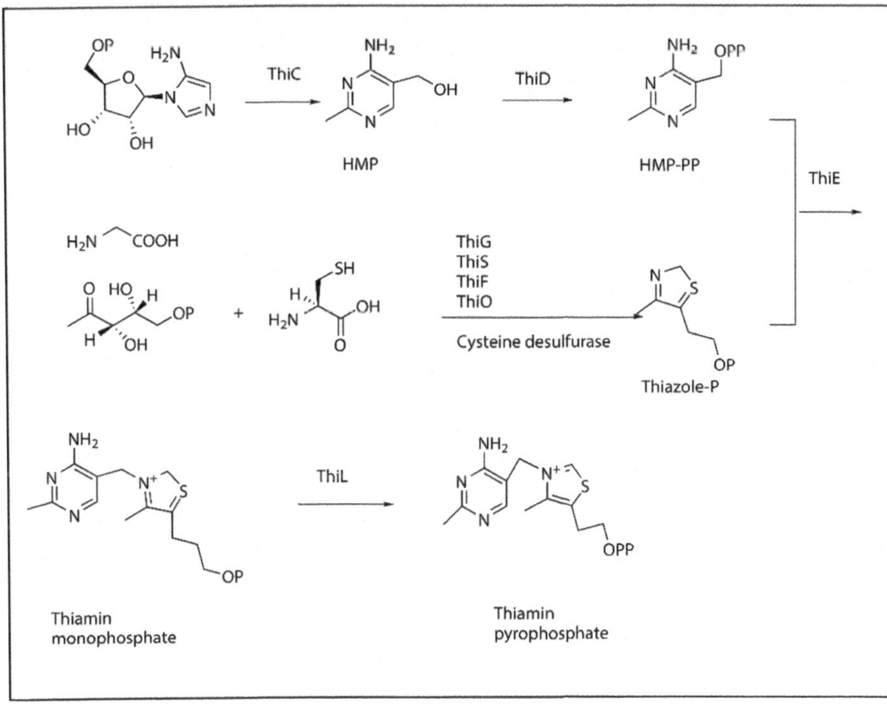

Fig. 7.8 Biosynthesis of vitamin B1 (modified from citation Jenkins et al. 2007). *ThiC* phosphomethylpyrimidine synthase, *ThiD* pyridoxine kinase, *ThiE* thiamine-phosphate synthase, *ThiL* thiamine-monophosphate kinase

and cysteine, requires at least The products of five different genes, the thiF, thiS, thiO and thiG genes undergo complex oxidative condensation reactions (Schyns et al. 2005). The thiamine phosphate pyrophosphorylase encoded by thiE catalyzes the coupling of HMP-PP and HET-P to produce thiamine monophosphate (TMP) (Begley et al. 1999). TMP is then phosphorylated by the action of thiamine-encoded thiamine monophosphate kinase (thiL) to form thiamine pyrophosphate (TPP).

Microbial fermentations have met limited successfully. Recently there are described many enzymes (such as thiC, theA, tenI and thiM) negatively regulated by the TPP-dependent riboswitch. Winkler et al. (2002) and their groups found that thiamine derivatives bind mRNA directly to regulate bacterial gene expression. The TPP-dependent riboswitch is present in the untranslated region (UTR) of mRNA encoding TPP synthase.

Vitamin B1 is mainly synthesized through chemical synthesis in present. While, due to the cost of chemical synthesis is higher and bring heavier environmental burden than biosynthesis, the price of vitamin B1 will more expensive. Nadia Drake reported that instead of chemically synthesizing thiamine to fortify foods, it may eventually be possible to employ modified microorganisms as primary vitamin factories - an advance that would greatly increase the efficiency of thiamine production while simultaneously decreasing the cost (Raschke et al. 2007).

7.2 Fat-Soluble Vitamin

Vitamins are an organic compound required for growth and health in very small quantity (Pilz et al. 2019). Vitamins are found in various food in minute amount and produced synthetically (Herbers 2003). Vitamins are classified according to their chemical and biological activity. Thus, each vitamin refers to some compounds that all exhibit the biological activity associated with a particular vitamin. To date, 13 vitamins are universally recognized. Among them, fat-soluble vitamins include: Vitamin K, Vitamin E, vitamin A, Vitamin D. The body must get it through various foods and supplements, because it cannot produce vitamins on its own. Vitamins are related with corresponding vitamin deficiency disease, such as vitamin D deficiency can lead to disease of bones (Savastano et al. 2017), vitamin A deficiency can lead to night blindness. Deficiency of vitamin E can cause the nerve damage uncommon and Vitamin K deficiency may result in spontaneous bleeding. Vitamins are extensively used as dietary supplements, meanwhile their usage in beverages and functional food have also increased tremendously (Tarento et al. 2018). In the elderly population, Alzheimer's disease (AD) is the mostly cause of dementia, affecting 46 million people worldwide currently. Vitamins A, D, E, and K are reported to affect the mechanisms of AD pathogenesis.

7.2.1 Vitamin A

7.2.1.1 Structure and Functions

Vitamin A is used in any compound having retinol biological activity (Fig. 7.9). Vitamin A is taken up as a retinyl ester or carotenoid and metabolized into an active compound, such as 11-cis-retinal, which is important for vision, while all-trans retinoic acid is a vitamin biological effect. All-trans retinoic acid binds to the retinoic acid receptor (RAR), which is heterodimerized with the retinoid X receptor. The RAR-retinoid X receptor heterodimer acts as a transcription factor, binding RAR response elements in promoters of different genes. Many cellular functions, including bone cell function, are mediated by vitamin A. The RAR-retinoid X receptor heterodimer acts as a transcription factor that binds to the RAR response element in the promoter of a different gene. Many cellular functions are mediated by vitamin A (Ye et al. 2000).

The International Union of Pure and Applied Chemistry (IUPAC) published a recommendation on vitamin nomenclature and suggested that the parent vitamin A should be called retinal, retinol, retinoic acid in 1960. These names summarize the

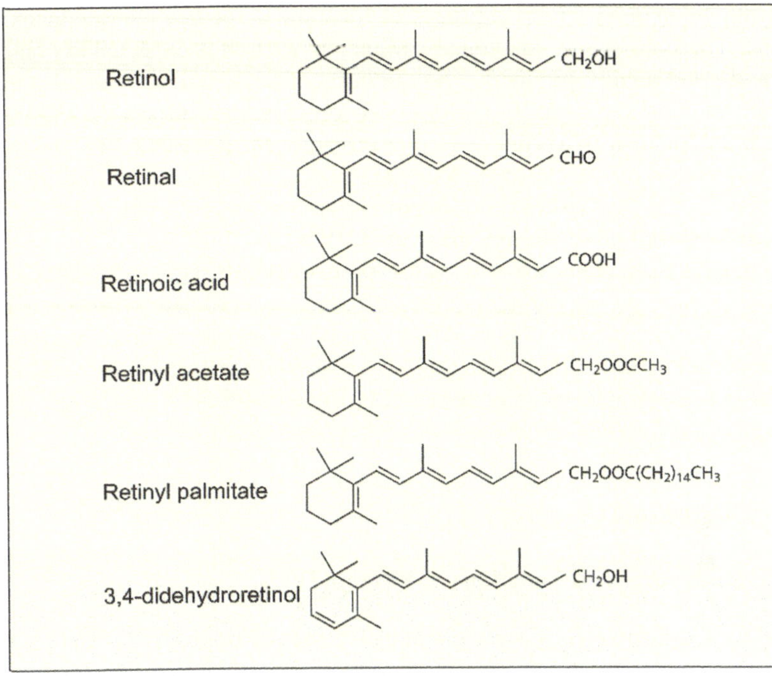

Fig. 7.9 Structure of retinoids

importance of these substances for retinal vision and also utilize the suffixes commonly used in organic chemistry to indicate the treatment of alcohol, aldehyde or carboxylic acid oxidation states at the polar ends of the molecule (Conaway et al. 2013).

7.2.1.2 Development of Vitamin A Synthesis

Vitamin A was first called fat soluble growth factor, a component of the liver oils of some marine animals. Although vitamin A can be extracted from animal tissues, most commercial vitamin A is chemically synthesized because of the distributed resources, cumbersome steps and high cost in extraction method. The main routes for the synthesis of vitamin A are Roche and BASF. Roche synthesis process was characterized by Grignard reaction with beta-ionone as starting material. Vitamin A acetate was synthesized by Darzens reaction, Grignard reaction, selective hydrogenation, hydroxyl bromination, dehydrobromination and six-step reaction (Stephensen 2001). BASF synthesis process is based on the Grignard reaction of beta-ionone with acetylene to produce acetylene-beta-ionol, which is selectively hydrogenated to ethylene-beta-ionol. After Witting reaction, Vitamin A acetate is synthesized by condensation with C5 aldehyde catalyzed by sodium alcohol (Tarento et al. 2018). Roche synthesis process is relatively mature, but research on new synthetic methods and new processes for some of the key intermediates is also ongoing. The synthesis of butenone by Mannich reaction is expected to obtain high yield and good quality products. The synthesis of hexacarbon alcohol by the new Grignard method can simplify the synthesis process, mild process conditions and improve the yield; it is worthy of further research. M. Rosenberger's new process for the synthesis of tetradecanal aldehyde, its process, process conditions, equipment conditions, etc., has more in-depth research value, in order to be applied in practical industrial production (Omenn et al. 1996).

There have been many ways to synthesize carotenoids. However, for symmetric carotenoids such as beta-carotene (Fig. 7.10), the most efficient synthesis approach has been a base-catalyzed double Wittig condensation. Furthermore, various modern metal-mediated coupling reactions have been described by using C-C bond forming processes in retinoid syntheses (Bernstein and Rando 1986).

7.2.2 *Vitamin D*

7.2.2.1 Structure and Functions of Vitamin D

Although vitamin D deficiency was originally thought to be a rickety/bone disease in the early seventeenth century, vitamin D was found to be a responsible nutritional factor only a century ago. This discovery has become difficult. In fact, vitamin D can be achieved by human skin and the diet of animal source (D3) and plant-derived form (D2) (Fig. 7.11). Vitamin D3 was first chemically synthesized and metabolized into the active form 1,25-dihydroxyvitamin D3 in the 1930s, and in the second half of the twentieth century, its mode of action in calcium and phosphate homeostasis was elucidated (Zittermann et al. 2014).

Synthetic vitamin D analogs that mimic the physiological effects of vitamin D are now available for diseases (Savastano et al. 2017). Vitamin D is a molecule that actively participates in a variety of metabolic pathways. It is primarily known for its implications for calcium metabolism. It actively participates in the cardiovascular system, affecting blood pressure, coronary artery disease and other vascular diseases (Beveridge et al. 2018). In addition, it has been determined that this vitamin is widely involved in the regulation of the immune system and the renin angiotensin aldosterone system (Perez-Hernandez et al. 2016).

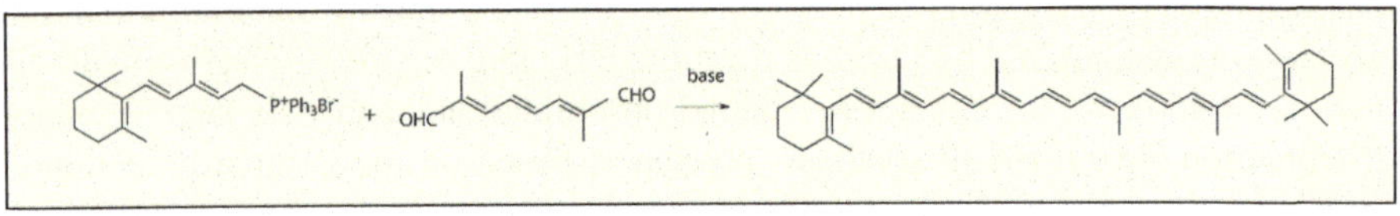

Fig. 7.10 Synthesis of β-Carotene via Wittig reaction. (Modified from Bernstein and Rando 1986)

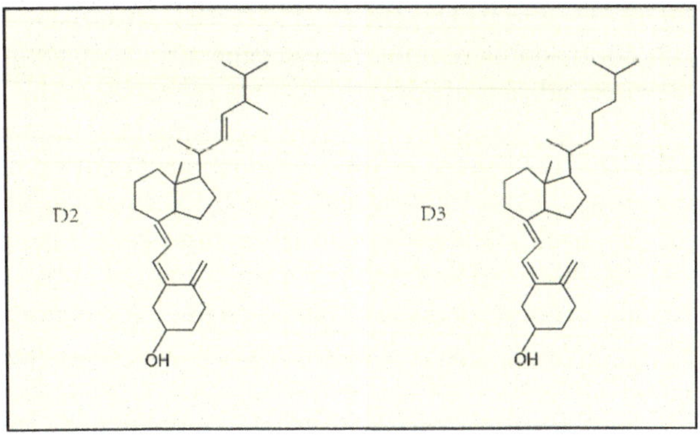

Fig. 7.11 Structure of vitamin D2 and D3

7.2.2.2 Development of Vitamin D

Vitamin D controls levels of phosphate and calcium in the body (Fig. 7.12) (Jones et al. 1998). Vitamin D in the diet or skin is an inactive precursor that requires two metabolic steps to become an active form of hormone. Hector DeLuca discovered the first step in these activation steps. The liver is an intermediate called

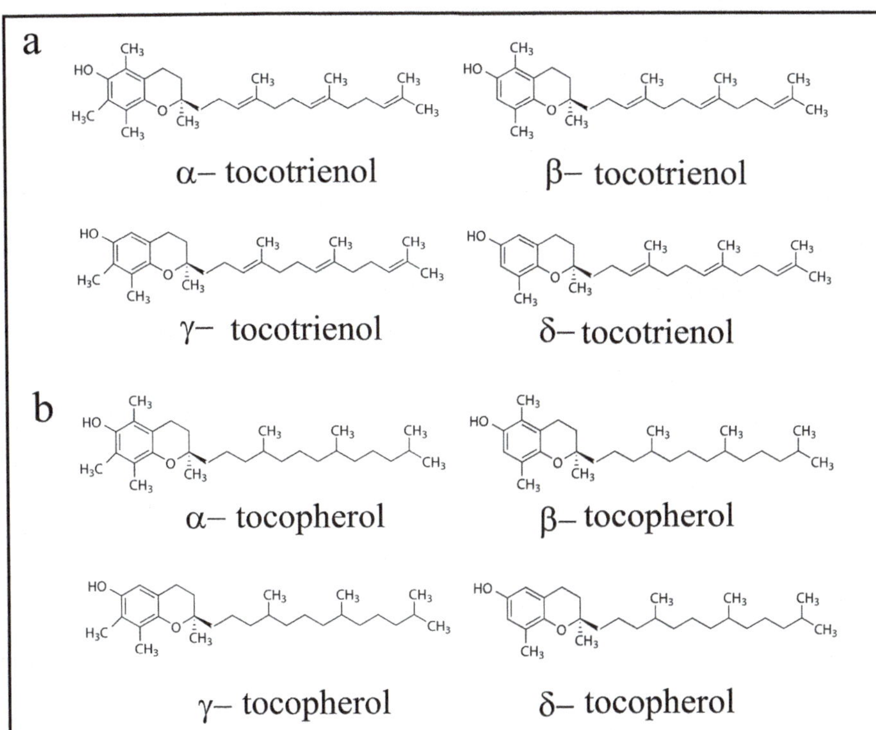

Fig. 7.12 Chemical structures of (**a**) tocotrienols and (**b**) tocopherols

25-hydroxyvitamin D3 (25-OH D3) (Blunt et al. 1968). 25-OH-D3 is the predominantly circulating form and its serum level is a generally accepted measure of vitamin D status. The second step of activation occurs primarily in the kidneys and in the resulting active form of vitamin D, called 1,25-dihydroxyvitamin D3. Several groups were considered to be the discovery and identification of calcitriol, including Mark Haussler, DeLuca, Egon Kodicek and Anthony Norman (Jones 2018). As the high potency of calcitriol, maintaining normal requires little calcium metabolism and metabolites to be produced and circulate a small amount of blood.

The traditional process includes three steps of photoreaction-column chromatography-recrystallization. In the new process, only two steps of photoreaction and recrystallization are needed. Photochemical synthesis of Vitamin D, the raw material ergosterol is mainly derived from yeast fermentation, from the production of penicillin and other drugs such as waste hyphae or vegetable oil, mushrooms and other products (Perez-Hernandez et al. 2016). Chemical Technology in Beijing University has developed a new route to extract ergosterol from penicillin mycelium with a yield of 50%, which greatly reduces production costs. Chemical Technology in Beijing University has also successfully developed a low-pressure mercury lamp and a new nitrogen agitated photoreactor, and established a new preparative liquid chromatography and computer control system, which was successfully applied to the separation and purification of VD2. It is unfavorable for scale production and cost reduction. Based on more than 20 years of research on photochemistry, the domestic research institute has proposed an innovative technical route for photochemical synthesis of vitamin D3. The technology has been transferred and implemented, and the manufacturer has a sales income of more than 500 million, showing great economic and social benefits (Pilz et al. 2019).

7.2.3 *Vitamin E*

7.2.3.1 Structure and Functions of Vitamin E

Vitamin E is found in many food including cereals, meat, vegetable oils, poultry, wheat germ oil, fruits, eggs and vegetables. Vitamin E has a group of eight lipophilic compounds, including four tocopherols designated as α-, β-, γ- and δ-tocopherols and four corresponding tocotrienols (Fig. 7.12) (Panfili et al. 2003).

Vitamin E is an antioxidant, which can maintain normal permeability, enhance skin capillary resistance, improve blood circulation and adjust fertility function, anti-aging effect (Dasgupta et al. 2015). Vitamin E can be used for treatment of coronary heart disease, arteriosclerosis, habitual abortion, muscular dystrophy, muscle spasm, neonatal scleroma, lupus, dermatomyositis, scleroderma, nodular vasculitis, etc. (Mohd Mutalip et al. 2018).

Vitamin E has important effects on humans and animals, but the synthetic pathway for vitamin E is limited in green photosynthetic plants, including lower single-celled cyanobacteria and higher plants (Cahoon et al. 2003). The pathway is not existence in humans or animals, vitamin E for daily nutrition is obtained from green plants, especially the seeds of various oil crops, and vegetable oils extracted from the seeds (Boukandoul et al. 2017).

7.2.3.2 Metabolic Pathways of Vitamin E

Vitamin E is synthesis by chemical synthesis or biosynthesis. The product of chemical synthesis is only α-tocopherol (Bonrath et al. 2007). While, α-, β-, γ- and δ-tocopherols and their four corresponding tocotrienols can be formed by biosynthesis (Saini and Keum 2016). There are some researchers report that compared with tocotrienols and the other three tocopherols, α-tocopherol has the highest bioactivity, but the antioxidant is opposite. Due to the exist of unsaturated bonds, the antioxidant of tocotrienols is higher than tocopherols.

As already reported, the strategy of vitamin E synthesis is based on the preparation of trimethylhydroquinone, isophytol, and the reaction (Fig. 7.13) followed by acylation to the commercial form (Bonrath et al. 2007).

In 2019, an analysis about natural vitamin E product market report that vitamin E in the world market is characterized by the five companies account for more than half of global demand, market patterns tend to be unified. Due to the low entry threshold for aspiring players, market competition is fierce. Besides, consumers are increasingly drawn toward natural products, which in turn, is one among the key factors favoring growth of natural vitamin E product market. Currently, manufacturers are spending effort and money to explore better production methods to meet the huge demand worldwide. They also attempted to assess the potential of alternative sources, such as rapeseed, palm oil deodorizer distillate and concentration characteristics of mixed sources of vitamin E (Abdul Kapor et al. 2017).

The biosynthetic pathways of tocotrienols and tocopherols in plastids have been widely studied. Tocotrienols are derived from the condensation reaction of HGA with plastidic geranylgeranyl diphosphate (GGDP), which is catalyzed by homogentisate geranylgeranyl transferase (HGGT). Tocopherols originates from the condensation reaction of phytyl diphosphate (PDP) with homogentisic acid (HGA), which is catalyzed by plastidic homogentisate phytyltransferase (HPT) (Fig. 7.14). Subsequent reactions are catalyzed by common enzymes, γ-tocopherol methyltransferase (γ-TMT) and 2-methyl-6-phytylbenzoquinone methyltransferase (MPBQMT) for tocotrienol and tocopherol biosynthesis. The tocopherol cyclase (TC) can not only product δ-tocopherol and γ-tocopherol with MPBQ and DMPBQ as substrates, but also can product δ- tocotrienols and γ- tocotrienols with MGGBQ and DMGGBQ as substrates.

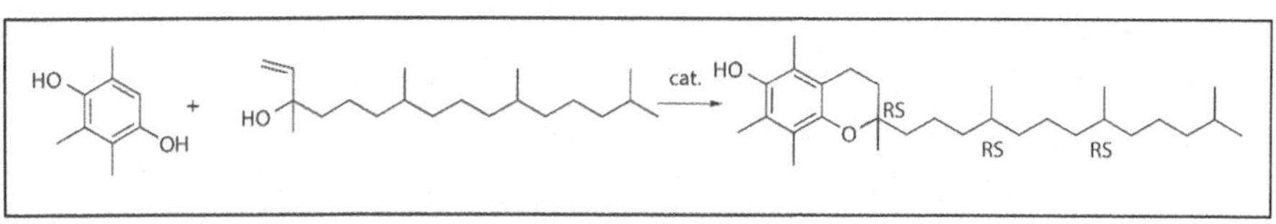

Fig. 7.13 Synthesis of α-tocopherol. (Modified from Bonrath et al. 2007)

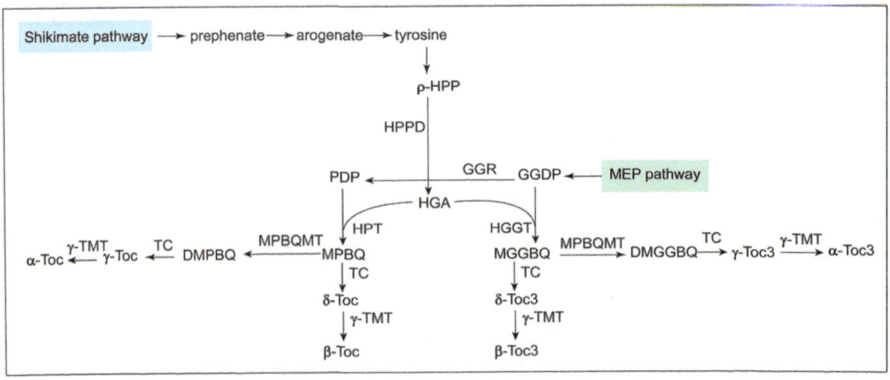

Fig. 7.14 Biosynthetic pathway for tocopherols and tocotrienols in plants (modified from Tanaka et al. 2015). *GGR* geranylgeranyl diphosphate reductase, *HPPD* p-hydroxyphenylpyruvate dioxygenase, *HPT* homogentisate phytyltransferase, *MPBQMT* 2-methyl-6-phytylbenzoquinone methyltransferase, *γ-TMT* γ-tocopherol methyltransferase, *MEP* 2-C-methylerythritol 4-phosphate, TC tocopherol cyclase, *GGDP* geranylgeranyl diphosphate, *PDP* phytyl diphosphate, *HGA* homogentisic acid, *p-HPP* p-hydroxyphenylpyruvate, *MGGBQ* 2-methyl-6-geranylgeranylbenzoquinone, *DMPBQ* 2,3-dimethyl-5-phytyl-1,4-benzoquinone, *MPBQ* 2-methyl-6-phytylbenzoquinone, *Toc* tocopherol, *DMGGBQ* 2,3-dimethyl-5-geranylgeranyl-1,4-benzoquinone, *Toc3* tocotrienol. (Tanaka et al. 2015)

Because of the key enzyme HPT found in experiments has limitations and the key enzyme HGGT mainly increases the content of tocopherol, which is not consistent with the purpose of increasing tocopherol, more studies hope to open a breakthrough in the upstream pathway of vitamin E biosynthesis. The biosynthetic pathway of vitamin E is also limited to some extent by the upstream synthetic pathway (Karunanandaa et al. 2005). The upstream of vitamin E biosynthesis pathway are shikimic acid pathway MEP pathway, respectively. Therefore, it is necessary to modify the synthesis pathways of vitamin E to increasing the output (Ajjawi and David 2004).

A lot of successful research on biosynthesis of vitamin E in plants, such as Cyanobacteria, Arabidopsis, tobacco, canola, corn and soybean. There are some reaction speed limit enzymes have been reported. In the Arabidopsis, or other host plants for overexpression the genes can increase the amount of vitamin E. In transgenic Arabidopsis and soybean seeds, vitamin E increase the maximum total of 1.8 times and 1.4 times. In transgenic Arabidopsis leaf, vitamin E content of up to 4.4 times that of wild-type (Savidge et al. 2002), further studies have shown that transgenic Arabidopsis thaliana in adversity after processing will raise a lot of vitamin E content in the leaves, can increase to up to 18 times (Collakova and DellaPenna 2003). Researchers explain that the activity of *hpt* is limited in the normal physiological, and depressed under the stress.

Tocopherols can accumulate in leaves of Arabidopsis and maize seeds by overexpressing *hggt* gene. There aren't tocotrienols synthesis in leaves of the wild-type Arabidopsis, the proportion of tocopherol can reach 85% of total vitamin E in the presence of the *hggt* (Cahoon et al. 2003). At the same time, tocopherols can be up to 4–6 times in the transgenic corn seed fertility when the *hggt* is present.

7.2.4 *Vitamin K*

7.2.4.1 Structure and Functions of Vitamin K

Vitamin K naturally exists in two forms, called vitamin K1 (phylloquinone) and vitamin K2 (menaquinone) (Shearer and Newman 2008), as shown in Fig. 7.15. All vitamin Ks are fat-soluble compounds with a common 2-methyl-1,4-naphthoquinone core but differ in the side chain structure at the 3-position. Vitamin K1 is present in the green leaf portion of the plant, which acts as an electron acceptor during photosynthesis. The purely extracted form of phylloquinone is a viscous pale yellow oil. Although chlorophyllin is a single compound, menaquinone is a series of vitamins having a polyisoprene unit at the 3-position of the naphthoquinone ring structure (Brodie and Ballantine 1960). Menaquinone is produced by Gram-positive and Gram-negative bacteria such as *Escherichia coli and Bacillus subtilis* as an electron carrier in the electron transport chain required for respiration (Tsukamoto et al. 2001).

Vitamin K is comprised of a 1,4-naphthoquinone group with an aliphatic side-chain in position 3 and a methyl formation in the 2-position (Fig. 7.15). To Vitamin K1, Doebel and Isler developed the most widely known method of Hoffman-La Roche. It involves the condensation of isophyol with menadione. First, during the condensation step, the menadione is reduced under acidic conditions and esterified at C1 to prevent undesirable reactions (Baker et al. 1942). Heat and acid catalyzed the condensation of the phytol with menadiol monoester. The product is de-protected by saponification and oxidized to form final vitamin K1. The process has two notable drawbacks. The first is that the solvents and catalysts are hazardous to both the environment and humans (Bonrath and Netscher 2005). The second drawback is

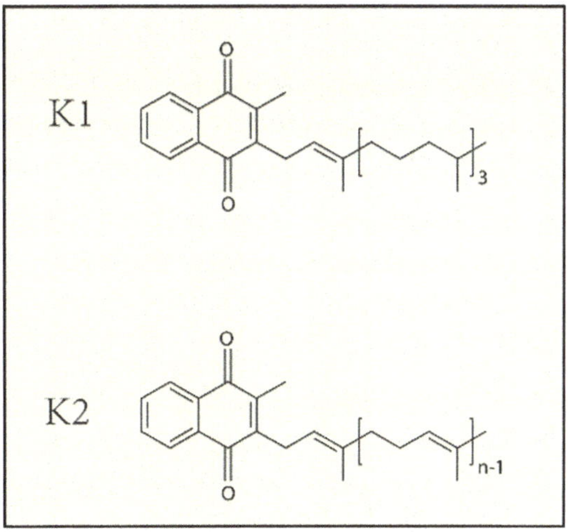

Fig. 7.15 Structure of K1 and K2

that the condensation of menadione with phytol results in the formation of the inactive Z-isomer (Daines et al. 2003). Studies on the chemical synthesis of vitamin K1 have focused on improving the stereoselectivity of the phytochemical step. Because the formation of the inactive isomer reduces the activity and yield of the final product (Daines et al. 2003). The method of Coman et al. overcomes difficulties to reduce or eliminate the use of toxic chemicals in the synthesis of vitamin K1 (Coman et al. 2010).

7.2.4.2 Metabolic Pathways of Vitamin K1

Metabolic engineering can be used to improve the output of the phylloquinone. Engineering work usually begins with determining the rate-influencing steps in the synthesis pathway. As shown in Fig. 7.16, phylloquinone and menaquinones in bacteria have some points of similarity in the biosynthetic pathway. In the process, the synthesis of the phytoside chain begins with isopentenyl pyrophosphate (IPP). By geranylgeranyl pyrophosphate synthase (GGPPS) or geranyl pyrophosphate synthase (GPPS), three IPP molecule and DMAPP are used to produce geranylgeranyl pyrophosphate (GGPP). Finally, phytyl pyrophosphate (phytyl-PP) can be produced in three sequential steps by geranyl pyrophosphate reductase (GGPR). Synthesis of the naphthoquinone skeleton begins with chorismate from the shikimate pathway. The chorismate is converted to o-succinyl benzoate (OSB), further converted to 1, 4-dihydroxynaphthalic acid (DHNA), cyclized to produce DHNA -CoA, and hydrolyzed to DHNA. 2-phytyl-1,4-naphthoquinone (PNQ) can be achieved by the DHNA and phytyl-PP. PNQ molecules are reduced and methylated to produce chlorophyll, which spontaneously oxidizes to produce phylloquinone.

7.2.4.3 Development of Vitamin K2

The MK-7 fermentation process can be carried out by solid or liquid state fermentation. Solid state fermentation (SSF) process can have a water content from 12% to 80% (Berenjian et al. 2013), while liquid fermentation (LSF), the water content is 90–95% (Berenjian et al. 2011). In addition, low productivity of MK-7 lead to an expensive process (Berenjian et al. 2015). Therefore, research has been conducted over the past decades to enhance the production of MK-7 (Berenjian et al. 2014).

In general, the main factors affecting the production of MK-7 by the SSF system are the selection of the water activity of the substrate, the size and type of the inoculum, the temperature, the fermentation time, the microbial strain, the appropriate substrate, the pretreatment and the particle size (Pandey 2003). The choice of matrix for MK-7 production in the SSF process is primarily dependent on cost and availability and therefore typically involves the screening of several solid matrices. Typically, raw matrices are used for SSF. Simultaneous substrate pretreatment and fermentation have been used to increase vitamin production. Among the bacterial strains, Bacillus licheniformis and Bacillus subtilis are well-studied strains for

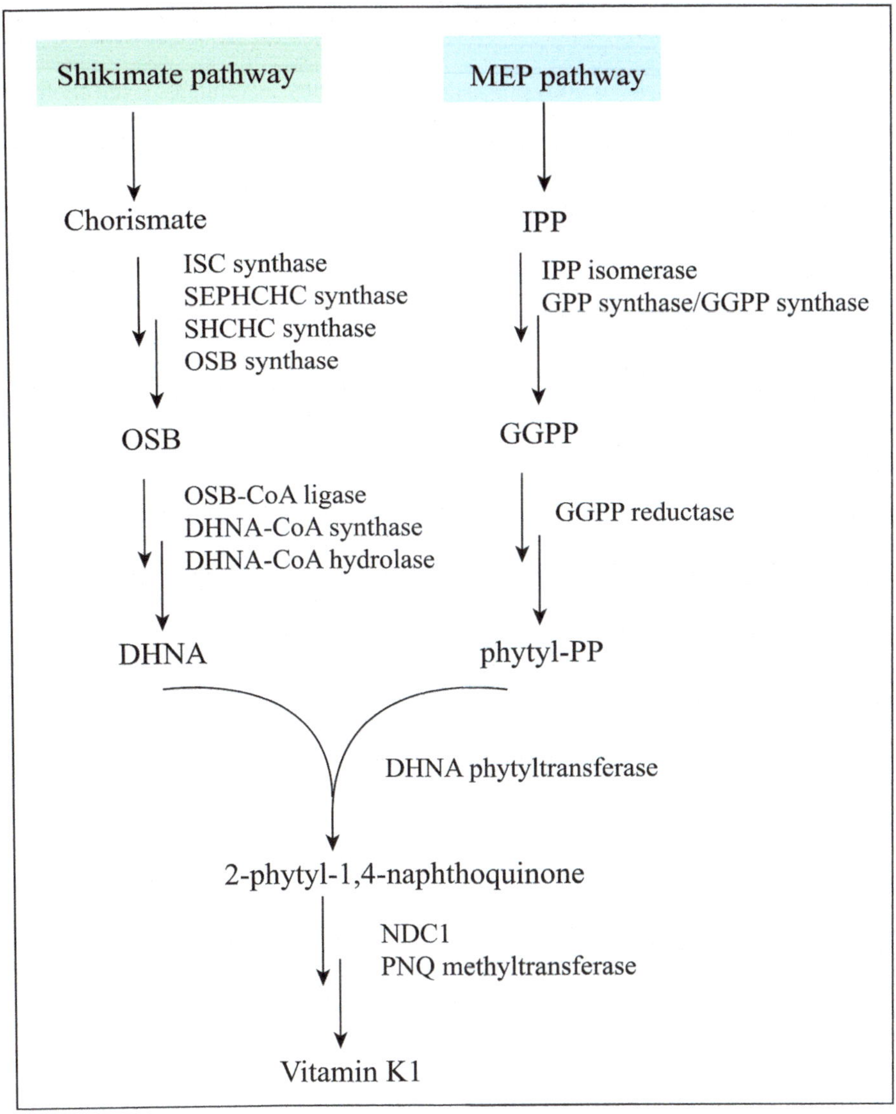

Fig. 7.16 The biosynthetic pathway of phylloquinone. (Modified from Tarento et al. 2018)

MK-7 (Berenjian et al. 2012). Bacillus licheniformis is an organism with a well characterized membrane containing menaquinone as the sole quinone and capable of anaerobic growth. MK-7 is present in wild-type and mutant strains with similar respiratory spasms. Goodman reported that the highest amount of MK-7 was 0.25 μg/mg dry weight of cells (Mahdinia et al. 2017).

7.3 Conclusions and Perspectives

Microbial processes for vitamin production have many advantages over chemical synthesis processes. The product from the chemical process is typically a racemic mixture, however, fermentation or bioconversion reaction produces the enantiomer compound required. Furthermore, advances in biochemistry, DNA technology and the genomic revolution have expanded the options for developing biotechnology in vitamin production. In addition, biotechnology processes and products often have a positive environmental impact.

In future studies, based on well-reported biosynthesis pathways, some lipid soluble vitamin metabolic pathways will be constructed to achieve microbiological product. Microbial synthetic vitamins face opportunities and challenges. In any case, there are many strategies for improving the productivity of secondary metabolites, including: selecting high-yielding organisms, metabolic engineering, and optimizing environmental conditions.

References

Abdul Kapor NZ, Maniam GP, Rahim MHA, Yusoff MM. Palm fatty acid distillate as a potential source for biodiesel production-a review. J Clean Prod. 2017;143:1–9.

Acevedo-Rocha CG, Gronenberg LS, Mack M, Commichau FM, Genee HJ. Microbial cell factories for the sustainable manufacturing of B vitamins. Curr Opin Biotechnol. 2019;56:18–29.

Ajjawi I, David S. Engineered plants with elevated vitamin E: a nutraceutical success story. Trends Biotechnol. 2004;22(3):99–100.

Attaluri P, Castillo A, Edriss H, Nugent K. Thiamine deficiency: an important consideration in critically ill patients. Am J Med Sci. 2018;356(4):382–90.

Avissar YJ, Ormerod JG, Beale SI. Distribution of δ-aminolevulinic acid biosynthetic pathways among phototrophic bacterial groups. Arch Microbiol. 1989;151(6):513–9.

Bailey SW, Ayling JE. The extremely slow and variable activity of dihydrofolate reductase in human liver and its implications for high folic acid intake. Proc Natl Acad Sci. 2009;106(36):15424–9.

Baker BR, Davies TH, McElroy L, et al. The antihemorrhagic activity of sulfonated derivatives of 2-methylnaphthalene. J Am Chem Soc. 1942;64(5):1096–101.

Banerjee RV, Matthews RG. Cobalamin-dependent methionine synthase. FASEB J. 1990;4(5):1450–9.

Basset GJC, Quinlivan EP, Gregory JF, Hanson AD. Folate synthesis and metabolism in plants and prospects for biofortification this work was supported in part by the Florida Agricultural Experiment Station, by an endowment from the C.V. Griffin, Sr. Foundation and by grant MCB-0129944 from the National Science Foundation. Journal Series no. R-09861. Crop Sci. 2005;45:449–53.

Begley TP, Downs DM, Ealick SE, McLafferty FW, Van Loon APGM, Taylor S, Campobasso N, Chiu H-J, Kinsland C, Reddick JJ, Xi J. Thiamin biosynthesis in prokaryotes. Arch Microbiol. 1999;171(5):293–300.

Berenjian A, Mahanama R, Talbot A, Biffin R, Regtop H, Valtchev P, Kavanagh J, Dehghani F. Efficient media for high menaquinone-7 production: response surface methodology approach. N Biotechnol. 2011;28(6):665–72.

Berenjian A, Mahanama R, Talbot A, Regtop H, John K, Fariba D. Advances in menaquinone-7 production by bacillus subtilis natto: fed-batch glycerol addition. Am J Biochem Biotechnol. 2012;8(2):105–10.

Berenjian A, Chan NL, Mahanama R, Talbot A, Regtop H, Kavanagh J, Dehghani F. Effect of biofilm formation by Bacillus subtilis natto on menaquinone-7 biosynthesis. Mol Biotechnol. 2013;54(2):371–8.

Berenjian A, Mahanama R, Talbot A, Regtop H, Kavanagh J, Dehghani F. Designing of an intensification process for biosynthesis and recovery of menaquinone-7. Appl Biochem Biotechnol. 2014;172(3):1347–57.

Berenjian A, Mahanama R, Kavanagh J, Dehghani F. Vitamin K series: current status and future prospects. Crit Rev Biotechnol. 2015;35(2):199–208.

Bernstein PS, Rando RR. In vivo isomerization of all-trans- to 11-cis-retinoids in the eye occurs at the alcohol oxidation-state. Biochemistry. 1986;25(21):6473–8.

Beveridge LA, Khan F, Struthers AD, Armitage J, Barchetta I, Bressendorff I, Cavallo MG, Clarke R, Dalan R, Dreyer G. Effect of vitamin D supplementation on markers of vascular function: a systematic review and individual participant meta-analysis. J Am Heart Assoc. 2018;7(11):1–20.

Blanche F, Debussche L, Thibaut D, Crouzet J, Cameron B. Purification and characterization of S-adenosyl-L-methionine: uroporphyrinogen III methyltransferase from Pseudomonas denitrificans. J Bacteriol. 1989;171(8):4222–31.

Blunt J, DeLuca H, Schnoes H. 25-hydroxycholecalciferol. A biologically active metabolite of vitamin D3. Biochemistry. 1968;7(10):3317–22.

Bocobza SE, Aharoni A. Switching the light on plant riboswitches. Trends Plant Sci. 2008;13(10):526–33.

Bonrath W, Netscher T. Catalytic processes in vitamins synthesis and production. Appl Catal A Gen. 2005;280(1):55–73.

Bonrath W, Dittel C, Giraudi L, Netscher T, Pabst T. Rare earth triflate catalysts in the synthesis of Vitamin E and its derivatives. Catal Today. 2007;121(1–2):65–70.

Botto LD, Yang Q. 5, 10-methylenetetrahydrofolate reductase gene variants and congenital anomalies: a HuGE review. Am J Epidemiol. 2000;151(9):862–77.

Boukandoul S, Casal S, Cruz R, Pinho C, Zaidi F. Algerian Moringa oleifera whole seeds and kernels oils: characterization, oxidative stability, and antioxidant capacity. Eur J Lipid Sci Technol. 2017;119(10):1–11.

Brodie AF, Ballantine J. Oxidative phosphorylation in fractional bacterial systems. III. Specificity of vitamin K reactivation. J Biol Chem. 1960;235(1):232–7.

Brushaber KR, O'Toole GA, Escalante-Semerena JC. CobD, a novel enzyme with L-threonine-O-3-phosphate decarboxylase activity, is responsible for the synthesis of (R)-1-amino-2-propanol O-2-phosphate, a proposed new intermediate in cobalamin biosynthesis in Salmonella typhimurium LT2. J Biol Chem. 1998;273(5):2684–91.

Cahoon EB, Hall SE, Ripp KG, Ganzke TS, Hitz WD, Coughlan SJ. Metabolic redesign of vitamin E biosynthesis in plants for tocotrienol production and increased antioxidant content. Nat Biotechnol. 2003;21(9):1082–7.

Capozzi V, Russo P, Dueñas MT, López P, Spano G. Lactic acid bacteria producing B-group vitamins: a great potential for functional cereals products. Appl Microbiol Biotechnol. 2012;96(6):1383–94.

Christensen KE, MacKenzie RE. Mitochondrial one-carbon metabolism is adapted to the specific needs of yeast, plants and mammals. Bioessays. 2006;28(6):595–605.

Collakova E, DellaPenna D. The role of homogentisate phytyltransferase and other tocopherol pathway enzymes in the regulation of tocopherol synthesis during abiotic stress. Plant Physiol. 2003;133(2):930–40.

Coman SM, Parvulescu VI, Wuttke S, Kemnitz E. Synthesis of vitamin K1and K1-Chromanol by Friedelâ crafts alkylation in heterogeneous catalysis. ChemCatChem. 2010;2(1):92–7.

Conaway HH, Henning P, Lerner UH. Vitamin a metabolism, action, and role in skeletal homeostasis. Endocr Rev. 2013;34(6):766–97.

Copp AJ, Stanier P, Greene NDE. Neural tube defects: recent advances, unsolved questions, and controversies. Lancet Neurol. 2013;12(8):799–810.

Crider KS, Yang TP, Berry RJ, Bailey LB. Folate and DNA methylation: a review of molecular mechanisms and the evidence for folate's role. Adv Nutr. 2012;3(1):21–38.

Czeizel AE, Dudas I, Vereczkey A, Banhidy F. Folate deficiency and folic acid supplementation: the prevention of neural-tube defects and congenital heart defects. Nutrients. 2013;5(11):4760–75.

Daines AM, Payne RJ, Humphries ME. The synthesis of naturally occurring vitamin K and vitamin K analogues. Curr Org Chem. 2003;7(16):1625–34.

Dasgupta N, Ranjan S, Mundra S, Ramalingam C, Kumar A. Fabrication of food grade Vitamin E Nanoemulsion by low energy approach, characterization and its application. Int J Food Prop. 2015;19(3):700–8.

Fan C, Bobik TA. The PduX enzyme of Salmonella enterica is an L-threonine kinase used for coenzyme B12 synthesis. J Biol Chem. 2008;283(17):11322–9.

Fang H, Kang J, Zhang D. Microbial production of vitamin B12: a review and future perspectives. Microb Cell Fact. 2017;16(1):1–14.

Fowler B, Leonard J, Baumgartner M. Causes of and diagnostic approach to methylmalonic acidurias. J Inherit Metab Dis. 2008;31(3):350–60.

Froese DS, Kochan G, Muniz JR, Wu X, Gileadi C, Ugochukwu E, Krysztofinska E, Gravel RA, Oppermann U, Yue WW. Structures of the human GTPase MMAA and vitamin B12-dependent methylmalonyl-CoA mutase and insight into their complex formation. J Biol Chem. 2010;285(49):38204–13.

Froese DS, Fowler B, Baumgartner MR. Vitamin B_{12}, folate, and the methionine remethylation cycle—biochemistry, pathways, and regulation. J Inherit Metab Dis. 2018;42:673–85.

Greenberg JA, Bell SJ, Guan Y, Yu Y-H. Folic acid supplementation and pregnancy: more than just neural tube defect prevention. Rev Obstet Gynecol. 2011;4(2):52–9.

Hanson AD, Gregory JF. Folate biosynthesis, turnover, and transport in plants. Annu Rev Plant Biol. 2011;62(1):105–25.

Herbers K. Vitamin production in transgenic plants. J Plant Physiol. 2003;160(7):821–9.

Hinterberger M, Fischer P. Folate and Alzheimer: when time matters. J Neural Transm. 2013;120(1):211–24.

Hoffbrand AV, Weir DG. The history of folic acid. Br J Haematol. 2001;113(3):579–89.

Iosue CL, Attanasio N, Shaik NF, Neal EM, Leone SG, Cali BJ, Peel MT, Grannas AM, Wykoff DD. Partial decay of thiamine signal transduction pathway alters growth properties of *Candida glabrata*. PLoS One. 2016;11(3):1–18.

Jenkins AH, Schyns G, Potot S, Sun G, Begley TP. A new thiamin salvage pathway. Nat Chem Biol. 2007;3(8):492–7.

Jones G. The discovery and synthesis of the nutritional factor vitamin D. Int J Paleopathol. 2018;23:96–9.

Jones G, Strugnell SA, Deluca HF. Current understanding of the molecular actions of Vitamin D. Physiol Rev. 1998;78(4):1193–231.

Kang Z, Wang Y, Gu P, Wang Q, Qi Q. Engineering *Escherichia coli* for efficient production of 5-aminolevulinic acid from glucose. Metab Eng. 2011;13(5):492–8.

Karunanandaa B, Qi Q, Hao M, Baszis SR, Jensen PK, Wong YH, Jiang J, Venkatramesh M, Gruys KJ, Moshiri F, Post-Beittenmiller D, Weiss JD, Valentin HE. Metabolically engineered oilseed crops with enhanced seed tocopherol. Metab Eng. 2005;7(5–6):384–400.

Lawrence MC, Iliades P, Fernley RT, Berglez J, Pilling PA, Macreadie IG. The three-dimensional structure of the bifunctional 6-hydroxymethyl-7,8-dihydropterin pyrophosphokinase/dihydropteroate synthase of Saccharomyces cerevisiae. J Mol Biol. 2005;348(3):655–70.

Létinois U, Schütz J, Härter R, Stoll R, Huffschmidt F, Bonrath W, Karge R. Lewis acid-catalyzed synthesis of 4-aminopyrimidines: a scalable industrial process. Org Process Res Dev. 2013;17(3):427–31.

Lonsdale D. A review of the biochemistry, metabolism and clinical benefits of thiamin(e) and its derivatives. Evid Based Complement Alternat Med. 2006;3(1):49–59.

Mahdinia E, Demirci A, Berenjian A. Production and application of menaquinone-7 (vitamin K2): a new perspective. World J Microbiol Biotechnol. 2017;33(2):1–7.

Martens J-H, Barg H, Warren M, Jahn D. Microbial production of vitamin B 12. Appl Microbiol Biotechnol. 2002a;58(3):275–85.

Martens J-H, Barg H, Warren M, Jahn D. Microbial production of vitamin B12. Appl Microbiol Biotechnol. 2002b;58(3):275–85.

Mohd Mutalip SS, Ab-Rahim S, Rajikin MH. Vitamin E as an antioxidant in female reproductive health. Antioxid Redox Signal. 2018;7(2):1–15.

Moretti D, Biebinger R, Bruins MJ, Hoeft B, Kraemer K. Bioavailability of iron, zinc, folic acid, and vitamin A from fortified maize. Ann N Y Acad Sci. 2014;1312(1):54–65.

Nahvi A, Sudarsan N, Ebert MS, Zou X, Brown KL, Breaker RR. Genetic control by a metabolite binding mRNA. Chem Biol. 2002;9(9):1043–9.

Nudler E. Flipping Riboswitches. Cell. 2006;126(1):19–22.

Omenn GS, Goodman GE, Thornquist MD, Balmes J, Cullen MR, Glass A, Keogh JP, Meyskens FL Jr, Valanis B, Williams JH Jr. Effects of a combination of beta carotene and vitamin A on lung cancer and cardiovascular disease. N Engl J Med. 1996;334(18):1150–5.

Pandey A. Solid-state fermentation. Biochem Eng J. 2003;13:81–4.

Panfili G, Fratianni A, Irano M. Normal phase high-performance liquid chromatography method for the determination of tocopherols and tocotrienols. J Agric Food Chem. 2003;51(14):3940–4.

Park JH, Burns K, Kinsland C, Begley TP. Characterization of two kinases involved in thiamine pyrophosphate and pyridoxal phosphate biosynthesis in Bacillus subtilis: 4-amino-5-hydroxymethyl-2methylpyrimidine kinase and pyridoxal kinase. J Bacteriol. 2004;186(5):1571–3.

Perez-Hernandez N, Aptilon-Duque G, Nostroza-Hernandez MC, Vargas-Alarcon G, Rodriguez-Perez JM, Blachman-Braun R. Vitamin D and its effects on cardiovascular diseases: a comprehensive review. Korean J Intern Med. 2016;31(6):1018–29.

Pfleger BF, Pitera DJ, Smolke CD, Keasling JD. Combinatorial engineering of intergenic regions in operons tunes expression of multiple genes. Nat Biotechnol. 2006;24(8):1027–32.

Pilz S, Zittermann A, Trummer C, Theiler-Schwetz V, Lerchbaum E, Keppel MH, Grubler MR, Marz W, Pandis M. Vitamin D testing and treatment: a narrative review of current evidence. Endocr Connect. 2019;8(2):27–43.

Poe M, Hoogsteen K, Matthews DA. Proton magnetic resonance studies on Escherichia coli dihydrofolate reductase. Assignment of histidine C-2 protons in binary complexes with folates on the basis of the crystal structure with methotrexate and on chemical modifications. J Biol Chem. 1979;254(17):8143–52.

Rahul. Vitamin B1 (thiamine mononitrate) market overview and forecast report. Market Research Gazette. 2019.

Raschke M, Bürkle L, Müller N, Nunes-Nesi A, Fernie AR, Arigoni D, Amrhein N, Fitzpatrick TB. Vitamin B1 biosynthesis in plants requires the essential iron–sulfur cluster protein, THIC. Proc Natl Acad Sci. 2007;104(49):19637–42.

Raux E, Lanois A, Levillayer F, Warren MJ, Brody E, Rambach A, Thermes C. Salmonella typhimurium cobalamin (vitamin B12) biosynthetic genes: functional studies in *S. typhimurium* and *Escherichia coli*. J Bacteriol. 1996;178(3):753–67.

Raux E, Schubert HL, Warren MJ. Biosynthesis of cobalamin (vitamin B12): a bacterial conundrum. Cell Mol Life Sci. 2000;57(13):1880–93.

Revuelta JL, Serrano-Amatriain C, Ledesma-Amaro R, Jiménez A. Formation of folates by microorganisms: towards the biotechnological production of this vitamin. Appl Microbiol Biotechnol. 2018;102(20):8613–20.

Rodionov DA, Vitreschak AG, Mironov AA, Gelfand MS. Comparative genomics of thiamin biosynthesis in procaryotes. New genes and regulatory mechanisms. J Biol Chem. 2002;277(50):48949–59.

Roth JR, Lawrence JG, Rubenfield M, Kieffer-Higgins S, Church GM. Characterization of the cobalamin (vitamin B12) biosynthetic genes of Salmonella typhimurium. J Bacteriol. 1993;175(11):3303–16.

Roth JR, Lawrence J, Bobik T. Cobalamin (coenzyme B12): synthesis and biological significance. Annu Rev Microbiol. 1996;50(1):137–81.

Saini RK, Keum Y-S. Tocopherols and tocotrienols in plants and their products: a review on methods of extraction, chromatographic separation, and detection. Food Res Int. 2016;82:59–70.

Saini RK, Nile SH, Keum Y-S. Folates: chemistry, analysis, occurrence, biofortification and bioavailability. Food Res Int. 2016;89:1–13.

Savastano S, Barrea L, Savanelli MC, Nappi F, Di Somma C, Orio F, Colao A. Low vitamin D status and obesity: role of nutritionist. Rev Endocr Metab Disord. 2017;18(2):215–25.

Savidge B, Weiss JD, Wong YH, Lassner MW, Mitsky TA, Shewmaker CK, Post-Beittenmiller D, Valentin HE. Isolation and characterization of homogentisate phytyltransferase genes from *Synechocystis* sp. PCC 6803 and Arabidopsis. Plant Physiol. 2002;129(1):321–32.

Schyns G, Potot S, Geng Y, Barbosa TM, Henriques A, Perkins JB. Isolation and characterization of new thiamine-deregulated mutants of *Bacillus subtilis*. J Bacteriol. 2005;187(23):8127–36.

Shane B. Folylpolyglutamate synthesis and role in the regulation of one-carbon metabolism. Vitam Horm, Elsevier. 1989;45:263–335.

Shearer MJ, Newman P. Metabolism and cell biology of vitamin K. Thromb Haemost. 2008;100(10):530–47.

Stephensen CB. Vitamin A, infection, and immune function. Annu Rev Nutr. 2001;21(1):167–92.

Sybesma W, Van Den Born E, Starrenburg M, Mierau I, Kleerebezem M, De Vos WM, Hugenholtz J. Controlled modulation of folate polyglutamyl tail length by metabolic engineering of Lactococcus lactis. Appl Environ Microbiol. 2003a;69(12):7101–7.

Sybesma W, Starrenburg M, Kleerebezem M, Mierau I, de Vos WM, Hugenholtz J. Increased production of folate by metabolic engineering of *Lactococcus lactis*. Appl Environ Microbiol. 2003b;69(6):3069–76.

Sybesma W, Burgess C, Starrenburg M, van Sinderen D, Hugenholtz J. Multivitamin production in Lactococcus lactis using metabolic engineering. Metab Eng. 2004;6(2):109–15.

Tanaka H, Yabuta Y, Tamoi M, Tanabe N, Shigeoka S. Generation of transgenic tobacco plants with enhanced tocotrienol levels through the ectopic expression of rice homogentisate geranylgeranyl transferase. Plant Biotechnol. 2015;32(3):233–8.

Tarento TDC, McClure DD, Talbot AM, Regtop HL, Biffin JR, Valtchev P, Dehghani F, Kavanagh JM. A potential biotechnological process for the sustainable production of vitamin K1. Crit Rev Biotechnol. 2018;39:1–19.

Tibbetts AS, Appling DR. Compartmentalization of mammalian folate-mediated one-carbon metabolism. Annu Rev Nutr. 2010;30(1):57–81.

Tsukamoto Y, Kasai M, Kakuda H. Construction of a Bacillus subtilis (natto) with high productivity of vitamin K2 (menaquinone-7) by analog resistance. Biosci Biotechnol Biochem. 2001;65(9):2007–15.

Vitreschak AG. Regulation of the vitamin B12 metabolism and transport in bacteria by a conserved RNA structural element. RNA. 2003;9(9):1084–97.

Vitreschak AG, Rodionov DA, Mironov AA, Gelfand MS. Regulation of the vitamin B12 metabolism and transport in bacteria by a conserved RNA structural element. RNA. 2003;9(9):1084–97.

Voelker AL, Miller J, Running CA, Taylor LS, Mauer LJ. Chemical stability and reaction kinetics of two thiamine salts (thiamine mononitrate and thiamine chloride hydrochloride) in solution. Food Res Int. 2018;112:443–56.

Williams J, Mai CT, Mulinare J, Isenburg J, Flood TJ, Ethen M, Frohnert B, Kirby RS. Centers for disease, C.; prevention. Updated estimates of neural tube defects prevented by mandatory folic acid fortification – United States, 1995–2011. MMWR Morb Mortal Wkly Rep. 2015;64(1):1–5.

Winkler W, Nahvi A, Breaker RR. Thiamine derivatives bind messenger RNAs directly to regulate bacterial gene expression. Nature. 2002;419(6910):952–6.

Wright A, Finglas P, Dainty J, Hart D, Wolfe C, Southon S, Gregory J. Single oral doses of ^{13}C forms of pteroylmonoglutamic acid and 5-formyltetrahydrofolic acid elicit differences in short-term kinetics of labelled and unlabelled folates in plasma: potential problems in interpretation of folate bioavailability studies. Br J Nutr. 2003;90(2):363–71.

Ye X, Al-Babili S, Klöti A, Zhang J, Lucca P, Beyer P, Potrykus I. Engineering the provitamin A (β-carotene) biosynthetic pathway into (carotenoid-free) rice endosperm. Science. 2000;287:303–5.

Zappa S, Li K, Bauer CE. The tetrapyrrole biosynthetic pathway and its regulation in Rhodobacter capsulatus. In: Recent advances in phototrophic prokaryotes. New York/London: Springer; 2010. p. 229–50.

Zhang J, Rana S, Srivastava RS, Misra RDK. On the chemical synthesis and drug delivery response of folate receptor-activated, polyethylene glycol-functionalized magnetite nanoparticles. Acta Biomater. 2008;4(1):40–8.

Zhao L, Ma X-D, Chen F-E. Development of two scalable syntheses of 4-Amino-5-aminomethyl-2-methylpyrimidine: key intermediate for vitamin B1. Org Process Res Dev. 2012;16(1):57–60.

Zhu T, Koepsel R, Domach MM, Ataai MM. Metabolic engineering of folic acid production. Ferment Biotechnol. 2003;862:207–19.

Zhu T, Pan Z, Domagalski N, Koepsel R, Ataai MM, Domach MM. Engineering of Bacillus subtilis for enhanced total synthesis of folic acid. Appl Environ Microbiol. 2005;71(11):7122–9.

Zittermann A, Ernst JB, Gummert JF, Borgermann J. Vitamin D supplementation, body weight and human serum 25-hydroxyvitamin D response: a systematic review. Eur J Nutr. 2014;53(2):367–74.

Chapter 8
Systems and Synthetic Biotechnology for the Production of Polyunsaturated Fatty Acids

Wei-Jian Wang, He Huang, and Xiao-Jun Ji

8.1 Introduction

Polyunsaturated fatty acids (PUFAs) have vital structural and functional roles in higher organisms including humans. As constituents of phospholipids, they confer flexibility, fluidity and selective permeability to membranes and consequently are of high physiological and therapeutic significance for human well-being (Bellou et al. 2016; Ji et al. 2014a, b, 2015a, b; Ji and Huang 2019; Ward and Singh 2005). The two key types of PUFAs are distinguished by the distance between their last double bond and the methyl-end of the acyl chain. Thus, omega-3/6 designates a PUFA whose last double bond is located at the third or sixth carbon from the omega-end of the carbon chain, respectively. Typical PUFAs include the omega-6 γ-linolenic acid (GLA) and arachidonic acid (ARA) (Sun et al. 2017; Wu et al. 2017), as well

W.-J. Wang
College of Biotechnology and Pharmaceutical Engineering, Nanjing Tech University, Nanjing, People's Republic of China

H. Huang
School of Pharmaceutical Sciences, Nanjing Tech University, Nanjing, People's Republic of China

State Key Laboratory of Materials-Oriented Chemical Engineering, Nanjing Tech University, Nanjing, People's Republic of China

Jiangsu National Synergetic Innovation Center for Advanced Materials (SICAM), Nanjing, People's Republic of China

X.-J. Ji (✉)
College of Biotechnology and Pharmaceutical Engineering, Nanjing Tech University, Nanjing, People's Republic of China

Jiangsu National Synergetic Innovation Center for Advanced Materials (SICAM), Nanjing, People's Republic of China
e-mail: xiaojunji@njtech.edu.cn

L. Liu, J. Chen (eds.), *Systems and Synthetic Biotechnology for Production of Nutraceuticals*, https://doi.org/10.1007/978-981-15-0446-4_8

as the omega-3 eicosapentaenoic acid (EPA) and docosahexaenoic acid (DHA) (Guo et al. 2017, 2018; Xue et al. 2013).

The omega-6 PUFAs are important precursors that are metabolized by various enzymes to produce a wide range of biologically and clinically important eicosanoid hormones, including prostaglandins, leukotrienes and thromboxanes, some of which play important roles in combating or preventing a number of human diseases (Ji et al. 2014a, b). While the omega-3 PUFAs are important for normal metabolism, they cannot be synthesized by mammals, which make them essential fatty acids which are of great importance for good health. They have many positive effects on human beings, such as anti-inflammatory and anti-blood clotting activity, lowering triglyceride (TAG) levels and reducing blood pressure, as well as reducing the risk of diabetes and certain types of cancer (Ji et al. 2015a, b). In addition, the omega-6 ARA and omega-3 DHA are important components of breast milk. These two PUFAs are present in breast milk but are absent from both cow's milk and infant formulas, which are frequently used as substitutes for breast milk. However, there is clinical evidence that these PUFAs can improve the memory and eyesight of babies because they are involved in the development of neural and retinal functions (Ratledge 2004).

Because of these benefits of PUFAs, there is a strong desire to develop an abundant, safe and economical source of edible PUFAs as an affordable food ingredient. However, the PUFA molecule is chemically unstable and prone to oxidation due to the unsaturated double bonds. Therefore, the use of PUFA-rich lipid, which is more stable than isolated PUFAs, as a food additive, has been suggested. Traditionally, GLA has been extracted from plants such as Oenothera, borage, etc., while ARA was derived from animal tissues such as adrenal glands, livers, etc. The most important traditional source of the omega-3 EPA and DHA are deep sea fish oils (Fig. 8.1). As awareness of the health benefits of these compounds has grown, traditional sources of PUFAs, including animals and plants, have not been able to keep up with demand. Driven by economic incentives, industrial microorganisms are increasingly being used for the production of PUFAs. These are mainly the so-called oleaginous microorganisms, including certain yeasts, molds, and algae, which can generally accumulate >20% lipids in their dry cell weight and can be cultured on a wide variety of substrates with relatively high growth rates, while the lipids they produce are of high purity (Ji and Huang 2019).

Importantly, some oleaginous species are known to naturally accumulate omega-3/6 PUFAs among their storage lipids (Table 8.1). However, the oil-production efficiency of these microbes is generally low, which makes it vital to improve their metabolic flux towards PUFA production, and thus enhance the competitiveness of the resulting biotechnological routes with the traditional extraction methods. With the emerging methods of systems and synthetic biotechnology in recent years, the microbial lipid and PUFA bio-accumulation mechanism has been elucidated and the microbial production of some of them has reached a high level. Engineering oleaginous strains in order to boost their PUFA accumulation is being

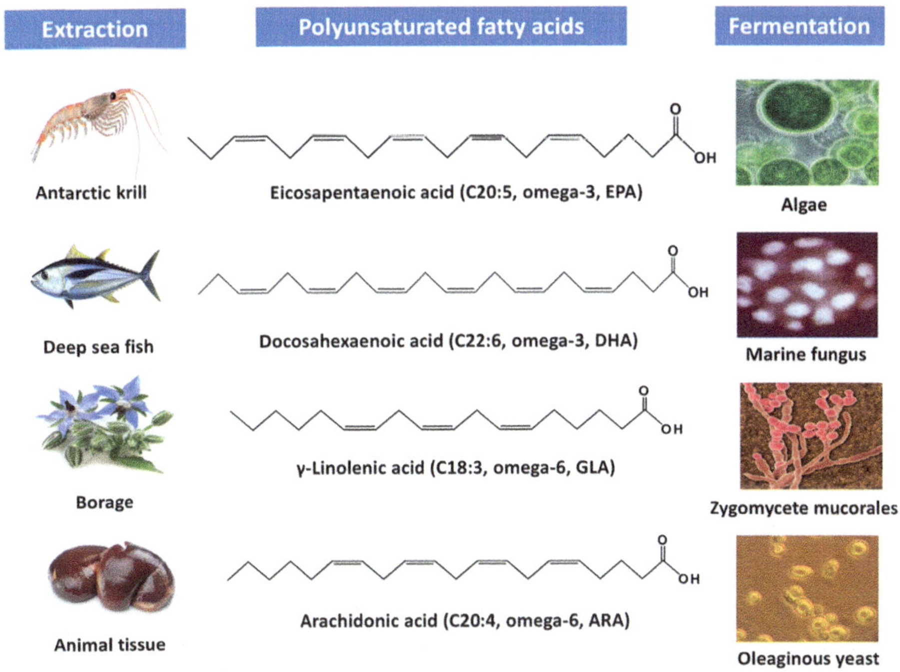

Fig. 8.1 Sources and structures of the omega-3 and -6 polyunsaturated fatty acids

attempted in both native PUFA producers and constructed microorganisms in which the ability to form PUFAs was introduced via genetic engineering. With advances in systems biotechnology, such as genomic and proteomic analysis, many studies revealed metabolic network shifts towards lipid accumulation under different kinds of stresses (Shi et al. 2017, 2018). Systems biotechnology has become a powerful tool to investigate the complex molecular mechanisms of lipid accumulation in oleaginous microorganisms. With the introduction of synthetic biotechnology, it has become possible to engineer microbial cell factories for the production of PUFAs. *Yarrowia lipolytica*, one of the most prominent non-conventional yeasts with potential biotechnological applications, has been engineered for PUFA production (Liu et al. 2017a, b; Xie et al. 2015; Xue et al. 2013). There are several reports on the reconstruction of heterologous fatty acids pathways in *Y. lipolytica*, which enabled the production of EPA, DHA, ARA and GLA. However, making cells into efficient factories is challenging because cells have evolved robust metabolic networks with hard-wired, tightly regulated lines of communication between molecular pathways, which resist efforts to divert resources. Developing new cell factories that meet the economic requirements for industrial-scale PUFA production is still challenging (Ji and Huang 2019; Nielsen and Keasling 2016). Here, we will focus on the recent progress in boosting the production of PUFAs using the aforementioned two kinds of emerging technologies and provide guidance for future research.

Table 8.1 The representative omega-3/6 PUFAs production using the oleaginous microorganisms

PUFAs	Conventional sources	Alternative microbial sources	References
GLA	Plants: Oenothera, Borage, Ornithogalum spp.	Mucor circinelloides	Zhao et al. (2016)
		Mortierella isabellina	Gao et al. (2014)
		Mortierella ramanniana	Dyal et al. (2005)
		Cunninghamella sp.	Al-Hawash et al. (2018)
ARA	Animal tissues: Adrenal glands, livers, egg yolks	Mortierella alpina	Ji et al. (2014a, b)
		Mortierella elongate	Yamada et al. (1987)
		Pythium irregulae	Cheng et al. (1999)
EPA	Fish oils: Brevoortia, Engraulis, Sardina, Scomber spp.	Pythium irregulae	O'Brien et al. (1993)
		Phaeodactylum tricornutum	Yongmanitchai and Ward (1991)
		Nitzschia laevis	Wen et al. (2002)
		Nannochloropsis sp.	Zou and Richmond (1999)
		Porphyridium cruentum	Hu et al. (2018)
		Shewanella sp.	Orikasa et al. (2004)
DHA	Fish oils: Brevoortia, Engraulis, Sardina, Scomber spp.	Crypthecodinium cohnii	Diao et al. (2019)
		Schizochytrium sp.	Ling et al. (2015)
		Aurantiochytrium sp.	Ma et al. (2017)
		Thraustochytrium sp.	Singh and Ward (1996)
		Ulkenia sp.	Kiy et al. (2005)
		Isochrysis galbana	Molina Grima et al. (1993)

Abbreviations: *PUFA* polyunsaturated fatty acid, *GLA* γ-linolenic acid, *ARA* arachidonic acid, *EPA* eicosapentaenoic acid, *DHA* docosahexaenoic acid

8.2 Systems Biotechnology for PUFA Production

The accumulation of PUFA-rich lipids is normally induced under various environmental stresses, such as high light, elevated salt concentrations, and deprivation of nutrients (Wang et al. 2004; Donot et al. 2014; Fan et al. 2014; Jeennor et al. 2006). Consequently, stress-based strategies are widely used as environmentally friendly approaches to induce lipid overproduction in cultured microorganisms, and a wide range of studies were carried out to identify and develop efficient induction techniques for lipid accumulation (Sharma et al. 2012; Shi et al. 2017). However, little is known about the signaling pathways linking these inducers to intracellular lipid synthesis.

The development of "-omics" analysis methods has illuminated the global reorganization of metabolic and transcriptional states and their integration between different stress regimens (Yu et al. 2017; Morano et al. 2012; Yang et al. 2014; Shi et al. 2017). The progress on elucidating the mechanisms for PUFA-rich lipid accumulation and boosting the production are benefiting from the advances in systems biotechnology.

8.2.1 Omega-6 PUFAs

The filamentous fungus *Mortierella alpina*, which can accumulate omega-6 ARA to over 60% of its lipid content, is considered to be the most prominent producer of ARA-rich lipids. Using a combination of high throughput sequencing and lipid profiling, Wang et al. (2011) first assembled the genomic sequence of *M. alpina* ATCC 32222, mapped its lipogenesis pathway and determined its major lipid species. This work laid the foundation for possible genetic engineering of *M. alpina* to produce higher levels and diverse contents of dietary lipids. Furthermore, Vongsangnak et al. (2013) constructed a genome-scale metabolic model based on the drafted genome map. However, it was just a refined network that could only be used to investigate genome annotation and metabolic routes, and was not detailed enough to analyze flux distributions or phenotypic behaviors. In order to overcome these limitations, Ye et al. (2015a) used the COBRA Toolbox to reconstruct a genome-scale metabolic model of *M. alpina*. The model was further used to investigate the roles of acetyl-CoA and NADPH in the regulation of ARA biosynthesis.

ARA accumulation in *M. alpina* may be induced under several stress conditions. For example, the aging approach, which entails culturing the fungal cells for several days without carbon source after regular fermentation, can significantly increase the ARA content in the final lipid product and total fatty acids (TFAs). However, the specific mechanisms involved in the crosstalk between these stresses and intracellular ARA synthesis are poorly understood. By using metabolomics and lipidomics analysis, Zhang et al. (2015) characterized the intracellular metabolic states and pathways closely associated with ARA biosynthesis in *M. alpina*, and revealed that the main reason for the increased ARA/TFA levels was not only at the expense of the degradation of other fatty acids, but also continued ARA biosynthesis during aging. Translocation may play a key role in ARA redistribution among the glycerol moieties of different triglycerides and membrane lipids. Several key pathways were activated for maintaining a relatively stable intracellular environment, which may give *M. alpina* a chance to survive during aging. Subcellular and whole-cell proteomics were also employed to systematically investigate the mechanism of aging (Yu et al. 2016a, 2017, 2018). An EC 4.2.1.17-hydratase was detected as a vital player in ARA accumulation during aging. Further pathway analysis suggested that reactive oxygen species (ROS) were accumulated and induced the malate/pyruvate cycle and isocitrate dehydrogenase, which might provide additional NADPH for ARA synthesis. In addition, the effects of some environmental factors such as dissolved oxygen on ARA accumulation in *M. alpina* were also investigated via comparative metabolomics (Zhang et al. 2017).

Another filamentous fungus, *Mucor circinelloides*, is of industrial interest as it can produce high levels of omega-6 GLA. There have been attempts to increase the production of GLA-rich lipids by applying different "omics" approaches (e.g. genomics and proteomics) (Klanchui et al. 2016; Vongsangnak et al. 2016). The genome of a *M. circinelloides* strain that produces high levels of GLA-rich lipids was sequenced and compared with that of the a low-producing strain, which elucidated the general features of the genome and the potential mechanism of high GLA-rich lipid accumulation (Tang et al. 2015). Furthermore, Tang et al. (2016) systematically analyzed the changes at the levels of protein expression during three growth stages (the balanced growth stage, the fast lipid accumulation stage, and the slow lipid accumulation stage) in the GLA-rich lipid high-producing strain. They found that coordinated regulation of central carbon metabolism upon nitrogen limitation may increase the carbon flux toward acetyl-CoA and NADPH for fatty acid biosynthesis. Additionally, ^{13}C-metabolic flux analysis was performed to gain insights into the mechanism and intracellular fluxes of lipid accumulation in *M. circinelloides* (Zhao et al. 2015).

8.2.2 *Omega-3 PUFAs*

DHA is the most representative of the omega-3 PUFAs. In nature, there are two pathways for DHA biosynthesis in the DHA accumulating microorganisms. The first is the traditional elongation/desaturation pathway, in which the fatty acid synthase (FAS) enzyme complex synthesizes the saturated fatty acids, mainly myristic acid (C14:0) and palmitic acid (C16:0), which are then transformed to DHA by a series of desaturases and elongases. The other pathway for DHA biosynthesis is the so-called alternative polyketide synthase (PKS) pathway, which is found often in marine microalgae. The PKS pathway was detected in *Schizochytrium limacinum* SR21 based on the developed genome-scale metabolic model (GSMM), iCY1170_DHA (Ye et al. 2015a, b). The reconstructed GSMM was then used to elucidate the mechanism by which DHA is synthesized in *S. limacinum* and predict the requirements for abundant acetyl-CoA and NADPH for DHA production. Similarly, in order to further identify which kind of DHA synthetic pathway is active in *Schizochytrium mangrovei*, transcriptome analysis based on next-generation sequencing was applied, which is a powerful method for discovering and identifying genes involved in the biosynthesis of various fatty acids. The results showed that the FAS, PKS, and elongation/desaturation pathways co-exist in *S. mangrovei* (Hoang et al. 2016). In a recent study, Pei et al. (2017) applied de novo transcriptome analysis to decipher the metabolic responses and determine a possible DHA biosynthesis mechanism in the microalgae *Crypthecodinium cohnii*. The results showed that *C. cohnii* might utilize a combination of PKS systems and elongation/desaturation steps for DHA biosynthesis, which was further confirmed by qRT-PCR and GC/MS-based metabolomic analyses.

DHA synthesis is sensitive to many factors such as the substrate, culture conditions and so on. Systems biotechnology contributes to a better understanding of the regulatory mechanisms controlling the synthesis of DHA-rich lipids. Compared to glucose, glycerol can increase DHA production in *Schizochytrium* sp., and the underlying mechanism was clarified by transcriptome analysis (Chen et al. 2016). In addition, some organisms significantly increase the DHA content under cold stress to maintain membrane fluidity and function. Ma et al. (2015) used Illumina's sequencing technology to examine global changes in the transcriptome of *Aurantiochytrium* sp. in response to low-temperature stress, contributing to a better understanding of the various functions of genes that are differentially expressed under cold stress. Furthermore, proteomics was used to interpret the molecular mechanisms underlying the increase in DHA contents at low temperature in *Aurantiochytrium sp.* The results showed that low temperature inhibits the cellular energy supply and leads to a significant upregulation of PUFA synthesis (Ma et al. 2017).

In recent years, many studies reported that phytohormones can regulate growth or lipid accumulation in microalgae, and the underlying mechanism was revealed by metabolomics. For example, Yu et al. (2016c) used GC/MS-based metabolomics methodology combined with multivariate analysis to reveal the metabolic mechanism by which 6-benzylaminopurine enhances the production of lipids and DHA in *Aurantiochytrium* sp. Similarly, metabolomics was used to reveal the mechanism by which the phytohormone gibberellin enhances the lipid and DHA biosynthesis in *Aurantiochytrium* sp. (Yu et al. 2016b). Moreover, this method was also applied in *Schizochytrium sp.* to investigate the metabolic profiles under different aeration conditions, and identified some key metabolites responsible for the responses to low oxygen supply (Li et al. 2013).

8.3 Synthetic Biotechnology for PUFA Production

8.3.1 Omega-6 PUFAs

ARA, a typical omega-6 PUFA, is synthesized via the aerobic Δ6 desaturase and Δ6 elongase pathway (Δ6 pathway) found in fungi and some other organisms, or the Δ8 desaturase and Δ9 elongase pathway (Δ8 pathway) found in euglenoids (Ji et al. 2014a, b). the metabolic pathway for ARA biosynthesis has been cloned and reconstructed in various organisms, such as *Arabidopsis thaliana* and *Brassica napus* (Qi et al. 2004; Petrie et al. 2012). An ARA titer corresponding to 0.25% of total fatty acids was first produced in engineered *Saccharomyces cerevisiae* via co-expression of a novel elongase, Δ6-desaturase and Δ5-desaturase, when linoleic acid was fed as substrate (Beaudoin et al. 2000). Recently, *Y. lipolytica* has been attracted increasing attention due to its special characteristics (Liu et al. 2015). The researchers at DuPont company developed an engineered *Y. lipolytica* for the production of ARA by traditional cloning and heterologous expression. In their work, more than 10% ARA in the total lipid fraction was obtained in the engineered strain, using either the Δ6 or Δ8 pathway (Damude et al. 2009). With the development of synthetic biotechnology, the abundant rDNA loci were used as genomic integration sites, and the Δ6 pathway for ARA biosynthesis was assembled and integrated into *Y. lipolytica* in one step, leading to a high level of ARA production reaching 0.4% of total fatty acids (Liu et al. 2017a). Moreover, it was found that the synthetic ARA biosynthesis pathway can redirect the carbon flux towards intracellular fatty acid accumulation at the expense of extracellular organic acid secretion in the engineered *Y. lipolytica* (Liu et al. 2017b).

8.3.2 Omega-3 PUFAs

EPA and DHA are typical omega-3 PUFAs with beneficial effects on human health. Presently, the main source of EPA and DHA is the oil of deep-sea fish. However, fishes do not naturally produce EPA and DHA. In fact, the EPA and DHA found in their tissues are synthesized *de novo* by marine microorganisms that form a part of their diet and the overall food chain. Generally, EPA and DHA can be synthesized via the anaerobic PKS pathway or the aerobic desaturase/elongase pathways, including the Δ6 and Δ8 pathways (Gong et al. 2014; Cao et al. 2012). Recent advances in synthetic biotechnology provide the possibility to heterologously assemble the EPA and DHA biosynthesis pathways in the non-native microbial hosts, such as *E. coli*, *S. cerevisiae* and *Y. lipolytica*.

Initially, the anaerobic PKS pathway was assembled in the model bacterium *E. coli* for the biosynthesis of EPA and DHA. However, a relatively low EPA yield of only 0.7% of total fatty acids was obtained when the EPA synthetic gene cluster from *Shewanella oneidensis* MR-1 was heterologously expressed in *E. coli* (Lee et al. 2006). Further substitution of promoter sequences of the EPA synthesis genes with strong promoters enhanced the EPA production to 7.5% of the total fatty acids (Lee et al. 2008). At the same time, a high DHA content was *de novo* synthesized by engineered *E. coli* via the co-expression of the PKS gene cluster and the *pfaE* gene from *Moritella marina* strain MP-1 (Orikasa et al. 2006). In addition, the EPA and DHA synthetic gene cluster (nearly 35 kb) was cloned from *Shewanella baltica* MAC1 and expressed in four different *E. coli* strains. In these engineered *E. coli* strains, the highest amount of both EPA and DHA was produced by *E. coli* EPI300T1, with EPA at 8–14% and DHA at 0.1–0.4% of the total fatty acids, respectively (Amiri-Jami and Griffiths 2010). In another study, a simplified 20-kb gene cluster for EPA and DHA synthesis was assembled, which led to the production of respectively 0.12 and 1.35 mg of EPA and DHA per g cell dry weight in the engineered *Lactococcus lactis* subsp. *cremoris* MG1363 (Amiri-Jami et al. 2014).

Due to the specific characteristics of yeasts such as *S. cerevisiae and Pichia pastoris,* they serve as arguably the most important model chassis apart from *E. coli* for the *de novo* production of EPA and DHA. Through the co-expression of the C_{18}-PUFA specific Δ6-elongase, Δ6-desaturase and Δ5-desaturase from *Caenorhabditis elegans,* the Δ6 pathway was assembled and EPA at 0.2% of total fatty acids was obtained in the engineered *S. cerevisiae* (Beaudoin et al. 2000). Domergue et al. (2015) constructed an engineered *S. cerevisiae* strain that carries the Δ6 desaturase from *Ostreococcus tauri*, the Δ6 elongase from *Physcomitrella patens* and the Δ5 desaturase from *Phaeodactylum tricornutum.* In their research, a high level of EPA corresponding to 4.5% of total fatty acids was produced under the optimized conditions. In another study, through evaluating different desaturases, an engineered *S. cerevisiae* was constructed and a high level of EPA up to 0.49% of the total fatty acids was obtained under the optimized conditions (Tavares et al. 2011). Similarly, Kajikawa et al. (2004) isolated and characterized three genes encoding Δ6 desaturases, an ELO-like elongase, and a Δ5 desaturase from *Marchantia polymorpha*, and assembled the pathway for EPA synthesis in *P. pastoris*. However, only an exceedingly low level of EPA (0.03% of total fatty acids) was produced by the engineered strain. In addition, DHA was first produced in engineered *S. cerevisiae* via the co-expression of a novel elongase from *Pavlova* sp. and a Δ4 desaturase from *Isochrysis galbana*, when EPA was fed as substrate (Pereira et al. 2004).

Due to its high content of linoleic acid (LA, C18:2), *Y. lipolytica* was investigated for the de novo synthesis and accumulation of different PUFAs, such as EPA and DHA (Xie et al. 2015; Damude et al. 2014; Xue et al. 2013). Through the overexpression of different heterologous desaturases and elongases within the EPA biosynthesis pathway, the engineered *Y. lipolytica* were capable of producing more than 25% EPA in the total lipid fraction using either the Δ6 or Δ8 pathway (Xie et al. 2015). Moreover, Xue et al. (2013) used a combinational approach and generated the engineered *Y. lipolytica* Y4305 that contained 30 copies of 9 different genes, which produced EPA at 56.6% of total fatty acids and accumulated lipids at up to 30% of dry cell weight. Specifically, the EPA biosynthetic and supporting pathways have been introduced into the oleaginous yeast to synthesize and accumulate EPA under fermentation conditions. The Yarrowia platform can also produce tailored omega-3 (EPA, DHA) and/or omega-6 (ARA, GLA) fatty acid mixtures in the cellular lipid profiles. Therefore, the DuPont researchers have combined the fundamental bioscience and industrial engineering skills to achieve large-scale production of Yarrowia biomass containing high amounts of EPA, which led to two commercial products, New Harvest™ EPA oil and Verlasso® salmon (Damude et al. 2011; Xie et al. 2015).

8.4 Conclusions and Future Outlook

The most challenging hurdle in the microbial production of PUFAs is to generate a large amount of PUFA on a comparatively lower budget and with greater efficiency. However, several factors, such as slow cell growth and low oleaginicity, still hinder the applications of the technology for microbial PUFA production. Therefore, it is very important to explore the potential oleaginous power of microbes as much as possible. Consequently, modifying the microbes, including approaches such as rewiring the microbial metabolism to adjust the composition of microbial PUFAs, is urgently needed.

In the future, the emerging systems and synthetic biotechnology approaches may solve the aforementioned bottlenecks and provide new insights and opportunities for microbial PUFA production. First, the potential genetic targets for microbial improvement could be identified by taking advantage of both two available systems biotechnology approaches – top-down (high-throughput "omics" tools) or bottom-up (mathematical modeling methods). Then, both systems biotechnology approaches can be used in unison to bridge the knowledge gap inherent in each, and thus better guide the identification of promising genetic targets. Furthermore, expanding the synthetic biotechnology toolbox for the PUFA accumulating microbes to include highly controllable and tunable expression cassettes can greatly speed up strain development. This would make rewiring the microbial metabolism for PUFA accumulation more feasible and efficient. Finally, based on the targets identified using systems biotechnology approaches, the synthetic biotechnology toolbox can be used to fine-tune or perturb the identified genes. Once these methods are efficiently applied, the goal of rationally engineering microbes for PUFA production will become attainable. Thus, the challenging obstacles standing in the way of widespread industrial production of microbial PUFAs would be resolved.

Acknowledgments This work was financially supported by the National Science Fund for Excellent Young Scholars of China (No. 21922806), the National Natural Science Foundation of China (No. 21776131), the National Key Research and Development Project of China, the Six Talent Peaks Project in Jiangsu Province of China (No. 2018-SWYY-047), and the Jiangsu Synergetic Innovation Center for Advanced Bio-Manufacture (No. XTD1814).

References

Al-Hawash AB, Li S, Zhang X, Zhang X, Ma F. Productivity of γ-Linoleic acid by oleaginous fungus Cunninghamella echinulata using a pulsed high magnetic field. Food Biosci. 2018;21:1–7.

Amiri-Jami M, Griffiths MW. Recombinant production of omega-3 fatty acids in Escherichia coli using a gene cluster isolated from Shewanellabaltica MAC1. J Appl Microbiol. 2010;109:1897–905.

Amiri-Jami M, Lapointe G, Griffiths MW. Engineering of EPA/DHA omega-3 fatty acid production by Lactococcuslactis subsp. cremoris MG1363. Appl Microbiol Biotechnol. 2014;98:3071–80.

Beaudoin F, Michaelson LV, Hey SJ, Lewis MJ, Shewry PR, Sayanova O, Napier JA. Heterologous reconstitution in yeast of the polyunsaturated fatty acid biosynthetic pathway. Proc Natl Acad Sci U S A. 2000;97:6421–6.

Bellou S, Triantaphyllidou IE, Aggeli D, Elazzazy AM, Baeshen MN, Aggelis G. Microbial oils as food additives: recent approaches for improving microbial oil production and its polyunsaturated fatty acid content. Curr Opin Biotechnol. 2016;37:24–35.

Cao YJ, Cao YG, Zhao MA. Biotechnological production of eicosapentaenoic acid: From a metabolic engineering point of view. Process Biochem. 2012;47:1320–6.

Chen W, Zhou PP, Zhang M, Zhu YM, Wang XP, Luo XA, Bao ZD, Yu LJ. Transcriptome analysis reveals that up-regulation of the fatty acid synthase gene promotes the accumulation of docosahexaenoic acid in Schizochytrium sp. S056 when glycerol is used. Algal Res. 2016;15:83–92.

Cheng MH, Walker TH, Hulbert GJ, Raman DR. Fungal production of eicosapentaenoic and arachidonic acids from industrial waste streams and crude soybean oil. Bioresour Technol. 1999;67:101–10.

Damude HG, Gillies PJ, Macool DJ, Picataggio SK, Pollak DMW, Ragghianti JJ, Xue Z, Yadav NS, Zhang H, Zhu QQ. High arachidonic acid producing strains of Yarrowia lipolytica. US Patent 7588931. 2009.

Damude HG, Gillies PJ, Macool DJ, Picataggio SK, Pollak DMW, Ragghianti JJ, Xue Z, Yadav NS, Zhang H, Zhu QQ. High eicosapentaenoic acid producing strains of Yarrowia lipolytica. US Patent 7932077. 2011.

Damude HG, Gillies PJ, Macool DJ, Picataggio SK, Ragghianti JJ, Seip JE, Xue Z, Yadav NS, Zhang H, Zhu QQ. Docosahexaenoic acid producing strains of Yarrowia lipolytica. US Patent 8685682. 2014.

Diao J, Song X, Cui J, Liu L, Shi M, Wang F, Zhang W. Rewiring metabolic network by chemical modulator based laboratory evolution doubles lipid production in *Crypthecodinium cohnii*. Metab Eng. 2019;51:88–98.

Donot F, Fontana A, Baccou JC, Strub C, Schorr-Galindo S. Single cell oils (SCOs) from oleaginous yeasts and moulds: production and genetics. Biomass Bioenerg. 2014;68:135–50.

Domergue F, Abbadi A, Zähringer U, Moreau H, Heinz E. In vivo characterization of the first acyl-CoA Δ6-desaturase from a member of the plant kingdom, the microalga Ostreococcus tauri. Biochem J. 2015;389:483–90.

Dyal SD, Bouzidi L, Narine SS. Maximizing the production of γ-linolenic acid in Mortierella ramanniana var. ramanniana as a function of pH, temperature and carbon source, nitrogen source, metal ions and oil supplementation. Food Res Int. 2005;38:815–29.

Fan JH, Cui YB, Wan MX, Wang WL, Li YG. Lipid accumulation and biosynthesis genes response of the oleaginous Chlorella pyrenoidosa under three nutrition stressors. Biotechnol Biofuels. 2014;7:17.

Gao D, Zeng J, Yu X, Dong T, Chen S. Improved lipid accumulation by morphology engineering of oleaginous fungus Mortierella isabellina. Biotechnol Bioeng. 2014;111:1758–66.

Gong YM, Wan X, Jiang ML, Hu CJ, Hu HH, Huang FH. Metabolic engineering of microorganisms to produce omega-3 very long-chain polyunsaturated fatty acids. Prog Lipid Res. 2014;56:19–35.

Guo DS, Ji XJ, Ren LJ, Li GL, Huang H. Improving docosahexaenoic acid production by Schizochytrium sp. using a newly designed high-oxygen-supply bioreactor. AIChE J. 2017;63:4278–86.

Guo DS, Ji XJ, Ren LJ, Li GL, Sun XM, Chen KQ, Gao S, Huang H. Development of a scale-up strategy for fermentative production of docosahexaenoic acid by Schizochytrium sp. Chem Eng Sci. 2018;176:600–8.

Hoang MH, Nguyen C, Pham HQ, Nguyen LV, Duc LH, Son LV, Hai TN, Ha CH, Nhan LD, Anh HTL, Thom LT, Quynh HTH, Ha NC, Nhat PV, Hong DD. Transcriptome sequencing and comparative analysis of Schizochytrium mangrovei PQ6 at different cultivation times. Biotechnol Lett. 2016;38:1781–9.

Hu H, Wang HF, Ma LL, Shen XF, Zeng RJ. Effects of nitrogen and phosphorous stress on the formation of high value LC-PUFAs in Porphyridium cruentum. Appl Microbiol Biotechnol. 2018;102:5763–73.

Jeennor S, Laoteng K, Tanticharoen M, Cheevadhanarak S. Comparative fatty acid profiling of Mucorrouxii under different stress conditions. FEMS Microbiol Lett. 2006;259:60–6.

Ji XJ, Huang H. Engineering microbes to produce polyunsaturated fatty acids. Trends Biotechnol. 2019;37:344–6.

Ji XJ, Ren LJ, Nie ZK, Huang H, Ouyang PK. Fungal arachidonic acid-rich oil: research, development and industrialization. Crit Rev Biotechnol. 2014a;34:197–214.

Ji XJ, Zhang AH, Nie ZK, Wu WJ, Ren LJ, Huang H. Efficient arachidonic acid-rich oil production by Mortierella alpina through a repeated fed-batch fermentation strategy. Bioresour Technol. 2014b;170:356–60.

Ji XJ, Mo KQ, Ren LJ, Li GL, Huang JZ, Huang H. Genome sequence of Schizochytrium sp. CCTCC M209059, an effective producer of docosahexaenoic acid-rich lipids. Genome Announc. 2015a;3:e00819–5.

Ji XJ, Ren LJ, Huang H. Omega-3 biotechnology: a green and sustainable process for omega-3 fatty acids production. Front Bioeng Biotechnol. 2015b;3:158.

Kajikawa M, Yamato KT, Kohzu Y, Nojiri M, Sakuradani E, Shimizu S, Sakai Y, Fukuzawa H, Ohyama K. Isolation and characterization of delta(6)-desaturase, an ELO-like enzyme and delta(5)-desaturase from the liverwort Marchantia polymorpha and production of arachidonic and eicosapentaenoic acids in the methylotrophic yeast Pichia pastoris. Plant Mol Biol. 2004;54:335–52.

Kiy T, Rüsing M, Fabritius D. Production of docosahexaenoic acid by the marine microalga, Ulkenia sp. In: Cohen Z, Ratledge C, editors. Single cell oils. Champaign: ACOS Press; 2005. p. 99–106.

Klanchui A, Vongsangnak W, Laoteng K, Meechai A. In silico analysis of mucor circinelloides genome-scale model for enhancing lipid production. In: International conference on computational systems-biology & bioinformatics. Macao: ACM; 2016. p. 14–8.

Lee SJ, Jeong YS, Kim DU, Seo JW, Hur BK. Eicosapentaenoic acid (EPA) biosynthetic gene cluster of Shewanella oneidensis MR-1: cloning, heterologous expression, and effects of temperature and glucose on the production of EPA in Escherichia coli. Biotechnol Bioprocess Eng. 2006;11:510–5.

Lee SJ, Kim CH, Seo PS, Kwon O, Hur BK, Seo JW. Enhancement of heterologous production of eicosapentaenoic acid in Escherichia coli by substitution of promoter sequences within the biosynthesis gene cluster. Biotechnol Lett. 2008;30:2139–42.

Li J, Ren LJ, Sun QN, Qu L, Huang H. Comparative metabolomics analysis of Docosahexaenoic acid fermentation processes by Schizochytrium sp. under different oxygen availability conditions. OMICS: A J Integr Bio. 2013;17:269–81.

Ling X, Guo J, Liu X, Zhang X, Wang N, Lu Y, Ng IS. Impact of carbon and nitrogen feeding strategy on high production of biomass and docosahexaenoic acid (DHA) by Schizochytrium sp. LU310. Bioresour Technol. 2015;184:139–47.

Liu HH, Ji XJ, Huang H. Biotechnological applications of Yarrowia lipolytica: past, present and future. Biotechnol Adv. 2015;33:1522–46.

Liu HH, Madzak C, Sun ML, Ren LJ, Song P, Huang H, Ji XJ. Engineering Yarrowialipolytica for arachidonic acid production through rapid assembly of metabolic pathway. Biochem Eng J. 2017a;119:52–8.

Liu HH, Zeng SY, Shi TQ, Ding Y, Ren LJ, Song P, Huang H, Madzak C, Ji XJ. A Yarrowialipolytica strain engineered for arachidonic acid production counteracts metabolic burden by redirecting carbon flux towards intracellular fatty acid accumulation at the expense of organic acids secretion. Biochem Eng J. 2017b;128:201–9.

Ma Z, Tan Y, Cui G, Feng Y, Cui Q, Song X. Transcriptome and gene expression analysis of DHA producer Aurantiochytrium under low temperature conditions. Sci Rep. 2015;5:14446.

Ma ZX, Tian MM, Tan YZ, Cui GZ, Feng YG, Cui Q, Song XJ. Response mechanism of the docosahexaenoic acid producer Aurantiochytrium under cold stress. Algal Res. 2017;25:191–9.

Molina Grima E, Sánchez Pérez JA, Garciá Camacho F, Garciá Sánchez JL, López Alonso D. n-3 PUFA productivity in chemostat cultures of microalgae. Appl Microbiol Biotechnol. 1993;38:599–605.

Morano KA, Grant CM, Moye-Rowley WS. The response to heat shock and oxidative stress in Saccharomyces cerevisiae. Genetics. 2012;190:1157–95.

Nielsen J, Keasling JD. Engineering cellular metabolism. Cell. 2016;164:1185–97.

O'Brien DJ, Kurantz MJ, Kwoczak R. Production of eicosapentaenoic acid by the filamentous fungus Pythium irregulare. Appl Microbiol Biotechnol. 1993;40:211–4.

Orikasa Y, Yamada A, Yu R, Ito Y, Nishida T, Yumoto I, Watanabe K, Okuyama H. Characterization of the eicosapentaenoic acid biosynthesis gene cluster from Shewanella sp. strain SCRC-2738. Cell Mol Biol. 2004;50:625–30.

Orikasa Y, Nishida T, Yamada A, Yu R, Watanabe K, Hase A, Morita N, Okuyama H. Recombinant production of docosahexaenoic acid in a polyketide biosynthesis mode in Escherichia coli. Biotechnol Lett. 2006;28:1841–7.

Pei GS, Li XR, Liu LS, Liu J, Wang FZ, Chen L, Zhang WW. De novo transcriptomic and metabolomic analysis of docosahexaenoic acid (DHA)-producing Crypthecodinium cohnii during fed-batch fermentation. Algal Res. 2017;26:380–91.

Pereira SL, Leonard AE, Huang Y, Chuang L, Mukerji P. Identification of two novel microalgal enzymes involved in the conversion of the omega3-fatty acid, eicosapentaenoic acid, into docosahexaenoic acid. Biochem J. 2004;384:357–66.

Petrie JR, Shrestha P, Belide S, Mansour MP, Liu Q, Horne J, Nichols PD, Singh SP. Transgenic production of arachidonic acid in oilseeds. Transgenic Res. 2012;21:139–47.

Qi BX, Fraser T, Mugford S, Dobson G, Sayanova O, Butler J, Napier JA, Stobart AK, Lazarus CM. Production of very long chain polyunsaturated omega-3 and omega-6 fatty acids in plants. Nature Biotechnol. 2004;22:739–45.

Ratledge C. Fatty acid biosynthesis in microorganisms being used for Single Cell Oil production. Biochimie. 2004;86:807–15.

Sharma KK, Schuhmann H, Schenk PM. High lipid induction in microalgae for biodiesel production. Energies. 2012;5:1532–53.

Shi K, Gao Z, Shi TQ, Song P, Ren LJ, Huang H, Ji XJ. Reactive oxygen species-mediated cellular stress response and lipid accumulation in oleaginous microorganisms: the state of the art and future perspectives. Front Microbiol. 2017;8:793.

Shi K, Gao Z, Lin L, Wang WJ, Shi XQ, Yu X, Song P, Ren LJ, Huang H, Ji XJ. Manipulating the generation of reactive oxygen species through intermittent hypoxic stress for enhanced accumulation of arachidonic acid-rich lipids. Chem Eng Sci. 2018;186:36–43.

Singh A, Ward OP. Production of high yields of docosahexaenoic acid by Traustochytrium roseum ATCC 20810. J Ind Microbiol. 1996;16:370–3.

Sun ML, Madzak C, Liu HH, Song P, Ren LJ, Huang H, Ji XJ. Engineering Yarrowia lipolytica for efficient γ-linolenic acid production. Biochem Eng J. 2017;117:172–80.

Tang X, Zhao L, Chen H, Chen YQ, Chen W, Song Y, Ratledge C. Complete genome sequence of a high lipid-producing strain of Mucor circinelloides WJ11 and comparative genome analysis with a low lipid-producing strain CBS 277.49. PLoS One. 2015;10:e0137543.

Tang X, Zan X, Zhao L, Chen H, Chen YQ, Chen W, Song Y, Ratledge C. Proteomics analysis of high lipid-producing strain Mucor circinelloides WJ11: an explanation for the mechanism of lipid accumulation at the proteomic level. Microb Cell Fact. 2016;15:35.

Tavares S, Grotkjaer T, Obsen T, Haslam RP, Napier JA, Gunnarsson N. Metabolic engineering of Saccharomyces cerevisiae for production of eicosapentaenoic acid, using a novel delta 5-desaturase from Paramecium tetraurelia. Appl Environ Microbiol. 2011;77:1854–61.

Vongsangnak W, Ruenwai R, Tang X, Hu X, Zhang H, Shen B, Song Y, Laoteng K. Genome-scale analysis of the metabolic networks of oleaginous Zygomycete fungi. Gene. 2013;521:180–90.

Vongsangnak W, Klanchui A, Tawornsamretkit I, Tatiyaborwornchai W, Laoteng K, Meechai A. Genome-scale metabolic modeling of Mucor circinelloides and comparative analysis with other oleaginous species. Gene. 2016;583:121–9.

Wang SB, Chen F, Sommerfeld M, Hu Q. Proteomic analysis of molecular response to oxidative stress by the green alga Haematococcuspluvialis (Chlorophyceae). Planta. 2004;220:17–29.

Wang L, Chen W, Feng Y, Ren Y, Gu ZN, Chen HQ, Wang HC, Thomas MJ, Zhang BX, Berquin IM, Li Y, Wu JS, Zhang HX, Song YD, Liu X, Norris JS, Wang S, Du P, Shen JG, Wang N, Yang YL, Wang W, Feng L, Ratledge C, Zhang H, Chen YQ. Genome characterization of the oleaginous fungus Mortierellaalpina. PLoS One. 2011;6:12.

Ward OP, Singh A. Omega-3/6 fatty acids: alternative sources of production. Process Biochem. 2005;40:3627–52.

Wen ZY, Jiang Y, Chen F. High cell density of the diatom Nitzschia laevis for eicosapentaenoic acid production: fed-batch development. Process Biochem. 2002;37:1447–53.

Wu WJ, Zhang AH, Peng C, Ren LJ, Song P, Yu YD, Huang H, Ji XJ. An efficient multi-stage fermentation strategy for the production of microbial oil rich in arachidonic acid in Mortierella alpina. Bioresour Bioprocess. 2017;4:8.

Xie D, Jackson EN, Zhu Q. Sustainable source of omega-3 eicosapentaenoic acid from metabolically engineered Yarrowialipolytica: from fundamental research to commercial production. Appl Microbiol Biotechnol. 2015;99:1599–610.

Xue Z, Sharpe PL, Hong SP, Yadav NS, Xie D, Short DR, Damude HG, Rupert RA, Seip JE, Wang J, Pollak DW, Bostick MW, Bosak MD, Macool DJ, Hollerbach DH, Zhang H, Arcilla DM, Bledsoe SA, Croker K, McCord EF, Tyreus BD, Jackson EN, Zhu Q. Production of omega-3 eicosapentaenoic acid by metabolic engineering of Yarrowia lipolytica. Nat Biotechnol. 2013;31:734–40.

Yamada H, Shimizu S, Shinmen Y. Production of arachidonic acid by Mortierella elongata 1S-5. Agric Biol Chem. 1987;51:785–90.

Yang ZK, Ma YH, Zheng JW, Yang WD, Liu JS, Li HY. Proteomics to reveal metabolic network shifts towards lipid accumulation following nitrogen deprivation in the diatom Phaeodactylum tricornutum. J Appl Phycol. 2014;26:73–82.

Ye C, Qiao W, Yu X, Ji XJ, Huang H, Collier JL, Liu L. Reconstruction and analysis of the genome-scale metabolic model of Schizochytrium limacinum SR21 for docosahexaenoic acid production. BMC Genomics. 2015a;16:799.

Ye C, Xu N, Chen H, Chen YQ, Chen W, Liu L. Reconstruction and analysis of a genome-scale metabolic model of the oleaginous fungus Mortierella alpina. BMC Sys Biol. 2015b;9:1.

Yongmanitchai W, Ward OP. Growth of and omega-3 fatty acid production by Phaeodactylum tricornutum under different culture conditions. Appl Environ Microbiol. 1991;57:419–25.

Yu Y, Li T, Wu N, Ren L, Jiang L, Ji XJ, Huang H. Mechanism of arachidonic acid accumulation during aging in Mortierella alpina: a large-scale label-free comparative proteomics study. J Agr Food Chem. 2016a;64:9124–34.

Yu XJ, Sun J, Sun YQ, Zheng JY, Wang Z. Metabolomics analysis of phytohormone gibberellin improving lipid and DHA accumulation in Aurantiochytrium sp. Biochem Engin J. 2016b;112:258–68.

Yu XJ, Sun J, Zheng JY, Sun YQ, Wang Z. Metabolomics analysis reveals 6-benzylaminopurine as a stimulator for improving lipid and DHA accumulation of Aurantiochytrium sp. J Chem Tech Biotechnol. 2016c;91:1199–207.

Yu Y, Li T, Wu N, Jiang L, Ji XJ, Huang H. The role of lipid droplets in Mortierella alpina aging revealed by integrative subcellular and whole-cell proteome analysis. Scientific Reports. 2017;7:43896.

Yu Y, Zhang L, Li T, Wu N, Jiang L, Ji XJ, Huang H. How nitrogen sources influence Mortierella alpina aging: From the lipid droplet proteome to the whole-cell proteome and metabolome. J Proteomics. 2018;179:140–9.

Zhang AH, Ji XJ, Wu WJ, Ren LJ, Yu YD, Huang H. Lipid fraction and intracellular metabolite analysis reveal the mechanism of arachidonic acid-rich oil accumulation in the aging process of Mortierella alpina. J Agr Food Chem. 2015;63:9812–9.

Zhang X, Jiang L, Zhu LY, Shen QK, Ji XJ, Huang H, Zhang HM. Effects of aeration on metabolic profiles of Mortierella alpina during the production of arachidonic acid. J Ind Microbiol Biotechnol. 2017;44:1225–35.

Zhao L, Zhang H, Wang L, Chen H, Chen YQ, Chen W, Song Y. (13)C-metabolic flux analysis of lipid accumulation in the oleaginous fungus Mucor circinelloides. Bioresour Technol. 2015;197:23–9.

Zhao L, Cánovas-Márquez JT, Tang X, Chen HQ, Chen YQ, Chen W, Garre V, Song Y, Ratledge C. Role of malate transporter in lipid accumulation of oleaginous fungus Mucor circinelloides. Appl Microbiol Biotechnol. 2016;100:1297–305.

Zou N, Richmond A. Effect of light-path length in outdoor flat plate reactors on output rate of cell mass and of EPA in Nannochloropsis sp. J Biotechnol. 1999;70:351–6.

Chapter 9
Microbial Production of Nutraceuticals: Challenges and Prospects

Ningzi Guan, Jianghua Li, Guocheng Du, Jian Chen, and Long Liu

Nutraceuticals are food substances with physiological functions and bioactivities that promote human health and prevent some diseases. The quality of human lives is threaten by nutrient deficiencies since dietary supplementation cannot meet the standard amount for nutraceuticals. With the improvement of living standards, people pay more and more attention to their health in recent years, which has caused a rapid increase in the market of nutraceuticals. Over $230 Billion of global market have been achieved in the field of nutraceuticals in 2018 (Yuan and Alper 2019).

Nutraceuticals on the market are of various forms such as nutrients, dietary supplements, and processed products. They are traditionally chemical synthesized or industrial isolated from plants, animals and microorganisms. However, owing to the limited resources, environmental pollution, toxic by-products and high cost on further processing, these processes have gradually become inadvisable and unsustainable for production. As an alternative, microbial synthesis is an attractive strategy for nutraceuticals production since the substrates are widespread and reproducible, bioprocesses are relatively environmentally friendly, and the products are considered

---

N. Guan
Synthetic Biology and Biomedical Engineering Laboratory, Biomedical Synthetic Biology Research Center, Shanghai Key Laboratory of Regulatory Biology, Institute of Biomedical Sciences and School of Life Sciences, East China Normal University, Shanghai, China

J. Li · G. Du · L. Liu (✉)
Key Laboratory of Carbohydrate Chemistry and Biotechnology, Ministry of Education, Jiangnan University, Wuxi, China

Key Laboratory of Industrial Biotechnology, Ministry of Education, Jiangnan University, Wuxi, China
e-mail: longliu@jiangnan.edu.cn

J. Chen
Key Laboratory of Industrial Biotechnology, Ministry of Education, Jiangnan University, Wuxi, China

L. Liu, J. Chen (eds.), *Systems and Synthetic Biotechnology for Production of Nutraceuticals*, https://doi.org/10.1007/978-981-15-0446-4_9

safe. Especially, the introduction of metabolic engineering provides a possibility for microbial production of nutraceuticals in industrial scale.

Applied in the food and pharmaceutical industries, strains must be safe enough and need to be characterized all sidedly. Therefore, GRAS (generally regarded as safe) strains were defined by the Food and Drug Administration (Burdock and Carabin 2004), which should be selected preferably for nutraceuticals biosynthesis. Some GRAS microbes have a long history of safe industrial use and typically selected for nutraceuticals production. Probiotics such as certain *Lactobacillus* and *Propionibacterium* are widely used to produce exopolysaccharides and vitamins as the quality and safety of products are guaranteed by the GRAS status (La China et al. 2018; Amiri et al. 2019). Lactic acid bacteria have attracted abundant interest as the hyaluronic acid producer. Propionibacteria are preferred to synthesize vitamin B_{12} and folate owing to the natural synthetic pathways as well as GRAS status (Hugenschmidt et al. 2011). *Lactobacillus* (Komatsuzaki et al. 2005), *Lactococcus* (Lu et al. 2009), *Bifidobacterium* (Park et al. 2005), and *Streptococcus* (Yang et al. 2008) have all been used for γ-aminobutyric acid biosynthesis.

As the eukaryotic model organism, *Saccharomyces cerevisiae* is selected and modified to synthesize glutathione (Lian et al. 2018). *Rhodotorula* and *Saccharomyces* are also considered as the ideal microbial producers of CoQ_{10} as the well-studied eukaryotes with CoQ biosynthetic pathways (Tran and Clarke 2007). *Corynebacterium glutamicum* is a traditional host for amino acids production and was introduced to synthesize non-proteinogenic amino acids such as β-alanine (Shen et al. 2014). *S. cerevisiae* and *Candida utilis* shows potential for carotenoids synthesis (Shimada et al. 1998). *Spirulina* is cultured commercially to collect polysaccharides, phycocyanins, and γ-linolenic acid (Suresh et al. 2014).

A variety of non-conventional GRAS strains are also explored for nutraceutical production. With the metabolic advantage that accumulating large amounts of organic acids, *Yarrowia lipolytica* is selected as a considerable producer of alpha-ketoglutaric acid. Besides, *Y. lipolytica* has been designed as a model to produce polyunsaturated fatty acids indcluding docosahexaenoic acid (DHA), eicosapentaenoic acid (EPA), α-linolenic acid (ALA) and arachidonic acid (ARA), the lipogenesis of which has been harnessed to create a platform for lipid and biofuel production (Blazeck et al. 2014). Owing to the complete genetic information, *Bacillus subtilis* has been metabolic engineered as a cell factory to synthesize hyaluronic acid, glucosamine, *N*-acetylglucosamine and so on (Liu et al. 2013).

In addition, a large amount of other valuable nutraceuticals are favorably received by the market, such as polyphenolic compounds (e.g. flavonoids, isoflavonoids), functional saccharides (e.g. trehalose), and so on. They are widely used in various industrial fields including foods, pharmaceuticals, and cosmetics, which appeal strict safety requirements and adequate quantities. The exposition of biosynthetic pathways and regulatory mechanisms of nutraceuticals in microorganisms provides huge opportunities for higher nutraceuticals microbial production (Fig. 9.1).

Constant improvements in genome sequencing and high-throughput detection technology have made systems and synthetic biology the most infusive technologies to build microbial cell factories (Otero et al. 2013). Systems biology aims at collect

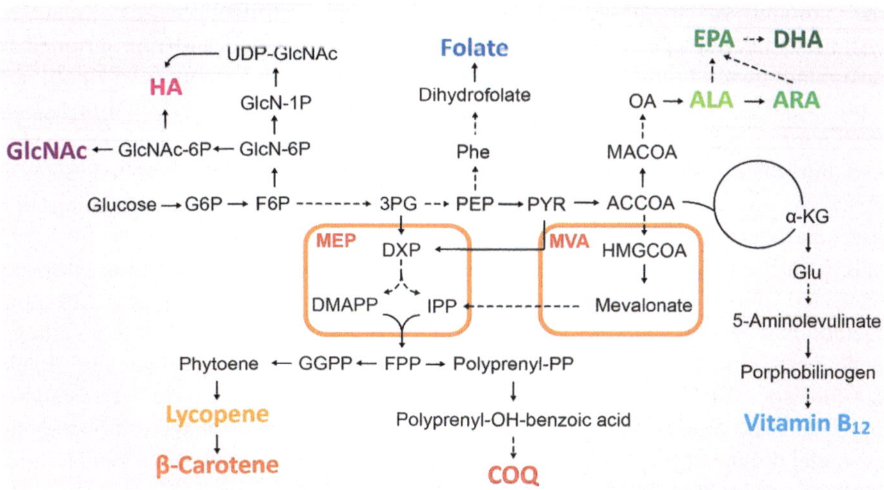

Fig. 9.1 A brief schematic diagram of the metabolic pathway for certain nutraceuticals. *3PG* 3-phospho-glycerate, *ACCOA* acetyl-CoA, *α-KG* alpha-ketoglutaric acid, *ALA* α-linolenic acid, *ARA* arachidonate, *COQ* coenzyme Q, *DXP* 1-deoxy-D-xylulose 5-phosphate, *F6P* fructose 6-phosphate, *FPP* farnesyl pyrophosphate, *G1P* glucose 1-phosphate, *G6P* glucose 6-phosphate, *GGPP* geranylgeranyl pyrophosphate, *GlcN-1P* glucosamine 1-phosphate, *GlcN-6P* glucosamine 6-phosphate, *GlcNAc* *N*-acetylglucosamine, *GlcNAc-6P* *N*-acetylglucosamine 6-phosphate, *Glu* glutamate, *HA* hyaluronic acid, *MEP* 2-C-methyl-D-erythritol 4-phosphate, *MVA* mevalonic acid, *OA* oleic acid, *PEP* phosphoenolpyruvate, *Phe* phenylalanine, *PYR* pyruvate, *UDP-GlcNAc* UDP-*N*-acetylglucosamine

integrated information of biological systems through analyzing all molecular elements quantificationally and comparatively and propose educated hypotheses about the underlying mechanisms of physiological regulation (Hood et al. 2004). As technologies of systems biology, omics are developing rapidly in recent years and have been widely employed in research and development for strain improvement, recognition of the bottlenecks in the biosynthetic pathways, and elevation of microbial compounds production (Liu et al. 2017).

However, a biological system is an extremely complex network of dependencies and interrelationships of genes, proteins, metabolites and even environment rather than a simple superposition. The characterization of their individual and interconnections provide merely a basis to comb the structure and dynamics of a system (Kitano 2002). The assemble principle, interactive model and regulative method of the subassembly are more crucial for the well-rounded cognition of a biological system, and also the research priorities. Systems biology is implemented to reveal molecular mechanisms that lurking beneath bioprocess phenotypes from all angles and provide foundations for construction of cell factories via synthetic biology tactics.

Synthetic biology is an integrated research field of biology and engineering that enables to reconstitute metabolic pathways, endow cells with new functions, alter the amount of metabolites, and develop new applications of organisms (Andrianantoandro et al. 2006). Organisms are investigated and reprogramed by designing novel biomolecule components, pathways and networks (Khalil and Collins 2010). Synthetic biology facilitates to construct multifunctional organisms in two major ways, invoking substitutable modules from another biological system and introducing artificial life with unnatural components (Benner and Sismour 2005). As a result, modified organisms are more appropriate for industrial applications and, hence, synthetic biology offers huge opportunities to construct efficient cell factories. Over the past decades, the development of synthetic biology has concentrated mainly on the design of biological devices and perfection of microbial metabolism, which gained remarkable achievements in *de novo* DNA synthesis, gene networks reconstruction and protein engineering (Heinemann and Panke 2006). Especially, multiple strategies of metabolic engineering have been designed on *E. coli* and *S. cerevisiae* to enhance the productive performances since they act as model organisms for bacteria and eukaryotic cells, respectively (Feist and Palsson 2008; Herrgård et al. 2008).

However, as programmable biological entities, cells are intricate that it is extremely urgent to exploit efficient approaches to assemble, implement and coordinate the biological devices, circuits and systems (Purnick and Weiss 2009). During nutraceuticals biosynthesis, for instance, the construction of feasible synthetic pathways is only the basic step. Although inserted into a GRAS strain successfully, a considerable yield depends on reasonable distribution of metabolic flux, calculating allocation of energy, and coordination between cell growth and product synthesis. What's more, the action mechanism and potential influence of exogenous components on cells must be investigated comprehensively and systematically combined with system biology analysis and computer-aided design (Nielsen and Keasling 2011). Then the development and application of system and synthetic biology could be improved increasingly, and rapid, reproducible, predictable building of platform cell factories could be done.

With the improvement of living standards, humans pay more attention to life quality and health. Amounts of studies have been devoted to discovering molecules advantageous to health—namely, nutraceuticals. Nutraceuticals have significant effect on disease prevention and are efficacious as active ingredients of drugs. Microbial synthesis of nutraceuticals has shown significant superiorities compared with traditional industrial processes. In particular, the identification of GRAS strains provides brilliant prospects for the microbial production of nutraceuticals. Although there is a large distance of most of these strains from industrial applications limited by low yields and certain legal restriction, they have shown tremendous potential for nutraceutical production. As the market demand for nutraceuticals is booming, and because this trend is likely to continue, more and more research attention has been focused on the biosynthesis by microbial cell factories. The utilization of systems biology strategies will bring new interpretation and intensive exploitation of GRAS strains, and then the pivotal role of synthetic biology approaches on nutraceuticals biosynthesis will gradually be borne out (Fig. 9.2). Although there may be some legal limitations on the utilization of genetic engineered microbes for the production

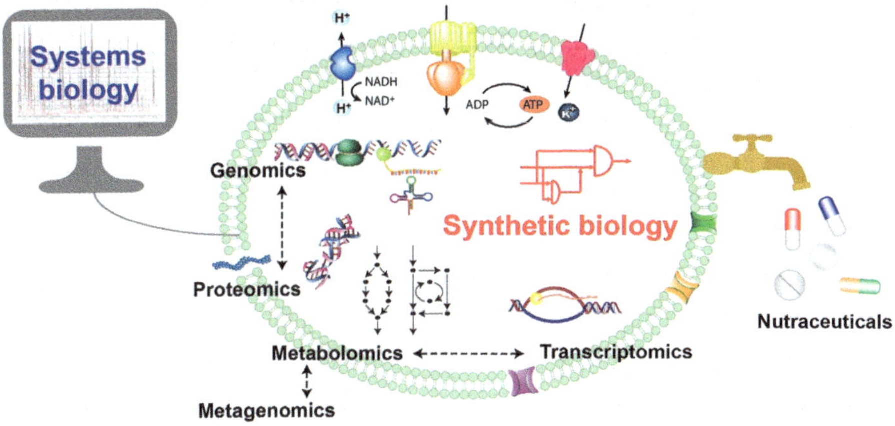

Fig. 9.2 Microbial synthesis of nutraceuticals by GRAS strains dissected and regulated through systems and synthetic biology approaches

of nutraceutical currently, appropriate extraction and purification methods could be adopted to conquer it. In conclusion, the difficulties on microbial nutraceutical production such as the high cost, low yield and legal restriction will be overcome, and substantial nutraceuticals will be harvested by constructing GRAS microbial cell factories.

References

Amiri S, Mokarram RR, Khiabani MS, Bari MR, Khaledabad MA. Exopolysaccharides production by Lactobacillus acidophilus LA5 and Bifidobacterium animalis subsp. lactis BB12: optimization of fermentation variables and characterization of structure and bioactivities. Int J Biol Macromol. 2019;123:752–65.

Andrianantoandro E, Basu S, Karig DK, Weiss R. Synthetic biology: new engineering rules for an emerging discipline. Mol Syst Biol. 2006;2(1):1.

Benner SA, Sismour AM. Synthetic biology. Nat Rev Genet. 2005;6:533.

Blazeck J, Hill A, Liu L, Knight R, Miller J, Pan A, Otoupal P, Alper HS. Harnessing Yarrowia lipolytica lipogenesis to create a platform for lipid and biofuel production. Nat Commun. 2014;5:3131.

Burdock GA, Carabin IG. Generally recognized as safe (GRAS): history and description. Toxicol Lett. 2004;150:3–18.

Feist AM, Palsson BØ. The growing scope of applications of genome-scale metabolic reconstructions using Escherichia coli. Nat Biotechnol. 2008;26:659.

Heinemann M, Panke S. Synthetic biology—putting engineering into biology. Bioinformatics. 2006;22:2790–9.

Herrgård MJ, Swainston N, Dobson P, Dunn WB, Arga KY, Arvas M, Blüthgen N, Borger S, Costenoble R, Heinemann M. A consensus yeast metabolic network reconstruction obtained from a community approach to systems biology. Nat Biotechnol. 2008;26:1155.

Hood L, Heath JR, Phelps ME, Lin B. Systems biology and new technologies enable predictive and preventative medicine. Science. 2004;306:640–3.

Hugenschmidt S, Schwenninger SM, Lacroix C. Concurrent high production of natural folate and vitamin B12 using a co-culture process with Lactobacillus plantarum SM39 and Propionibacterium freudenreichii DF13. Process Biochem. 2011;46:1063–70.

Khalil AS, Collins JJ. Synthetic biology: applications come of age. Nat Rev Genet. 2010;11:367.

Kitano H. Systems biology: a brief overview. Science. 2002;295:1662–4.

Komatsuzaki N, Shima J, Kawamoto S, Momose H, Kimura T. Production of γ-aminobutyric acid (GABA) by lactobacillus paracasei isolated from traditional fermented foods. Food Microbiol. 2005;22:497–504.

La China S, Zanichelli G, De Vero L, Gullo M. Oxidative fermentations and exopolysaccharides production by acetic acid bacteria: a mini review. Biotechnol Lett. 2018;40:1289–302.

Lian J, Mishra S, Zhao H. Recent advances in metabolic engineering of Saccharomyces cerevisiae: new tools and their applications. Metab Eng. 2018;50:85–108.

Liu L, Liu Y, Shin H-d, Chen R, Li J, Du G, Chen J. Microbial production of glucosamine and N-acetylglucosamine: advances and perspectives. Appl Microbiol Biotechnol. 2013;97:6149–58.

Liu L, Guan N, Li J, Shin H-d, Du G, Chen J. Development of GRAS strains for nutraceutical production using systems and synthetic biology approaches: advances and prospects. Crit Rev Biotechnol. 2017;37:139–50.

Lu X, Xie C, Gu Z. Optimisation of fermentative parameters for GABA. Czech J Food Sci. 2009;27:433–42.

Nielsen J, Keasling JD. Synergies between synthetic biology and metabolic engineering. Nat Biotechnol. 2011;29:693.

Otero JM, Cimini D, Patil KR, Poulsen SG, Olsson L, Nielsen J. Industrial systems biology of Saccharomyces cerevisiae enables novel succinic acid cell factory. PLoS One. 2013;8:e54144.

Park KB, Ji GE, Park MS, Oh SH. Expression of rice glutamate decarboxylase in Bifidobacterium longum enhances γ-aminobutyric acid production. Biotechnol Lett. 2005;27:1681–4.

Purnick PEM, Weiss R. The second wave of synthetic biology: from modules to systems. Nat Rev Mol Cell Biol. 2009;10:410–22.

Shen Y, Zhao L, Li Y, Zhang L, Shi G. Synthesis of β-alanine from L-aspartate using L-aspartate-α-decarboxylase from Corynebacterium glutamicum. Biotechnol Lett. 2014;36:1681–6.

Shimada H, Kondo K, Fraser PD, Miura Y, Saito T, Misawa N. Increased carotenoid production by the food yeast Candida utilis through metabolic engineering of the isoprenoid pathway. Appl Environ Microbiol. 1998;64:2676–80.

Suresh D, Madhu M, Saritha C, Shankaraiah P. Antidepressant activity of spirulina platensis in experimentally induced depressed mice. Int J Pharm Arch. 2014;3(3):1.. ISSN:2319-7226

Tran UPC, Clarke CF. Endogenous synthesis of coenzyme Q in eukaryotes. Mitochondrion. 2007;7:S62–71.

Yang SY, Lü FX, Lu ZX, Bie XM, Jiao Y, Sun LJ, Yu B. Production of γ-aminobutyric acid by Streptococcus salivarius subsp. thermophilus Y2 under submerged fermentation. Amino Acids. 2008;34:473–8.

Yuan S-F, Alper HS. Metabolic engineering of microbial cell factories for production of nutraceuticals. Microb Cell Factories. 2019;18:46.

Printed by Printforce, the Netherlands